Outsider Scientists

Outsider Scientists

ROUTES TO
INNOVATION
IN BIOLOGY

EDITED BY

OREN HARMAN AND

MICHAEL R. DIETRICH

THE UNIVERSITY OF CHICAGO PRESS

Chicago and London

Oren Harman is the chair of the Graduate Program in Science, Technology, and Society at Bar-Ilan University, Ramat Gan, Israel. **Michael R. Dietrich** is professor in the Department of Biological Sciences at Dartmouth College.

The University of Chicago Press, Chicago 60637
The University of Chicago Press, Ltd., London
© 2013 by The University of Chicago
All rights reserved. Published 2013.
Printed in the United States of America

22 21 20 19 18 17 16 15 14 13   1 2 3 4 5

ISBN-13: 978-0-226-07837-3 (cloth)
ISBN-13: 978-0-226-07840-3 (paper)
ISBN-13: 978-0-226-07854-0 (e-book)

Library of Congress
Cataloging-in-Publication Data
Outsider scientists : routes to innovation in biology /
edited by Oren Harman and Michael R. Dietrich.
      pages cm
   Includes bibliographical references and index.
   ISBN 978-0-226-07837-3 (cloth : alkaline paper) —
ISBN 978-0-226-07840-3 (paperback : alkaline paper) —
ISBN 978-0-226-07854-0 (e-book)
1. Biologists—Biography.   2. Biology—History.
I. Harman, Oren Solomon, editor.   II. Dietrich, Michael R., editor.
   QH26.O98 2013
   570.92′2—dc23   2013016559

DOI: 10.7208/chicago/9780226078540.001.0001

♾ This paper meets the requirements of
ANSI/NISO Z39.48-1992 (Permanence of Paper).

To outsiders everywhere
who are still looking in

CONTENTS

PREFACE

No one likes an outsider. They know it all, haven't paid their dues, and often think little of the rules everyone else has been required to play by—except that outsiders are also sometimes godsends, blowing in like a felicitous wind, carrying new energy and whispering new truths. Outsiders often see things differently than those who have been gazing at a problem for a long time, and it is this perspective that makes them so valuable.

As students of biology and its history, we have both long felt that outsiders have played a special role within our field. The questions asked in the life sciences are both so fundamental and so broad that it stands to reason that tools of many kinds, not just biological, must be employed in order to crack outstanding mysteries. Certainly, this has been the case historically. And so we present to the reader in this volume what we hope is a thought-provoking sample of some of the remarkable boundary crossers of the modern era who have come into biology. They hail from physics, chemistry, mathematics, and computer science; and linguistics, philosophy, engineering, mathematics, and popular writing, too. Each, in their distinctive way, has brought new tools and methods from his or her home disciplines, new ways of looking at and thinking about old problems. Along the way, they have revolutionized biology, but they have also taught us something valuable about the importance of the external eye, the transformative ways in which boundary crossing plays a role in creative thinking and innovation.

It is our hope that students of biology, as well as its history and philosophy, will be able to look differently at their fields after having read *Outsider Scientists*. There are great advantages, we feel, to teaching the growth of the life sciences in the modern era by looking through the prism of those who have come to enrich it from the outside. So, too, do we hope that people interested in how novelty and inventiveness come about find enlightenment in the essays presented here: taken together and considered as a whole, they present a strong argument for ingenuity borne of emigration, and a call to arms for a more fluid boundary between the different divisions of human knowledge and research, as well as for stronger ties of cooperation between them. Outsiders matter because they are often the fuel for all that is new. Gazing into an unknown future, perhaps we can learn more about this by carefully considering their storied past.

ACKNOWLEDGMENTS

We thank our brave contributors—Sandy, Michael, Fred, Jay, Erika, Luis, Tim, Ehud, Maya, Rob, Tecumseh, Sahotra, Hallam, Jonathan, Greg, Bill, and Michel, who with great dedication (and patience with our editorial harassment) have given flesh and body to a skeletal intuition about the ubiquity and importance of outsiders in biology. Your fine scholarship helps to turn the notion of boundary crossing from a disparate description of idiosyncratic events into an important category in the study of biology and science more generally. Special thanks, too, to Richard Lewontin, who for the second time around, concludes our edited volume with his thoughtful remarks.

Karen Merikangas Darling, our editor, has been a wonderful supporter and guide throughout the entire process. We thank her for her wise council, her encouragement, and above all for her genuine belief in the project. We benefited from a fine set of comments from our manuscript reviewers that helped us clarify and communicate our central themes. Thank you, too, Kristie Reilly, for your eagle eye and talent in bringing the manuscript to final polish.

Finally, to all those outsiders, young and old, timidly contemplating a boundary crossing, a word of professional advice: Go for it!

OREN HARMAN & MICHAEL R. DIETRICH

INTRODUCTION
OUTSIDERS AS INNOVATORS
IN THE LIFE SCIENCES

INTRODUCTION

Both intellectually and institutionally, the life sciences occupy a fascinating middle ground between the physical and exact sciences, on the one hand, and the social sciences and humanities on the other. If biology were an animal, it would be a duck-billed platypus—something that appears chimeric, yet is fully rooted in its own historical lineage of accumulating adaptations, tinkering, and change.

Like that strange aquatic mammal, "half bird, half beast,"[1] its features point to its origins and ecology. Biology as a science has come into being as a patchwork, assuming its present visage as a consequence of myriad interactions between different traditions of knowledge, method, and philosophy while maintaining an overarching quest for understanding of the natural world. Indeed, historically, many researchers have come from outside biology to ask fundamentally *biological* questions. These outsiders have played a crucial and defining role in the growth of modern biology; they have brought new skills and ideas to the "inside" and have thus added something new to biology. As a consequence, biology can feel sometimes as if it is a strange hybrid—with a bill, a flat tail, fur, and webbed feet. After all, biologists include among their number men and women who sit before computers crunching numbers, as well as cavers who crawl through subterranean spaces in search of lizards; and biology counts among its tools patch-clamps and test tubes and microchip arrays and bird-snares. Its worldviews range from reductionism to dualism, idealism to emergence. It can often seem confusing: is biology really just one thing? Like that "highly interesting novelty," as the beguiling Australian bird-and-reptile-like mammal was once called, it is indeed one thing. And like the platypus, biology has been formed by adapting forces coming from outside, from the environment of other disciplines and practices. The platypus may seem like a paradox, because it appears to be chimeric. Biology likewise appears chimeric, but has attained an internal integrity and innovative potential from those external forces.

The molecular revolution of the late twentieth century, for example, was to a large degree stimulated by the influx of physicists into biology, applying as they did both a different style and approach to the problem of heredity. Ecology and population biology, too, have been determinatively shaped by the arrival of mathematicians to these fields, using tools from their own discipline to resolve biological problems with unfamiliar instruments. Linguists have applied their training and tools to investigate problems in cognition, social scientists to attack the puzzles of animal behavior, philosophers to probe conceptual foundations, writers to sharpen their pens on evolution, computer scientists and engineers to try to crack the mystery of life. As such, these "outsiders" have supplied important sources of innovation in biology and, each in his or her way, contributed to its patchwork design. What is of interest to us here is the manner in which scientists recruited from different disciplines have helped, and continue to help, produce novel approaches, concepts, theories, experiments, practices, insights, and—ultimately—novel scientific understanding.

This book seeks to provide historical descriptions and analyses for the ways in which researchers from the "outside" have been sources of significant innovation. The collection of cases assembled here critically examines these sources of innovation by considering how different researchers were able to integrate ideas, techniques, and methods across divergent scientific communities. As will become apparent, these innovations were NOT idiosyncratic accidents, but the result of the careful work of making intellectual connections, translating idioms, creating languages, and fostering new forms of collaboration that bridged training and experience in the biological sciences with a rich array of fields, disciplines, and perspectives. In the end, outsider interventions have given biology its peculiar form.

THE PROBLEM OF INNOVATION

As early as 1667, Thomas Sprat, historian of the Royal Society of London, noted a connection between being an outsider to a trade and inventiveness. A glance from an angle, Sprat argued, might well reveal a new aspect of nature. More recently, sociologists Joseph Ben-David and Robert Merton have shown the importance of disciplinary immigrants for the development of a particular science.[2] Merton, though, problematized a strict dichotomy or divide between those considered insiders and outsiders. As a result, later thinking about disciplinary boundaries reflected a more dynamic perspective regarding disciplinary identity. Lynn Nyhart's discussion of the birth of the discipline of physiology from the older anatomy in nineteenth-century German universities, on the other hand, considered the role played by different institutions in erecting boundaries between old and incipient fields.[3]

In a broader, more theoretical manner, Peter Galison has applied the anthropological notion of the "trading zone" to scientific practice, analyzing a number of examples from physics in which scientists of different subfields have met—creating common pidgin idioms, and then creoles—in order to jointly attack conundrums.[4]

But if the "outsider" and the "outsider as innovator" have been recognized as important categories in the history of science more generally, the treatment of the "outsider" in the history of biology has been focused more narrowly on specific instances. A number of histories of the molecular revolution, for example, highlight the role of physicists, such as Max Delbrück and Francis Crick, who became biologists and played a foundational role in the creation of molecular biology.[5] Evelyn Fox Keller, in her book *Making Sense of Life*, features some of the cyberneticists and artificial lifers who used metaphors from the world of computing to help probe deep problems in development and embryology.[6] A comparative treatment of the range of "outsiders" that have shaped the course of biology more broadly is sorely missing. Bringing together a diverse set of examples allows us to explore the various conditions that fostered both their movement into biology and their innovative contributions to biological understanding.

WHAT MAKES AN "OUTSIDER"?

In *Outsider Scientists* we conceive of outsiders in terms of academic disciplines. We are interested in scholars trained or practicing in a nonbiological discipline who moved into some branch of biology. These disciplinary newcomers or outsiders bring with them perspectives, skill, and training that are often not shared by insiders—those trained within biology. The fundamental question we are considering asks how moving from a field outside of biology to a field within biology has served as a significant source of scientific innovation.[7] We have asked our authors to consider what features of their subjects' original scientific training and research experience in a nonbiological context allowed them to make innovative contributions to the field of biology that they eventually joined. But a word of caution: we do not wish to hang too much on the category of *discipline*, because we do not think that the question of training and innovation depends strictly on moving from one discipline to another, nor do we believe that disciplines, as such, are hard and fixed categories. Rather we are interested in considering movement between communities of scientists with divergent practices, paradigms, or habitus and the role that this intellectual movement plays in innovation within biology.

Movement between communities occurs not just between disciplines but also within them. Increasingly, recognized subdisciplines have developed almost insurmountable barriers, as specializations divide the landscape and render movement more difficult within. This is true for biology as much if not more than for physics, chemistry, and computing. For that reason, we also consider a number of examples in which researchers from one subdiscipline within biology crossed into a second subdiscipline to make contributions there. An exemplary case would be Ilya Metchnikoff moving from developmental biology to immunology, or Francois Jacob, moving from work on bacteria to mice. Such cases are similar to those of nonbiologists crossing into biology because here too, researchers bring with them completely new skills, perspectives and training. These particular outsiders we term "insider-outsiders."

Our definition of the outsider, then, is restricted. Excluded from it are outsiders on account of religion, ethnicity, gender, and character—though for all of these, to be sure, fascinating examples abound. The sole and guiding principle for *Outsider Scientists* is that the individual in question should have moved from one intellectual community, with its distinctive practices and established conceptions, into an area of biology new to that individual. Because these migrating scholars often bring with them tools, techniques, theories, and practices, we could have chosen to follow these instrumentalities into new areas, but we chose instead to follow individuals into new communities and institutions. The biographical focus of each of the following chapters is not intended to portray scientists as lone knowers, but as members of new disciplinary communities—members who significantly alter the practices of those communities.

Making judgments as to who is an outsider and who isn't, however, necessarily remains a complicated affair. To begin with, one needs to assume that there is an "inside" outsiders must enter, and this was not always true in biology. Lamarck may have coined the term in 1802, but biology as a coherent field and well-defined community, with institutions and academic programs, particular subdisciplines, research agenda, and journals, took time to establish, and of course remains in flux. When does one mark the inception of a field: When its name is coined? When the first society of practitioners is founded? When the subject is included as a field of study in the universities? However one approaches this problem, it is clear that the trajectory and growth of biology was unique in different historical contexts, such as in the German-, French-, and English-speaking worlds.[8]

Wary of the slipperiness—and to a degree the arbitrariness—of defining a hard and fast historical date for the birth of biology as a discipline, we have chosen to include in this volume a first section that will treat a number of early examples of interesting nineteenth-century practitioners whose engagement with problems of a living nature illustrates the very difficulty involved in speaking about "outsiders" with any confidence before the late nineteenth century. Gregor Mendel was a clergyman who had little or no formal training in anything called "biology." The worlds that he uniquely united—experimental physics gleaned at the University of Vienna, the middle-European business of practical plant and animal breeding, and the local scientific society at Brünn—gave birth to a research program that would play a crucial role in the establishment of genetics and the establishment of biology as an identifiable field years later. Mendel helps us understand, both intellectually and in terms of earlier local traditions, what the creation of an "inside" for modern biology entails. Similarly, the role of Pasteur the businessman and chemist, moving into what was rapidly becoming an institutional *biologie* in France, helps put a finger on the process of the birth of the disciplinary divides that defined a distinct *biology*, as does Felix d'Herelle's uniquely self-taught (and fascinatingly international) trajectory in microbiology. Finally, to round off the early examples, the contributions of Samuel Butler, the Victorian novelist, serve to trigger a discussion of the ways in which literary engagement with the idea of evolution challenged a number of crucial divides: the science-philosophy divide via the teleology and causality debate, and the public-private divide via the debate concerning the proper forum for negotiating scientific disputes. These four individuals play an important role in allowing us more carefully to consider the criteria for "inside" and "outside" in biology as they developed historically.

Thus, the first part of the book, "Outsiders before the Inside," treats the dichotomies of teleology–efficient causality, amateur-professional, local-international, industry-academia, and public-private, each of which played a role in the birth of modern biology. Other examples could have served us here, but we have chosen these early individuals in order to create a meaningful set of contrasting cases to later figures who were involved in the creation of innovations following explicit acts of boundary crossing into areas of modern biology.

Recognizing "outsiders" in biology becomes more straightforward as we consider the development of the biological sciences in the twentieth century, and it is this century that is the main focus of *Outsider Scientists*. Here the challenge of understanding outsiders and their innovations in biology pertains less to the ambiguity of describing an "inside" or an "outside," and more to a problem of selecting a range of both diverse and representative outsiders. There have been many practitioners in biology who can be thought of as "outsiders," and we have had to think long and hard about whom to include. We would have liked to include more than eventually made it in—there are, in other words, outsiders who "got away." One thinks in particular of Max Delbrück, Herbert Simon, Isaac Asimov, Gerald Edelman, Seymour Benzer and, reaching further back, Goethe as an early "outsider before the inside," who as an artist and morphologist attempted to reconcile his divergent pursuits. It is our hope that this particular collection will spur others to examine such figures in the mode we suggest. The figures that have made it into our book have been chosen to illustrate the myriad ways in which "outsider science" comes about and functions. For each case we have chosen expert contributors, each with a broad and deep knowledge of the relevant history and context.

Before we continue any further, we'd like to address a quick word to the skeptic. The category of the "outsider" in science, the objection might run, is too diffuse to be of any value. After all, there are many ways to be an "outsider," and the dynamic of insider-outsider interactions will necessarily take many forms. Our reply to the skeptic is meant to disarm: we agree. Our goal is not to define exhaustively what it has meant to be an "outsider" to biology. Given the shifting nature of biology as a discipline, that would be a Herculean task. But we do not shy away from this diversity. Very much to the contrary, we are consciously setting out to present it in as full a fashion as possible rather than unthinkingly "lumping" all the disparate histories into a conceptual straitjacket. Clearly, the contingencies matter, as do the myriad facets of the outsider incursions—that is the point of the historical narratives that follow. Our goal is to offer a range of historical cases that allow us to comparatively understand the elements of discipline crossing that contribute to processes of scientific innovation.

Our analysis does assume that "discipline" is a legitimate historical category. While most scholars would agree that biology is and has been a discipline, they can differ on what exactly constitutes a discipline.[9] Specialization and institutionalization through markers such as professional societies, journals, and designated funding streams have typically been recognized as elements of discipline formation. More recently, epistemic criteria of problem definition and practice have been added. Minding and maintaining the boundaries of scientific disciplines has also been the object of scholarly research, especially as biology itself has emerged as a dynamic enterprise. While contemporary biology, especially since the rise of the molecular revolution, is widely recognized as a mosaic or hybrid of many diverse subfields, in the early twentieth century there was a distinct movement to seek a unified biology. Historians, such as Betty Smocovitis, have written eloquently about the desire to articulate a common core to the discipline of biology and about the challenges to this unification. While consensus on a unifying theory, even within the so-called evolutionary synthesis, proved elusive, scientific societies were formed, journals established, and the social and cultural definition of biology was perpetuated, even if it was always in motion. The fluidity of the disciplining of biology does make it a moving target for historians. However, we do not need an entity etched in stone. We need an entity that is sufficiently different from neighboring areas of inquiry that we can say that chemistry as a discipline, for instance, differs from biology as a discipline in terms of imparting to its members distinctive concepts, theories, methods, practices, and approaches. The various sub-branches of biology will differ among themselves, but the general pattern of the whole will yet distinguish it from other major areas of inquiry, such as chemistry.

Discipline crossing draws our critical attention to forms of epistemic difference that may be rooted in the style of thought an outsider brings with her, a particular set of intellectual tools, an experimental apparatus or design, or that may involve more broadly (and deeply) a general vision or specific motivation.[10] Discipline crossing may relate to the way that the reception of outsiders is determined by sociological as opposed to intellectual reasons, and how this varies depending on the particular "outside" one is coming from. The salient objective is that the cast of "outsiders" illustrate, as a group, a wide spectrum of the different facets of the phenomena.

THE OUTSIDERS WHO MADE IT IN

To help the reader, and in order to provide an organizing framework, we have divided the book into six parts. They are 1) Outsiders before the Inside, 2) Outsiders from the Physical Sciences, 3) Outsiders from Mathematics, 4) Outsiders from the Human Sciences, 5) Insider-Outsiders, and 6) Outsiders from Informatics.

As we have mentioned, the category "Outsiders before the Inside" includes accounts of Mendel, Pasteur, d'Herelle, and Butler. The histories of these figures will introduce a perspective on innovation through the integration of diverse interests, approaches, and practices before there was a clearly demarcated discipline identified as biology. Importantly, they provide a contrast to the stories of the later periods in which disciplinary markers are more easily discerned, since those markers were more actively enforced after the turn of the twentieth century. Indeed, many of the dichotomies these early examples highlight—such as teleology vs. efficient causality, amateur vs. professional, industry vs. academia—provided the definitional distinctions that later biologists used to create and enforce disciplinary boundaries. All four cases, authored by Sander Gliboff, Jonathan Simon, William Summers, and Michael Ruse, respectively, speak to the power of movement across the intellectual terrain as a means to foster new insights.

In part 2, essays by Sahotra Sarkar, Gregory Morgan, and Hallam Stevens introduce us to a sample of the many physicists who crossed into biology in the twentieth century. Erwin Schrödinger and Linus Pauling may be familiar subjects, but Sarkar and Morgan provide careful new consideration of how these two Nobel laureates translated their knowledge of physics into biological idiom, and in so doing helped create the foundations of molecular biology. Stevens describes the work of Walter Goad, a less well-known figure, who used his understanding of computational physics acquired in atomic bomb work to reshape the genetic databases and algorithms that now form the basis of bioinformatics.

In our third part, Michael Dietrich and Robert Skipper Jr., Maya Shmailov, and Jay Odenbough each consider scientists at the interface of mathematics and biology. R. A. Fisher, Nicolas Rashevsky, and Robert MacArthur all brought mathematical and statistical insights to bear on biological phenomena in ways that transformed biological practice from its earlier naturalist tradition. The statistical tools developed by Fisher alone have become completely commonplace in all branches of biology as a result.

Part 4 considers outsiders from the human sciences, with essays by W. Tecumseh Fitch on the linguist Noam Chomsky, T. J. Horder on the philosopher David Hull, and Erika Lorraine Milam on the writer Elaine Morgan. These cases do not represent equally influential incursions into biology: Chomsky's attempt to pry open the brain by exploring the rules of grammar helped bring about a revolution in the cognitive sciences, while Hull's and Morgan's grappling with particular theories of systematics and evolution, respectively, produced more of a glancing blow toward their discipline of evolutionary biology. Still, taken together, the three examples illustrate salient features of biology's intersection with the humanities.

In part 5 we meet two "insider-outsiders": Ilya Metchnikoff and François Jacob. In their essays on these internal migrants, Fred Tauber and Michel Morange show how movement across subfield boundaries can be both difficult and transformative. Drawing from his background in embryology and development, Metchnikoff challenged the prevailing theories of immunity of his day, while Jacob took principles he had learned working on bacteria and phage in molecular biology and applied them to the mouse in the study of disease.

The final section of the book deals with the influence of informaticians on the life sciences. Chapters by Ehud Lamm on John von Neumann and Norbert Weiner, Oren Harman on George Price, and Luis Campos on Drew Endy reveal how biological systems have been reimagined in sometimes radical ways by outsiders redesigning their new disciplinary homes using the theoretical frameworks and idioms of computer science and informatics.

WHAT HAVE WE LEARNED FROM OUR OUTSIDERS?

The essays that follow shed light on three elements of the relationship that is our focus: the outsider, the process of coming in to biology, and the process of innovation. Concentrating on these three elements allows us to explore the roles of features of personalities, institutions, and prior training that have shaped the wide range of scientific novelties described in the chapters. We start with the outsider.

On the Outsider

The outsiders described here are not your typical scientists. When it comes to "outsiderness" as an aspect of character, many of these individuals reveal traits that rendered their crossing of boundaries almost natural. They are *bona fide* transgressors. They see little point in respecting conventional boundaries, either because they view them as inherently ridiculous or because they don't see them at all. A quintessential example is George Price. He was nothing if not an intellectual scavenger. Trained in nuclear chemistry, he switched from the Manhattan Project to work at Bell Labs on transistors and informatics, then to cancer research at a Minnesota hospital, then to magazine writing on current affairs, then to computer problems at IBM, then finally to mathematical evolution, all the while sending unsolicited letters to Nobel laureates that claimed breakthroughs in fields as disparate as neurophysiology and economics. Price saw problems, not disciplines, and, fueled by a cocksure attitude and dismissiveness toward convention—for better and for worse—acted accordingly. Linus Pauling shared with Price a similar disposition. Fiercely independent of mind, Pauling used the occasion of his 1954 chemistry Nobel lecture to admonish young scientists never to take anything on authority and always to think for themselves, respecting no boundaries; eight years later, he was in Stockholm again receiving a second Nobel Prize, this time for Peace. Erwin Schrödinger, too, possessed an aspect of character that made him a natural outsider: the confidence of a man who thought—together with Einstein, it must be admitted, but against the better judgment of the rest of the physics community—that the apparent paradoxes of quantum mechanics would eventually disappear. It was this confidence, no doubt, that helped to stoke his pretension to explain heredity by means of quantum mechanics when he attempted an answer to the question "What is life?" in a series of lectures delivered in 1943.

"Cocksure," "arrogant," "confident"—these are appellations we find applied again and again to our outsiders, and not by insiders alone. Nicolas Rashevky stormed into the life sciences from mathematics seeking to shake its very core ("You name it, he had a theory on it," one commentator quipped); "hot-tempered" Ilya Metchnikoff humiliated Nobel laureates in a field he had never studied but wished to transform; Drew Endy, hyperconfident and extolling a culture of "cool," sought to revolutionize biology by using engineering principles to synthesize life itself. The diminutive Elaine Morgan, Erika Milam tells us, "had sass," marshaling wit and humor to take her male-chauvinist targets to task. Earlier in the century, Felix d'Herelle, marshaled the autodidact's bold self-possession to revolutionize microbiology, and R. A. Fisher, like a terrier hound (which he incidentally resembled), showed incorrigible persistence against opponents in applying statistics to evolution and heredity. Of course, outsiders' personalities were nevertheless by no means static.

Often outsiders possessed a broad "vision" which they actively pursued: Rashevsky and Fisher and MacArthur sought to mathematize biology and Pauling sought to bring physical chemistry to biology, for example. Fisher believed that this kind of intervention from the outside was most difficult for insiders to accept: "A new subject for investigation," Dietrich and Skipper quote him as saying, "will find itself opposed by indifference, by inertia, and usually by ridicule. A new point of view, however, affecting thought on a wide range of topics may expect a much fiercer antagonism." Sometimes, as with Price, there is no more than a kind of problem-specific intellectual opportunism. Sometimes, as with Endy, perhaps both are present.

But incursions from the outside are not always the result of a particular aspect of personality; sometimes they simply describe the act of crossing an unseen, or alternatively a closely patrolled, divide to solve a particular problem. Our "Outsiders before the Inside" are examples of the former. Each, in his own way, moved from one métier to another without necessarily exercising the muscles of overbearing confidence, or expressing hatred of authority, or indulging in contempt for convention (think of the gentle curate, Mendel)—though Butler probably imagined himself a Renaissance man. Louis Pasteur, to the medical establishment, might have been insufferable, but microbiology, at any rate, had yet to define its boundaries. Walter Goad, by contrast, is a modern example of a man who didn't possess the fiery "outsider" character, but nevertheless recognized a void, entering the field, with the help of a long-standing institutional interest at Los Alamos in biology and medicine, to apply numerical data management tools to genetic databasing. It was his tool—the computer—rather than his temperament that led Goad into biology, allowing him to import ready-made ways of thinking, doing, and organizing with little resistance.

Regardless of personality, the outsider's training was always of the utmost importance. Perhaps we should not be surprised to find that no small fraction of our outsiders actually had a prior connection to biology before trying to enter the field. Drew Endy may have received a D in high school biology for failing to recite the Latin names of 200 insects, but Robert MacArthur, whose dad was a geneticist, actually got his PhD in ecology. Norbert Weiner, too, studied biology before becoming a mathematician, showing particular interest in physiology and teleology. Fisher, from the outset, had been hooked by eugenics and biometry, alongside mathematics, and Schrödinger, though most people don't know this, was an international authority on the physiology and biophysics of color vision. Still, it is the prior training in the nonbiological discipline that we are most interested in, since the training usually lays the foundation for the incursion to begin with. We'll address this particular issue when we turn to innovation, but first let's take a look at the process of the outsider moving in.

On Moving In

The process of crossing a divide entails a number of elements that we find recurring in one form or another in many of the outsiders considered here. These are the role of patrons and forward-looking funding bodies; the role of institutions; collaboration both with insiders and fellow outsiders; courting; and—closer to the content of innovation itself—processes of translation, simplification (especially with theoreticians), and sometimes popularization (especially with outsiders from the humanities). Each of these features represents an aspect of institutions and the social context that supported the outsider seeking to bring an original result, method, or perspective to the life sciences.

To begin with, a patron, it would seem, is a wonderful thing to have for an outsider. A number of our outsiders manifestly benefited from having enthusiastic supporters, though others neither sought nor were offered assistance. Perhaps the starkest example of patronage here is that of Major Leonard Darwin, Charles Darwin's fourth son and an avid eugenicist, who, from quite early on, decided that he was going to do everything he could to help advance R. A. Fisher's career. This meant helping him publish his famous 1918 paper on the correlation of relatives on the supposition of Mendelian inheritance. The paper played a historical role in wedding Mendelian genetics to Darwinian selection. It was, however, initially rejected by Fisher's great nemesis, Karl Pearson, for publication in his journal *Biometrika*. Thanks to Leonard Darwin's intervention, the paper was published in the *Transactions of the Royal Society of Edinburgh*. Moreover, Darwin arranged to have Fisher supported by monthly stipends, enabling him to develop his synthesizing insights, which culminated in what became his magnum opus, *The Genetical Theory of Natural Selection*, in 1930. George Price, too, enjoyed patronage from John Maynard Smith and Bill Hamilton, collaborators both; but their aid went beyond the usual bounds. It was Hamilton, after all, who by way of a ruse cajoled the editor of *Nature* into publishing Price's path-breaking covariance paper, which no doubt otherwise would not have seen the light of day. Similarly, without the generosity and encouragement of John Maynard Smith, their historic joint paper, which applied game theory to animal conflict, would most probably have never been written.

Felix d'Herelle provides perhaps the starkest counterexample, a man who made it decidedly on his own. Having left school at seventeen, and working from the periphery, d'Herelle was neither a member nor even known to either the Koch or Pasteur school of microbiology, at least for quite some time. Relying on his own reason and confidence, and all the while moving from place to place (Canada to Guatemala to Mexico to Argentina to Columbia to Algeria to Tunisia to Cyprus to France, and more), he never enjoyed any form of patronage, except for a short-term commission tendered by the Argentine Minister of Agriculture to exterminate locusts in his country. D'Herelle was a lone maverick.

Individual patrons may not be necessary, but some form of support or acceptance is often crucial, such as forward-looking funding bodies. Warren Weaver of the Rockefeller Foundation is a celebrated example of a supportive administrator, one who had the foresight to offer critical aid to both Pauling and Rashevsky at Cal Tech and at the University of Chicago, respectively.[11] In both cases, Weaver saw what many who were unequivocally insiders didn't notice, that outside tools—structural chemistry and mathematics, in these cases—could go a long way to help solve important "insider" problems. Drew Endy, on the other hand, at least when he began, was rather impeded by the main funding taps: one agency threw his grant request out the window citing irrelevance, and worse, complete lack of believability.

Indeed, outsiders don't always find institutional homes that are willing to take a chance on projects that seem to many unimportant or even sinister. Pauling, however good a chemist he was known to be, ended up creating his own institute, the Institute of Orthomolecular Medicine, later renamed the Linus Pauling Institute of Science and Medicine. He used the institute to pursue his vitamin C research, which does indicate that buying a home for "free thinking" doesn't always lead to the best results. Rashevsky, too, found the going rather rough at the University of Chicago, Weaver's support notwithstanding. Moving (or rather being moved) from the Department of Psychology to the Department of Physiology and back again, he found his work continually falling between the cracks. He was too mathematical for the biologists and too biological for the physicists and mathematicians. Finally, he solved the problem by creating his own *Journal of Mathematical Biophysics*, which had more success than the Institute of Orthomolecular Medicine.

Outsider incursions always occur within some institutional context. Some outsiders were fortunate to find the "right" institution, one that provided support for newcomers, encouraged collaboration, and sought interdisciplinary connections to address biological problems. Fisher, for example, was free at the Rothamsted Experimental Station to pursue both practical and theoretical work integrating statistics, biology, and eugenics that might very well have been impossible elsewhere (including Cambridge). Price, too, years later, enjoyed the backing of a kind institution: when he walked off the street into the University College London biostatistics department with his covariance equation written on a piece of paper—the ultimate outsider act if ever there was one—he was afforded an honorary research position within the hour and summarily helped to secure a grant for further research; University College London, mind you, was a world-leading center of genetics at the time. Goad's career, too, makes the point of institutional importance. Indeed, Los Alamos's wartime successes rendered it continually crucial to national security, which meant greater latitude for senior scientists in following curiosity-driven research. As Hallam Stevens shows, the "exigencies of wartime work" also promoted interdisciplinary collaboration. This meant for Goad that he could play a leading role in convincing the National Institutes of Health to fund the GenBank project. Institutional backing, then, seems to be a relative quantity when it comes to outsiders. Some, like Fisher, Price, and Goad, were lucky to be spurred on and provided the means by their institutions; others, like Rashevsky and Pauling, fought within until they found external solutions; still others, like Butler, d'Herelle, and Morgan, didn't even try.

But if institutional support has a variable influence, collaborations, almost across the board, seem crucial. Pauling (with Alfred Mirsky, Karl Landsteiner, and then Emile Zuckerkandl), Price (with John Maynard Smith and Bill Hamilton), and Weiner (with Arturo Rosenblueth) all prove how important work with bona fide insiders possessing complementary (rather than identical) skills can be. None of these outsiders would have been able to get very far without their collaborators. Their work on protein stabilization by hydrogen bonds, antibody specificity, evolutionary molecular clocks, the evolutionary stable strategy, multilevel selection, and negative feedback, respectively, would have been the worse for it. Pauling explained the cooperative dynamic nicely: "Landsteiner would ask, 'What do these experimental observations force us to believe about the world?' and I would ask, 'What is the most simple, general, and intellectually satisfying picture of the world that encompasses these observations and is not incompatible with them?'" The collaborators' methodological departures and differences in perspective, as well as their help in the more mundane technicalities—such as learning correct notations and suitable experimental designs—proved crucial to these outsiders for solving important problems. In some cases, as in Price's, affixing one's name beside that of a well-known insider also made a great difference.

Outsiders turn to fellow outsiders for collaborations as much as they do to insiders. Pauling, for instance, worked with the biochemist Robert Corey on determining structures of amino acids and on models of protein chains; Corey was as much an outsider in his way to biochemistry as Pauling was to molecular biology. Weiner joined hands with an electrical engineer collaborator, Julian Bigelow, to work together with the insider Rosenblueth. Goad worked with the physical chemist John Camm; together they examined transport processes in biological systems on IBM 704 and 7094 machines. Francois Jacob sent his own bacteria and phage men, Hubert Condamine and Charles Babinet, to study mammalian embryology and to return to his lab; Jacob himself declined to learn from the insiders directly. Schrödinger may have wanted to shake up biologists, but it was the physicists—Seymour Benzer, Crick, George Gamow, Salvador Luria—who heeded his call more than anyone else. But Endy is perhaps the ultimate example of an outsider who knew he would need to turn to like-minded outsiders to get anywhere at all: he extended his hand to Tom Knight and his fellow electrical engineers at MIT rather than engage true insiders in biology. When he found himself at a disciplinary crossroads—should he study more molecular biology from the inside or think as an outsider engineer?—Endy determined to "screw it," since the complications and details of biology seemed "of little interest."

Courting the inside is sometimes a requisite for outsiders, even if looking at Endy's path doesn't immediately divulge this. The biological world into which both Weiner and von Neumann were attempting to enter, for example, was anything but hospitable. Ehud Lamm quotes E. B. Wilson, at the 1934 Cold Spring Harbor Symposia on Quantitative Biology, offering a number of axioms to the initiated, the first of which was "science need not be mathematical," and the second, "simply because a subject is mathematical it need not therefore be scientific." The inside, clearly, was less than inviting. Incidentally, Weiner and von Neumann ended up choosing different approaches to engaging biologists—the former seeking out collaboration, the latter going it alone.

The writers Samuel Butler and Elaine Morgan, too, understood full well that they needed to court their readership—whether by gripping drama, scathing wit, gentle humor, or all of these—and directed their talents inward as much as out. Still, popularization was an issue: the way to succeed, both authors knew, entailed capturing the hearts and minds of the public. As Michael Ruse explains, when it came to evolution, before 1859 the subject was considered a pseudoscience, after 1930 it was professionalized, and in between its status was ambiguous (though it may be objected that early figures as central as George Cuvier and Karl Ernst von Baer took evolution seriously enough to go to great lengths to dispute it, and that evolution was taught in many German universities by 1860, and even before).[12] Focusing on the English-speaking world, Ruse argues that evolution was a popular science during this period, and the popular book or novel as legitimate and influential a venue as the scientific paper. Insiders like T. H. Huxley and George Romanes grasped this, which is why they themselves attempted to speak to the public alongside their more professional writings, understanding full well, if reluctantly, that this meant the door had been pushed wide open. Butler capitalized, building a successful career as a popular writer on evolution. After him, Morgan did too, though in her day she was required to mount a tighter argument, based on a careful reading of the scientific evidence. And while Noam Chomsky may have awaited his Steven Pinker, he has nevertheless used the television to great effect in making himself known as a public intellectual, as much for his linguistic theories as for his politics.

On Innovation

Outsiders bring with them new language as well as designs. In order to express their innovative ideas, then, they must engage in a process of translation.[13] Schrödinger, for example, transported terminology such as "isomers" from organic chemistry to describe different stable states of genes, and "tunneling" from physics to speak of the process of translation between such states to help explain, among other things, mutations. "Negative entropy" was another concept he used to translate a concept from thermodynamics into one in molecular biology, a translation that may well have given birth to more confusion than clarity.

Price, too, as Harman shows, went about the business of translating concepts, in his case from Claude Shannon's channel capacity informatics to selection dynamics more broadly. It was the precision and beauty of the formalization of the theory of communication that Price sought to translate into biology. Von Neumann and Weiner were very much engaged in a kind of translation enterprise as well, using "self-reproducing automata" and "negative feedback" as central concepts otherwise unheard of within biology. And Endy fashioned repressible promoters as transistors, the biological and genetic equivalents to toggle switches and oscillators.

Translating, or rather getting insiders to understand translations, isn't always easy, as the correspondence between Joshua Lederberg and Jon von Neumann attests. Lederberg wanted to know how intracellular components correspond to the elements of the cellular automata model, but von Neumann's conceptual model had no relation to biological realities. Lionel Penrose, too, found it difficult to find answers to specific biological questions in von Neumann's model, in particular those having to do with the physical and chemical aspects of self-reproduction. For both biologists, the engagement with the mathematician proved frustrating.

Indeed, biology is a messy science, full of details and exceptions. It is for this reason that many of the outsiders coming in—in particular, theoreticians and those with mathematical and physical skills—sought to simplify matters in order, as it were, to see the forest independently of the trees. Rashevsky, to be sure, wanted to transform biology into a deductive science rather than an empirically based one, where theory based on oversimplified—often grossly oversimplified—scenarios (of cell division, growth, nerve conduction, brain function, etc.) could be used to predict trends rather than exact values, and help direct avenues of research. In looking at island biogeography, species abundance distribution, and optimal foraging strategies, MacArthur preferred "patterns" to trends. Indeed, science itself was to him essentially a matter of detecting patterns, an approach that allowed him to transform ecology from a descriptive science to a structured and predictive one. But integrating genetics, ecology, biogeography, and ethology was no small order, and by necessity it called for simplifying. Indeed, the members of the Marlboro Circle to which he belonged—Egbert Leigh, E. O. Wilson, Richard Levins, and Richard Lewontin—all agreed that only two of any three goals— generality, realism, and precision—could ever be maximized, and MacArthur preferred the first two. They called it the "simple theory" approach. "Our truth is the intersection of independent lies," was how Levins put it, with commendable honesty.

Price, too, sought to simplify by generalizing. His tautological covariance equation had exactly zero biological assumptions in it, which is precisely why it was difficult for many biologists to appreciate its import (many still don't). Endy's synthetic biology, almost by definition, was a science of standardization, hence simplification. And von Neumann, looking at self-reproduction, was interested in the internal functional organization of the system, an exercise accomplished by axiomatizing the behavior of the system's components. This again was an idealization, and hence a simplification, rather than an attempt to describe real biological phenomena.

Simplifications offered by outsiders invariably ruffle the feathers of insiders to the point of fury. Theodosius Dobzhansky was being gentle when he intoned, as Lamm quotes him: "Experience has shown that, at least in biology, generalization and integration can best be made by scientists who are also fact-gatherers, rather than by specialists in biological speculation." The future Nobel laureate neurophysiologist John Eccles was harsher, letting the community know that he thought his own field would be strictly impeded, rather than advanced, by superficial analogies to automata. And Endy's detractors so resisted the idealization behind his "Standard Parts List for Biological Circuitry" that Endy invited disgruntled biologists to send complaints to the "Office of Biological Disenchantment, MIT 68-580" (in other words, his office in the electrical engineering department). Still, despite its detractors, simplification has been a real motor for innovation.

At the base of the ability to simplify, generalize, and translate stands the particular training and intellectual territory from whence the outsider arrives. Take for example Pauling's background in crystallography, which solidified his unshakeable belief in the idea that properties of all substances depend on their structure. It was this "methodological structural reductionism," as Gregory Morgan calls it, that pushed him to seek the explanation of cellular behavior at a "deeper" level, leading to the solution of the alpha helix, among other problems. Take Weiner's work on target-tracking machines for the American Air Force during World War II: it was here that he first encountered the oscillatory movement for which he found an analogue in the intention tremors of human cerebral patients, and which, via the wedding of feedback to intentionality, became the conceptual centerpiece of the science of cybernetics. Or take Price's work at Bell Labs with Shockley and Bardeen, which familiarized him with Claude Shannon's theory of communication, a theory he then sought to apply to the process of biological selection. Or Goad's earlier work in physics on the hydrogen bomb, which, Hallam Stevens claims, is responsible for the introduction of computers into mainstream biology. Goad recognized that the kind of numerical and statistical methods he had used to solve data-intensive problems in fission and fusion could be applied, using digital electronic computers, to genetics. GenBank, the database he created, was premised on the notion that biological problems could be framed as pattern-matching and data-management problems. Ultimately, the system of sharing, communication, organizing, and distributing DNA sequence data that it produced allowed for the birth of the Human Genome Project. In all these cases, there is a direct connection, even a direct analogy, between prior work in one field and the later work in biology.

Endy, too, in calling for an "open source biology" based on "tools of mass construction," deployed an analogy between Boolean electrical gates that can perform simple operation such as AND, OR, NOT, NAND, NOR and genetic components that could be construed as "BioBricks." In so doing, he swept away with one blow the model organism approach, since, to his mind at least, standardized biological parts rendered such lab-based practices unnecessary. As Luis Campos argues, this was the ultimate outsider biology— outside of the confines of even the organism itself or the species—but it was based on a functional analogy and a confident assurance of its validity. Pasteur, too, saw an analogy between fermentation and infectious diseases, as Jonathan Simon elucidates. This vision drew a direct line for him between his chemical work, on the one hand, and his medical-biological work on the other. The Frenchman's eye for the utility of organisms oriented microbiology first to industrial production and then to the treatment of disease, two directions that would be enhanced by d'Herelle's international work in the twentieth century.

Analogy is not the only route for the introduction of a divergent disciplinary understanding into biology. Sometimes a general approach, or a particular skill, will suffice. Take, for instance, Metchnikoff. It was the Russian's embryological preoccupations and Darwinian framework that informed his challenge, both methodological and theoretical, to the prevailing immunological theories of his time. These proved of little interest to microbiologists and immunochemists, who had no background in development to speak of and who were focused on defining the mediators of immune response rather than looking at the etiology of the entire system. But Metchnikoff had the vision to perceive the connection between one set of problems—evolutionary and developmental—and the mystery of identity as presented by immunity. As Fred Tauber argues, it may very well be the case that only an outsider, aloof from the immediate concerns of the dominant scientific community, could have posed the question of identity in the face of dynamic change so starkly.

Elaine Morgan, too, crossed a divide by bringing with her something of value. In her case, it was her pen. An English major at Oxford and a mother of three, Morgan had spent a life writing screenplays and dramas for the BBC from a quiet base in Wales. But with an irascible wit and a fortunate turn of phrase, she was able to address that most basic of a readership's faculties directly: its common sense. "Learn to trust the evidence of your own senses over that of the written word," she wrote with just a pinch of disingenuousness, knowing full well that the written word was her best weapon. Indeed, Morgan could afford to give a full and forceful treatment to a theory that had been proposed thirty years earlier by an insider, Alister Hardy, who, not yet having "Sir" affixed to his forename, had been wary to publish anything that might damage his nascent scientific career. When it came to advocating the "aquatic ape" theory, Morgan, by her own admission, had "nothing to lose, no high academic position to think of." What she did have, however, was the skills of a dramaturge.

Coming from the outside in and of itself may allow for exercising greater imagination alongside, or even in some cases as an alternative to directly transporting particular tools, skills, or methods from home disciplines. Schrödinger's precise combinatorial model, Sahotra Sarkar argues, and his truly revolutionary and insightful notion of a genetic code, had less to do with his background as a physicist and more with an unrestrained inventiveness. George Price, too, in translating game theory from economics and international affairs to animal behavior, wasn't necessarily applying hard-won skills to a new setting, but rather exercising the kind of imaginative associative thinking which is so often stifled by internalist training and worldview. To be sure, outsiders have often been lambasted for being unqualified: such was Elaine Morgan's fate, as well as that of Rashevsky's, at least up to a certain point. But Samuel Butler, who had the distinction of being attacked by both T. H. Huxley and his grandson Julian, rather saw being an amateur as an advantage. (Huxley the grandson regarded Butler, alongside George Bernard Shaw, as "literary men," whose views are "based not on scientific fact and method, but on wish-fulfillment"). As Michael Ruse shows, Butler remained an outsider with respect to clubs and scientific societies and the like, but thought that being an amateur actually gave him a fresher perspective over the professionals (including and especially Darwin!).

A fresh perspective, then, may be the lot, or luck, of a different kind of professional even when he thinks of himself, or is termed, an "amateur." Prior training in a different discipline or intellectual community may afford special access to associative or imaginative thinking and the ability to analogize. Specific methods and tools from home disciplines can serve as keys with which to enter from the outside and unlock particular questions. And the fact that the outsider may have little to lose may pose as an advantage, too.

But so can something even more prosaic. When François Jacob decided that he was going to try to apply the principles he had learned from studying bacteria and the viruses that attack them to the mouse and the diseases that attack it (and other mammals), he was following a considered personal philosophy of science that regarded theories and models as nothing more than the recombination of elements present in previous theories and models. This to him was not only how "Nature the tinkerer" worked, but how science itself advances. And outsiders, he believed, are often in the best position to introduce new combinations. But, as Michel Morange aptly recognizes, this may very well be due simply to the fact that outsiders are much less conscious of obstacles, while being prone to "transgressions" because they are not at all familiar with the "rules of the game." In Jacob's case, there were massive obstacles to climb in moving from the simplest of model organisms to the most complex. Had these obstacles been known to him more precisely, they might have dissuaded him from trying. In the end, due to his work and others', the mouse became the choice model organism for studying mechanisms of disease in humans. In Schrödinger's case there were massive biochemical and molecular obstacles to scale as well, and much relevant biological knowledge remained unmastered. But as Francis Crick made clear, the main point of *What Is Life?*—that biology needs the stability of chemical bonds—"was one that only a physicist would feel it necessary to make." And, of course, it made a difference. Lack of knowledge, or naiveté, may be as important to innovation as highly specific forms of know-how and sophistication.

The Platypus
Outsiders don't always leave a mark, even when they try hard to do so. In this book David Hull is a stark example, a man as well placed as any on both sides of a disciplinary divide (in his case philosophy and systematics). Hull himself believed that a philosopher could "uncover, explicate and possibly solve problems in biological theory and methodology," but, as T. J. Horder shows, he ended up contributing more to "studying the science of science scientifically"—a title of one of his papers—than to any debate within systematics. Indeed, Hull was very much aware that the majority of scientists invariably find the work of philosophers, at least when it comes to their own field, superfluous. This is not to say that there is an unbreachable chasm between philosophy and the "hard core" of scientific practice, but that some disciplinary divides are harder to negotiate than others. Persuading biologists of the deep relevance of philosophy to their work continues to be a challenge.

What is clear, however, is the extent to which modern biology has been constituted as a pastiche, a conglomeration of different methods and tools and points of view and approaches. The term "genetics" was coined by William Bateson in a private letter in 1905, but as a glance at Gregor Mendel's story makes obvious, the modern theory of heredity had come into being as a blend between experimental physics, commercial plant breeding, botany, animal husbandry, local natural history, and even law. Microbiology, too, was later forged in a disciplinary furnace, to which chemistry, agriculture, medicine, and economics all contributed. And molecular biology was constituted like a tassel of disparate strands, as researchers from different fields led by particular problems found themselves obliged to master a host of tools and methods hitherto unlisted on the "how to" menu of the biologist.

Indeed, when Noam Chomsky challenged behaviorism, wielding the sword of generative linguistics, he was functioning as an outsider storming the gates, whereas when he championed animal behavior–influenced nativist biolinguistic theories, as W. Tecumseh Fitch shows, he was more like the insider importing "outsider" ideas. The overall result of these interventions was the refashioning of something called cognitive sciences that expanded the purview of biology. Whether or not "synthetic biology" will become a mainstay of the life sciences remains to be seen, but if Endy's gradual refashioning of the engineering vision from a revolutionary agenda into "nothing new here" is an indication of anything, it is that "biology" as a discipline is an ever-changing quantity.

The point, as Heraclitus might have appreciated, is that what is construed as "in" and "out" is and always has been in constant flux. The biological traveler can never step twice into the same river; he will find that not only have the waters changed—their color, temperature, and speed—but the banks, too, have changed, laying out a new topography. Indeed, the very organisms swimming about have morphed and been hunted and restocked from neighboring rivers. In biology—especially in biology—"outsiders" have been transformative.

In a very real sense, as Richard Lewontin stresses in the epilogue to this volume, the extreme dynamism of the life sciences problematizes the concept of biology "outside the box." Indeed, it problematizes the very notion of a "box." An example comes from a 1997 paper from the *Proceedings of the National Academy of Sciences* on resistance to phosphate insecticide in a sheep blowfly. The paper—the fruit of a collaboration between botanists, zoologists, and chemists—makes the point starkly: here, the biological effect of resistance is unrobed step by step all the way down to a single atom effect, moving the analysis from biology to chemistry to physics. Does speaking of an epistemic biological "box" help to understand how science is done in this case, Lewontin asks? Indeed, is this really an epistemic box at all? The answer, it seems, is to a great degree a reflection of the history of modern biology: the gradual erasure, made possible both by methodological and theoretical advances, of the boundary between life and non-life, as well as the growing ability to look at systems as constituted by components, amenable to mathematical and physical analysis. This is not merely, or simply, a story of reduction, but rather more accurately of accumulating more tools deemed relevant to the solution of mysteries provided by the natural world. As the tools multiply, so the epistemic "box" is enlarged to accommodate them. Yes, asking how a blowfly has become resistant to phosphate insecticide would be considered by most to fall under the purview of biology, but the way we attempt to answer such a question today as opposed to forty or sixty or one hundred years ago renders "biology" an ever-changing constant.

■

Looking at the "duck-bill mole" and other oddities of Australian wildlife, Governor John Hunter offered in 1793 that "a promiscuous intercourse between the different sexes of all these different animals might account for their unlikely forms."[14] Indeed, so strange was the beast that the great German comparative anatomist Johann Blumenbach christened it in 1800 *Ornithorhynchus paradoxus*. It would take time and careful scrutiny before the evolutionary lineage of the platypus was better understood, but even today, with all that we know, it remains a wondrous vision.

And so does biology, that most hodgepodge of all sciences. Still, as we hope readers will agree, the chimeric character of biology has often served it well. The cases in *Outsider Scientists* reveal how personal features such as persistence, institutional features such as the presence of willing patrons, mentors, and collaborators, and intellectual features such as the ability to create useful analogies and translations between fields fostered and promoted innovation in biology by newcomers as they constantly shaped and reshaped its form. The outsiders profiled in these pages, never content to merely "think differently," engaged themselves in the hard work of articulating connections between ideas, practices, people, and institutions that allowed their work to get a hearing among biologists and to gain a measure of influence. The success of these outsiders speaks not merely to their persuasiveness, but to their ability to understand key features of disparate fields and then build the bridges that connect and ultimately transform them.

NOTES

1. Ann Moyal, *Platypus: The Extraordinary Story of How a Curious Creature Baffled the World* (Washington: Smithsonian Institution Press, 2001), 5.

2. Joseph Ben-David, "Role and Innovations in Science," *American Journal of Sociology* 65 (1960): 557–568; Robert Merton, "The Perspectives of Insiders and Outsiders," in *The Sociology of Science* (Chicago: University of Chicago Press, 1973), 99–136.

3. Lynn Nyhart, *Biology Takes Form: Animal Morphology and the German Universities, 1800–1900* (Chicago: University of Chicago Press, 1995).

4. Peter Galison, *Image and Logic: A Material Culture of Microphysics* (Chicago: University of Chicago Press, 1997).

5. Robert Olby, *Francis Crick: Hunter of Life's Secrets* (Cold Spring Harbor: Cold Spring Harbor Press, 2009); Horace Judson, *The Eighth Day of Creation* (New York: Penguin, 1979).

6. Evelyn Fox Keller, *Making Sense of Life* (Cambridge: Harvard University Press, 2002).

7. On the connections between interdisciplinarity and innovation, see Jonathon N. Cummings and Sara Kiesler, "Collaborative Research Across Disciplinary and Organizational Boundaries," *Social Studies of Science* 35 (2005): 703–722.

8. Compare, for example, Lynn K. Nyhart, *Modern Nature: The Rise of the Biological Perspective in Germany* (Chicago: University of Chicago Press, 2009); and Jane Maienschein, *Transforming Traditions in American Biology, 1880–1915* (Baltimore: Johns Hopkins University Press, 1991).

9. Robert Kohler, *From Medical Chemistry to Biochemistry: The Making of a Biomedical Discipline* (Cambridge: Cambridge University Press, 1986); Jean Gayon, "Is Molecular Biology a Discipline?" *History and Epistemology of Molecular Biology and Beyond*, Preprint 310 (Berlin: Max Plank Institute for the History of Science, 2006), 249–252; Alexander Powell et al. "Disciplinary Baptisms: A Comparison of the Naming Stories of Genetics, Molecular Biology, Genomics, and Systems Biology," *History and Philosophy of the Life Sciences* 29 (2007): 5–32; Evelyn Fox Keller, "Physics and the Emergence of Molecular Biology: A History of Cognitive and Political Synergy," *Journal of the History of Biology* 23 (1990): 389–409; Vassiliki B. Smocovitis, "Unifying Biology: The Evolutionary Synthesis and Evolutionary Biology," *Journal of the History of Biology* 25 (1992): 1–65; Tim Lenoir, *Instituting Science* (Stanford: Stanford University Press, 1997).

10. Jonathan Harwood, *Styles of Scientific Thought: The German Genetics Community, 1900–1933* (Chicago: University of Chicago Press, 1993).

11. See J. Rogers Hollingsworth's analysis of scientific discovery at the Rockefeller University in *Creating a Tradition of Biomedical Research*, ed. Darwin Stapleton (New York: Rockefeller University Press, 2004).

12. See Robert Richards, *The Tragic Sense of Life: Ernst Haeckel and the Struggle Over Evolutionary Thought* (Chicago: University of Chicago Press, 2009).

13. See also Galison, *Image and Logic*, on trading zones and the creation of creoles.

14. Moyal, *Platypus*, 11.

OUTSIDERS BEFORE THE INSIDE

SANDER GLIBOFF

THE MANY SIDES
OF GREGOR MENDEL

INTRODUCTION

Gregor Mendel (1822–1884) was long viewed as the ultimate scientific outsider. After all, he was not celebrated as the founder of genetics until sixteen years after his death. He was not a professional researcher by any definition, but a monk and a schoolteacher, and later an abbot. He lived and worked far from the great European intellectual centers of his day, in Brünn, provincial capital of Moravia, then part of the Austrian Empire (Brno, in Czech; since 1993, part of the Czech Republic).[1] He presented the results of his experimental crosses of pea plants, along with the foundational ideas of genetics, only at his local scientific society in 1865. His now-famous article on the subject came out in the *Proceedings* of the society in 1866,[2] not the most conspicuous place for it, and, indeed, little notice was taken of it at first. It took until 1900 for it to be "rediscovered" and recognized as a cornerstone of the emerging science of genetics. Only then did Mendel find a place among the scientific insiders, or so it seemed to early geneticists.

The geneticists quickly embraced Mendel as one of their own. His name was given to "Mendel's laws," traits that "Mendelize," "Mendelism" as a theory of heredity, and "Mendelism" as a theory of evolution. Geneticists from all over the world donated money for a Mendel monument in Brünn. Mendel's work seemed to fit in so well in twentieth-century science that it was hard to imagine him ever belonging anywhere else, and the image developed of Mendel as a man so far ahead of his time that he had no intellectual company, no peers, no teachers. Unencumbered by old-fashioned preconceptions or intellectual commitments, he could see ahead to future problems and solve them, while his shorter-sighted contemporaries would not have seen the point of his work, had they even read it.

The blinkered view of Mendel as a figure without peers is reflected in one of the best-known photographs of him (figure 1.1), which is actually a detail from a group portrait. The corresponding narrative of the lone and unrecognized genius is told most eloquently by Loren Eisley. Here is how he describes Mendel's 1865 presentation to the Brünn Society of Naturalists:

Figure 1.1: Gregor Mendel. Curt Stern Papers.
American Philosophical Society, Philadelphia, PA.

Stolidly the audience had listened. Just as stolidly it had risen and dispersed down the cold, moonlit streets of Brünn. No one had ventured a question, not a single heartbeat had quickened. In the little schoolroom one of the greatest scientific discoveries of the nineteenth century had just been enunciated by a professional teacher with an elaborate array of evidence. Not a solitary soul had understood him.

Thirty-five years were to flow by and the grass on the discoverer's grave would be green before the world of science comprehended that tremendous moment. Aged survivors from the little audience would then be importuned for their memories. Few would have any.

A few pages later, Eisley offers this judgment about Mendel's place in history:

Mendel is a curious wraith in history. His associates, his followers, are all in the next century. That is when his influence began. . . . Gregor Mendel had a strange fate: he was destined to live one life painfully in the flesh at Brünn and another, the intellectual life of which he dreamed, in the following century.[3]

It is a tragic tale indeed. But cheer up: I am going to tell it a little differently by drawing on more recent historical research that has found some company for Mendel—some associates, audiences, and influences.

Geneticists and historians alike have spun the Mendel story for a variety of rhetorical purposes.[4] Versions like Eisley's, which emphasize long neglect and isolation, tend to aggrandize twentieth-century Mendelians for their greater openness to new ideas and their superior understanding of Mendel's paper, while the idea of a rediscovery reassures us that science is self-correcting and sooner or later will give credit where credit is due. But we should also make allowances for the lack of information about Mendel's community and context available to the rediscoverers. For them it was hardly unreasonable to infer that his monastic life and teaching duties precluded full-time research and close contacts to the international scientific community, or that the long neglect of his paper was good prima facie evidence that he had been misunderstood in Moravia and ignored everywhere else.

But more information is now available, and it has become clear that, for all the modernity of his scientific thinking, Mendel was also rooted in the nineteenth century: in the intellectual, economic, and religious life of Brünn, in the pure and applied sciences of the Austrian Empire (especially meteorology, biogeography, plant breeding, and evolution) and in European science generally. Several separate lines of post-Eisley research have recovered sides of the historical Mendel other than just the misplaced geneticist.

I will try here to bring these lines together into a more complete portrait and to show how his many sides also enrich our understanding of his famous paper.

Recent historical research also recognizes multiple nineteenth-century sources for later ideas about heredity. Far from springing fully formed from Mendel's head, genetics is beginning to look like a synthesis of many lines of thinking, not only in biology, but also in medicine, agriculture, law, and other spheres.[5] From this point of view, the many-sided Mendel to be developed here, with his multiple affiliations, audiences, and intellectual resources, makes a much more plausible founding figure than the lone outsider. His achievement then emerges as less a crossing of existing disciplinary boundaries than as a merger of disciplines into something new.

HISTORICAL AND POLITICAL BACKGROUND

In the eighteenth century, some eighty years before Mendel's birth, the Habsburg dynasty ruled over a sprawling multinational, multicultural realm in Central and Eastern Europe, comprising what is now Austria, Hungary, the Czech Republic, Slovakia, Croatia, Slovenia, Bosnia and Herzegovina, Romania, and parts of Poland and Italy. Along with Russia, France, and Britain (Germany was still fragmented into many separate kingdoms and principalities), the Habsburg Empire[6] was one of the "great powers" of Europe, militarily and politically.

Its rulers were shocked, however, in 1740, when the much-smaller Kingdom of Prussia attacked and captured their northern province of Silesia, their most highly industrialized province, with its textile factories and iron and coal mines. The loss was an economic blow as well as a military embarrassment. The young empress, Maria Theresa (ruled 1740–1780), spent over twenty years trying to get Silesia back, fighting the War of the Austrian Succession (1740–1748) and the Seven Years' War (1756–1763) against Prussia and a changing constellation of allies. She never did reclaim it, but held on only to a sliver, where the Mendel family happened to live, and which became part of Moravia, administratively.

Maria Theresa's efforts to compete with Prussia and maintain the status of the empire as a great power were not exclusively military and diplomatic in nature. She, and especially her son Joseph II (ruled 1780–1790, but co-regent with his mother from 1765 on), also modernized and centralized their administration. They applied the eighteenth-century ideals of "enlightened absolutism," a political and economic theory that emphasized efficiency and rational organization, and they instituted reforms that aimed to weaken the competing, decentralized, redundant, and inefficient feudal powers, such as the noble landlords, high church officials, and religious orders. Some of these Theresian-Josephine reforms were to have direct effects on Mendel's life and career: partially emancipating the serfs, modernizing agricultural practices, making primary and secondary education more widely available, encouraging the study of the natural sciences, and forcing the monasteries to be of service to the Empire and the economy.

MENDEL'S EARLY LIFE AND SCHOOLING

Mendel's father, Anton Mendel, was a peasant in a little village called Heinzendorf (Czech: Hynčice).[7] The family was ethnically German, and the region majority-German as well. Anton Mendel took a special interest in cultivating and grafting fruit trees, and taught young Mendel how to do it, too. He was encouraged in this by the local countess, an enlightened ruler in the Josephine tradition, who made an effort to promote scientific agriculture in her territory. She imported and distributed fruit trees and had natural science taught at village schools.

Our scientist was born Johann Mendel on July 22, 1822, the second of three children. (He took the name Gregor as an adult, upon entering the monastery.) Only limited schooling was available nearby, but because he was considered exceptional, the priests who taught the village children arranged for him to continue his education in town, and helped him talk his parents into it. They had little means of financing his studies, and they would miss the help of their only boy on the farm, for peasants like Anton Mendel had been only partially emancipated. He owned his own farm, but still had to work three days a week for the countess, an obligation known as the *Robot*.[8]

In 1834, at the age of thirteen, Mendel began attending *Gymnasium* (the academically oriented secondary school that opened the door to university education and the elite professions) some twenty-five miles away. His parents could only afford room and half board, and he had to tutor slower but wealthier pupils to earn his lunch money. Mendel got along like that until 1838, when his father was crushed in a logging accident while performing his *Robot*. He survived, but never recovered fully, had trouble maintaining the farm, and could not support his son at school as before. Mendel in the meantime had earned a teaching certificate that qualified him as a private tutor and was able to eke out a living at school, but at times, the pressure of school and work and worries about his family became too much for him. He suffered some kind of breakdown in 1839 and returned home for several months to recuperate. Several more of these breakdowns are recorded, both during and after his schooling, but no precise medical information about them is available.

In 1840, Mendel completed *Gymnasium* and moved on to Olmütz (Czech: Olomouc) in southern Moravia for a two-year course of university-level study (all that was offered; there was no full-fledged university in Moravia). Because of another breakdown and an extended stay on the farm, it took him three years to finish at Ölmütz, after which he decided not to continue his struggle to get a university degree. Instead, he followed his family's wishes and the advice of his Olmütz professors (one of whom was a monk himself), and took holy orders. Against stiff competition, he was accepted into the Augustinian order, at the monastery of St. Thomas in Brünn in 1843. There he received training in theology and was prepared for the priesthood. He was ordained in 1847.

The monastery offered Mendel not only some much-needed security, but also an intellectual community and an opportunity to do good works and improve the lives of Moravians through education and applied science. The monastery had been influenced by the Josephine reforms, which had eliminated the more contemplative (or uncooperative) monastic orders and confiscated their property. The surviving ones had to be active and productive in worldly affairs.

The monks of Brünn served as highly qualified instructors at several *Gymnasien* as well as at Olmütz. Some were experts in scientific agriculture, managed the monastery's extensive landholdings, and made an effort to share their knowledge with farmers and businesses in the region. Several, among them the abbot, were interested in pure science, too, and they had experimental gardens, a herbarium, a mineralogical collection, meteorological instruments, and a big, up-to-date library. In short, Mendel found himself in learned company, who thought the study of nature was important and useful for their work and their community. It was, admittedly, not a major European research center, but the isolation of the monastery should not be exaggerated, either. Had Mendel been seeking an opportunity to get involved in scientific research, he could hardly have made a more practical choice.

MENDEL'S NINETEENTH-CENTURY SOURCES (1):
PLANT AND ANIMAL BREEDING

The work of the monks and the economic interests of the monastery and the surrounding community were tied to a large extent to agriculture and its improvement, and Mendel clearly was motivated to keep the practical goals of his research in sight. His work promised to explain, among other things, how traits combined and interacted in plant hybrids, and it suggested ways of rationalizing hybridization methods to produce desired combinations of traits at will.[9]

An important line of research into Mendel's agricultural connections has focused on the theory and practice of animal breeding, particularly sheep breeding. Vítězslav Orel and Roger Wood have sought out the writings of sheep breeders who were active in Brünn from the 1810s through the 1830s and have analyzed their methods and their conceptions of heredity. The Moravians, like many breeders throughout Europe, built upon the work of British breeder Robert Bakewell (1725–1795) and others, who had explored ways of making breeding more systematic and scientific. They analyzed the animal into checklists of simple characteristics, measured those characteristics objectively and quantitatively, and devised procedures for comparing, selecting, and inbreeding the animals.

In Brünn, there were also significant discussions of theoretical matters, and it was asserted that heredity must be a law-abiding and predictable process, amenable to scientific study. The breeders there inquired into the effects of inbreeding and outcrossing, and into the stability of varieties. Why did inbred varieties not always breed true? What made them revert to their ancestral condition or become more average? When two varieties were crossed, what determined whether maternal or paternal characteristics would predominate in the offspring? Did it depend on the inner constitution or organization of the parent, or on environmental conditions?[10]

Since Mendel only arrived in Brünn in the 1840s, when sheep breeding was in decline there and few of the older breeders were still around, it is difficult to document a direct influence of their methods and ideas on Mendel's later work. Nonetheless, it seems safe to assume that Mendel was aware of the earlier Moravian ideas and practices, either through his abbot, who had been a breeder himself, from the published breeding literature, or from younger plant and animal breeders who were active, along with Mendel, in the Brünn scientific society.

And there are several important things that Mendel does seem likely to have learned from the old sheep breeders and from his own efforts at plant breeding. First and foremost was the "breeder's gaze," the ability to look at the animal and analyze the overall impression into individual characteristics— the "points of the breed," or, much later, the "unit characters" of the geneticist. The kinds of questions the breeders had asked also recurred in Mendel's work: when does one trait from one parent get inherited in preference to a contrasting trait from the other parent? How can a breeder reliably get desirable traits from two different strains to combine in a hybrid? Also of great importance was the conviction that heredity would turn out to be a predictable and repeatable process, amenable to scientific study and the formulation of general laws.

For most of his early years at the monastery, Mendel was assigned to teach at the local *Gymnasium*. He taught Latin and Greek, German literature, math, and science, and was found to be very good at it. The abbot sent him to Vienna in 1850 to take the licensing examinations, which were very demanding and dragged on for several weeks. Although Mendel barely passed the written parts and failed the orals, at least one Viennese professor thought he had some potential and advised the abbot to send him to the University for further training, which he did, for two years (1851–1853).

Much had changed in the educational system since Mendel had left it for monastic life, with several of the changes resulting from the Revolutions of 1848. Politically, the Revolutions are considered failures. In Austria, the Habsburgs beat them down and emerged more powerful than before, but some reform efforts did ensue. One of them was to modernize the universities and build them into research centers, an effort that also placed new pressures on secondary education to better prepare students for the reformed universities. That is what created the sudden demand at the *Gymnasien* for teachers like Mendel. Meanwhile, at the university level, Vienna had recruited a number of prominent researchers, and Mendel had the opportunity to take courses with some of the best-known scientists of his day, including the physicist Christian Doppler (1803–1853) and the botanist Franz Unger (1800–1870).

MENDEL'S NINETEENTH-CENTURY SOURCES (2):
FRANZ UNGER AND ACADEMIC BOTANY IN VIENNA

Mendel's university studies provide historians several more ways of linking him to nineteenth-century scientific thought and methodology. Much has been made of his physics, math, and meteorology coursework as the sources of his quantitative thinking, his conception of a scientific law, and the value he ascribed to experimentation. His exposure to issues in biology, especially pre-Darwinian evolutionary thought, have also been noted.[11]

The old view of Mendel as an outsider in biology, along with the unsafe assumption that biologists were unfamiliar with quantitative and experimental methods, has led most authors to focus on physics, math, and meteorology as the important academic influences from Vienna. But I am much more impressed by Mendel's apparent debts to his botany professor Franz Unger, whose multi-faceted research program covered everything from cell theory to microscopic plant anatomy and plant pathology, as well as plant biogeography, paleobotany, and evolutionary theory.

Central to Unger's approach was the assumption that there were special forces at work in living creatures that made them grow, develop, reproduce,

and evolve, and that these forces obeyed quantifiable laws. In his paleobo-
tanical work, for instance, Unger counted fossil species in different geological
periods, broke the counts down by taxonomic group, calculated the ratios
between the groups, and traced how the ratios shifted over time. That enabled him to show, quantitatively, that the flora of the earliest period was
dominated by algae; that there followed an age of ferns; then horsetails
and club mosses; conifers; and finally flowering plants. To him, the pattern of changing ratios indicated that a quantifiable law of nature was
at work, a law of evolution or development. (Unger used the same word—
"*Entwicklung*"—for all kinds of progressive, organic change, whether in
paleontology or embryology.)[12]

Classifying and counting plants, calculating ratios, and searching for
numerical relationships and laws were not peculiar to Unger, but were common practices in mid-nineteenth-century botany and zoology, particularly
in the study of the geographic distribution of species.[13] As we shall see,
Mendel classified and counted plants in a similar way, drew conclusions from
the resulting ratios, and used the language of *Entwicklung* to describe the
changes in the ratios over successive generations. It was a novel application
of the method, but it is hardly surprising to find a student of Unger's using
it, and it can hardly have been incomprehensible to the botanists of the day.
This use of counts and ratios in deriving laws of change marks Mendel as a
member of yet another nineteenth-century intellectual community.

SCIENTIFIC RESEARCH IN BRÜNN

Mendel never did pass his exams and earn proper teaching credentials, but
that was apparently because of one of his breakdowns during his last try in
1856, which caused him to give up before finishing the written portion. But
back home in Brünn, there were no doubts about either his teaching ability
or his knowledge of math and science. During the post-1848 reform period,
the *Gymnasium* employed quite a few unlicensed adjuncts, and they had no
trouble keeping Mendel on until he was elected abbot in 1868.

In the years after his return to Brünn in 1853, Mendel appears to have engaged in many and varied scientific activities in his spare time. He was a co-
founder and active member of the Brünn scientific society. He ran a weather
station at the monastery and wrote about weather forecasting and the possibilities for communicating weather reports by semaphore and telegraph. He
studied sunspots. He analyzed epidemiological data for correlations with
changes in the water table. He tried his hand at beekeeping and became one
of the very first to breed bees systematically. He was a chess player and inven-
tor of chess problems. And, of course, he continued his gardening and plant

breeding, with special interests in ornamental plants (he bred prizewinning fuchsias), fruit trees, and the peas and beans with which he did his famous experimental crosses. He began his published experiments with the peas in 1856, right after his last attempt at the licensing exams in Vienna.

MENDEL'S NINETEENTH-CENTURY SOURCES (3): THE HYBRIDIZING TRADITION

In the published article on those "Experiments in Plant Hybridization," Mendel himself suggests some additional nineteenth-century (or earlier) sources and inspirations for his research. Most prominently, he cites the scholarly literature on plant hybridization, particularly the work of Joseph Gottlieb Koelreuter (1733–1806) and Carl Friedrich Gaertner (1772–1850). Several authors, most notably Robert Olby, have looked into these pre-Mendelian hybridizers, and have seen Mendel as using their methods and following up some of their open questions, especially concerning the nature and stability of plant species and of hybrids, the possibility of species transformation, whether—and how—plants reproduce sexually, and the relative importance of the pollen and the germ or egg cell in determining the appearance of the offspring.[14]

Although Mendel does not discuss him explicitly, the great taxonomist Linnaeus (1707–1778) also had taken an interest in plant hybridization as a mechanism by which new species could be produced from old, for example, if hybrids were stable and remained distinct from the parent stocks. So it would seem that Mendel, at least indirectly, was addressing questions about evolution and classification that had been quite central in European science since the eighteenth century.[15]

THE HYBRIDIZATION PAPER

Mendel's published account of his experiments systematically addresses each of his communities and audiences: "Artificial pollinations, done with ornamental plants with the aim of producing new color variations, occasioned the experiments to be discussed here."[16] So begins the famous paper, with a reference to practical breeding—maybe his own experience cross-pollinating fuchsias—as a resource and motivation for the research.

Immediately after this overture to the breeders, Mendel engages the Unger-style biogeographers, developmentalists, and evolutionists, with their interests in formulating laws of organic change:

> The conspicuous regularity with which the same hybrid forms always reappeared, as long as the pollination occurred between the same strains,

provided the motivation for further experiments, whose task it was to follow the development [*Entwicklung*] of the hybrids in their descendents.[17]

As has also been noted by Floyd Monaghan and Alain Corcos, Mendel writes repeatedly about developmental laws and the changing composition of the entire experimental population as he lets the hybrids and their offspring self-pollinate for six generations. This is not quite the approach of a modern geneticist, who would focus instead on the mechanisms of transmission from one pair of parents to its offspring.[18]

But Mendel is still not finished introducing his paper and drawing his communities and audiences together. The second paragraph is devoted to the academic plant hybridizers, such as Gaertner and Koelreuter. Those authors are on the right track, Mendel finds, yet have not quite been able to satisfy the demand for "a generally valid law for the formation and development [*Entwicklung*] of the hybrids,"[19] that is, the kind of law that Unger would want him to look for.

Mendel weaves these approaches together. With his breeder's gaze, he analyzes the pea plant into individual characteristics of interest and chooses pairs of parental plants that contrast in each, for example, having green or yellow as the pea color, round or wrinkled as the pea shape, or dwarf or tall as the plant height.[20] But unlike the breeder, his goal is not so much to improve these features or create new combinations of them, but rather to ask general questions about their interactions and changing ratios. Like the academic plant hybridizers, he wants to know about the nature of hybrids, the extent to which each of a pair of contrasting parental characteristics gets expressed, and what rules determine whether hybrid characteristics persist over multiple generations. Will, for example, yellow and green blend to an intermediate color? Interact somehow to produce something wildly different? No, he observes that one trait wins out consistently over the other, and defines "dominance" and "recessiveness" of traits accordingly.[21]

Next, Mendel pursues the developmental/evolutionary question by letting the hybrids and their offspring propagate by self-pollination, and by classifying and counting the progeny. He observes the recessive trait reappearing in predictable ratios for six generations before he stops the experiments. He can then provide a formula that predicts the composition of the experimental population after any number of generations of self-pollination following a cross. This fulfills Franz Unger's ideal of a developmental law.[22]

Finally, and very tentatively, Mendel also provides something destined to catch the eye of the twentieth-century geneticist. He hypothesizes that there are physical differences in the pollen and germ cells that correspond to the differing parental characteristics,[23] and begins to speak of material "elements"[24] that segregate into the reproductive cells and come together in new pairings after fertilization. He is exceedingly cautious and vague about what these elements might be, physically, but it would be enough to inspire later conceptions of the "gene" as a hereditary unit, locatable on a chromosome.[25]

Also of great importance to the early geneticists were the ideas that these elements occurred in pairs, one from each parent, and that the entire plant could be analyzed into individual characteristics, each governed by such a pair of elements. The geneticists also adopted and expanded upon the rules of dominance and recessiveness and looked for other kinds of interactions between genes. They took over the rules of "segregation" and "independent assortment," which governed the way in which the individual elements (later "genes") were divvied up into the pollen (or sperm) and egg cells, to be paired up in new ways after fertilization. And they soon made the connection between these rules and the movements of chromosomes during cell division, and modified Mendel's rules by allowing for linkage of multiple genes on the same chromosome. More generally, they prized Mendel's paper for its use of experimentation, its quantitative approach, its search for mechanisms, and its emphasis on making and testing precise predictions.

CONCLUSIONS

Mendel did indeed have much to say to the early geneticists, and he gave them good reason to claim him posthumously as a founder and a long-lost insider, but he was hardly such a tragic and isolated figure as some have imagined,. He was a successful breeder, teacher, monk, and abbot, and a member of multiple communities and intellectual traditions.

To be sure, the image of the isolated genius was never universally accepted among geneticists. Thomas Hunt Morgan, for example, one of the founders of classical *Drosophila* genetics, was pretty sure that "The genial abbot's work was not entirely heaven-born, but had a background of one hundred years of substantial progress that made it possible for his genius to develop to its full measure."[26]

Writing at the time of the Mendel centennial of 1965, geneticist L. C. Dunn, too, complained of "the aura of isolation which has clung" to Mendel and how "Even some biologists of today tend to think of him as though he had been a visitor from outer space whose brief transit through European Science was unobserved at the time."[27] But they both had only limited means of correcting the picture and putting Mendel into context. Morgan referred to the literature on plant hybridization for pre-Mendelian hints about dominance and segregation. Dunn was able to add a bit more, by pointing to some of Mendel's more worldly activities and connections to local scientific societies, but even he had to admit that "[Mendel] does seem rather an outsider in European botany."[28]

It has taken much longer for historians to piece together the more complete picture that is symbolized nicely by figure 1.2, the group portrait that was the source of figure 1.1. It shows Mendel not only in his most important community, but also choosing to hold a fuchsia as his attribute, thus connecting himself to the plant breeders as well as the monastery.

As a plant breeder, he was an heir to a local tradition of practical breeding that had already developed methods of delimiting and analyzing individual traits and had begun to investigate theoretical questions about heredity as early as the 1820s. He could hardly have done his research at all were it not for his membership in the Abbey of St. Thomas in Brünn and the

Figure 1.2: Gregor Mendel in a group. George Mendel, third from right, with friars of the Augustinian Abbey in Brno. Mendel holds a fuchsia plant. Courtesy of the Mendel Museum, Brno, Czech Republic.

transformations of monastic life in the wake of enlightened absolutism and the Josephine reforms. Mendel's order fulfilled its obligations to the state in part by promoting science, especially scientific agriculture, and by sending out experts like Mendel to teach in the schools. Further, the monastery enabled him to go to Vienna for exposure to new ideas and methods from several fields of scholarship, such as experimental physics and especially quantitative botany and biogeography. And either his studies or his own reading led him to cutting-edge questions about the nature of hybridization and its role in evolution.

In Mendel's Moravia, the lines between pure and applied science, religious and scientific institutions, and professional scientists and amateur naturalists were not drawn as sharply as in the twentieth century. Neither were the fields of heredity, development, and evolution as strongly demarcated. Mendel could participate in various fields, communities, and institutions. Perhaps he was a modern geneticist, too, in some sense, but that would not make him an outsider everywhere else. On the contrary, it would underscore his multi-sidedness.

With the many sides of Mendel in mind, we can see how his celebrated paper on hybridization addressed the interests of contemporary breeders, plant hybridizers, Mendel's teachers in Vienna, brothers at the monastery, and colleagues at the Brünn Society, while also engaging with themes of the international scholarly literature. Contemporaries would easily have seen where Mendel was getting his ideas and methods from, and where he wanted to go with his laws and hypothetical elements, but did not yet have much reason to draw general conclusions from Mendel's pea data. The breeders, especially, would have known of too many cases that did not follow Mendel's laws, as would Mendel himself. In fact, his paper freely discusses counterexamples such as hybrids that breed true instead of segregating out into dominants and recessives in the proper ratios.[29]

In contrast, early twentieth-century geneticists and historians underestimated the importance of practical breeding, hybridization, and development as resources for Mendel. The laws and hypothetical mechanisms looked to them like strokes of genius, without significant input from earlier studies. They were also willing to take up the challenge of explaining the exceptions, working out additional rules and mechanisms, and building Mendel's pea rules into a general body of theory.

Instead of an "outsider" or a "boundary crosser," Mendel is better described as a synthesizer of multiple approaches, one whose synthesis has been mistaken for an abrupt origin de novo. Early twentieth-century geneticists, eager to distinguish their new field from older lines of research, overemphasized the novelty and exaggerated the divisions between the first geneticist and his contemporaries. As genetics became a discipline and defined its boundaries, it drew Mendel in, while banishing the breeders, monks, Linnaeans, and old-fashioned hybridizers to the outside and obscuring their presence in the historical picture.

FURTHER READING

Allen, Garland E. "Mendel and Modern Genetics: The Legacy for Today." *Endeavour* 27, no. 2 (2003): 63–68.

Gliboff, Sander. "Gregor Mendel and the Laws of Evolution." *History of Science* 37 (1999): 217–235.

Mendel, Gregor. "Experiments in Plant Hybridization." MendelWeb, http://www .mendelweb.org/MWpaptoc.html (accessed 05/26/2011).

Müller-Wille, Staffan and Hans-Jörg Rheinberger, editors. *Heredity Produced: At the Crossroads of Biology, Politics, and Culture, 1500–1870.* Cambridge, MA: MIT Press, 2007.

Olby, Robert C. *Origins of Mendelism.* 2nd ed. Chicago: University of Chicago Press, 1985.

Orel, Vítězslav. *Gregor Mendel: The First Geneticist.* Translated by Stephen Finn. Oxford: Oxford University Press, 1996.

Wood, Roger J. and Vítězslav Orel. *Genetic Prehistory in Selective Breeding: A Prelude to Mendel.* Oxford: Oxford University Press, 2001.

NOTES

Work on this paper was supported by a grant from the NSF (award no. 0843297). Figure 1.1 was supplied by the library of the American Philosophical Society and is reprinted with their permission. For figure 1.2, I thank Michaela Jarkovska of the Mendel Museum for the reproduction, and Abbot Lukas Evzen Martinec of the Augustinian Abbey in Old Brno for permission to use it in this publication.

Thanks also to Andy Fiss for research assistance and a critique of an early draft. Garland Allen, Anne Mylott, and participants in the Biology Studies Reading Group at Indiana University also read the manuscript at various stages and provided helpful commentary.

1. Throughout this paper, I use the German place names, as Mendel would have done.

2. Gregor Mendel, "Versuche über Pflanzenhybriden," *Verhandlungen des naturforschenden Vereines in Brünn* 4 (1866): 3–47; also available in English and online: Curt Stern and Eva R. Sherwood, eds., *The Origin of Genetics: A Mendel Source Book* (San Francisco: W. H. Freeman, 1966); Gregor Mendel, "Experiments in Plant Hybridization," MendelWeb, http://www.mendelweb.org/MWpaptoc.html (accessed 05/26/2011).

3. Loren Eisley, *Darwin's Century: Evolution and the Men Who Discovered It* (Garden City, NY: Anchor Books, 1961), 206, 211.

4. For a more thorough survey, see Jan Sapp, "The Nine Lives of Gregor Mendel," in *Experimental Inquiries*, ed. H. E. Le Grand (Dordrecht: Kluwer Academic Publishers, 1990).

5. Staffan Müller-Wille and Hans-Jörg Rheinberger, eds., *Heredity Produced: At the Crossroads of Biology, Politics, and Culture, 1500–1870* (Cambridge, MA: MIT Press, 2007), especially the editors' introduction, "Heredity—The Formation of an Epistemic Space."

6. After 1804 also called the Austrian Empire; after 1867, Austria-Hungary.

7. Biographical details are from Hugo Iltis, *The Life of Mendel* (New York: W. W. Norton, 1932); and Vítězslav Orel, *Gregor Mendel: The First Geneticist*, trans. Stephen Finn (Oxford and New York: Oxford University Press, 1996).

8. From the Slavic *"robota,"* meaning the corvée or, figuratively, "drudgery"; the word has entered the English language, by way of Czech science fiction, to mean an artificial human.

9. For a concise overview of Mendel's life, work, and legacy, emphasizing the interconnections with agriculture, see Garland E. Allen, "Mendel and Modern Genetics: The Legacy for Today," *Endeavour* 27, no. 2 (2003): 63–68.

10. Vítězslav Orel and Roger J. Wood, "Empirical Genetic Laws Published in Brno before Mendel Was Born," *Journal of Heredity* 89 (1998): 79–82; Roger J. Wood and Vítězslav Orel, "Scientific Breeding in Central Europe during the Early Nineteenth Century: Background to Mendel's Later Work," *Journal of the History of Biology* 38, no. 2 (2005): 239–272; Roger J. Wood, "The Sheep Breeders' View of Heredity Before and After 1800," in Müller-Wille and Rheinberger, *Heredity Produced*; Roger J. Wood and Vítězslav Orel, *Genetic Prehistory in Selective Breeding: A Prelude to Mendel* (Oxford and New York: Oxford University Press, 2001).

11. Robert C. Olby, *Origins of Mendelism*, 2nd ed. (Chicago: University of Chicago Press, 1985), ch. 5; Vítězslav Orel, "Mendel and New Scientific Ideas at the Vienna University," *Folia Mendeliana Musei Moraviae Brno* 7 (1972): 27–36; Franz Weiling, "J. G. Mendels Wiener Studienaufenthalt 1851–1853," *Sudhoffs Archiv für Geschichte der Medizin und der Naturwissenschaften* 51 (1966): 260–266; Franz Weiling, "J. G. Mendel als Statistiker und Biometriker: Sowie die Quellen seiner statistischen Kenntnisse," in *Biometrische Vorträge, Deutsche Region der Internationalen Biometrischen Gesellschaft, 15. Biometrischen Kolloquium* (Hannover, 1968); Franz Weiling, "Das Wiener Universitätsstudium 1851–1853 des Entdeckers der Vererbungsregeln Johann Gregor Mendel," *Folia Mendeliana Musei Moraviae Brno* 21 (1986): 9–40.

12. Sander Gliboff, "Evolution, Revolution, and Reform in Vienna: Franz Unger's Ideas on Descent and Their Post-1848 Reception," *Journal of the History of Biology* 31, no. 2 (1998): 179–209; Sander Gliboff, "Gregor Mendel and the Laws of Evolution," *History of Science* 37 (1999): 217–235; Sander Gliboff, "Franz Unger and Developing Concepts of *Entwicklung*," in *"Einheit in der Diversität": Franz Ungers Naturfoschung im internationalen Kontext*, ed. Marianne Klemun (Vienna: Austrian Academy of Sciences, in preparation).

13. Janet Browne, *The Secular Ark: Studies in the History of Biogeography* (New Haven: Yale University Press, 1983); Susan Faye Cannon, "Humboldtian Science," in *Science*

in Culture: The Early Victorian Period (New York: Science History Publications, 1978); Sander Gliboff, "H. G. Bronn and the History of Nature," *Journal of the History of Biology* 40 (2007): 259–294; Malcolm Nicolson, "Alexander von Humboldt, Humboldtian Science and the Origins of the Study of Vegetation," *History of Science* 25 (1987): 167–194.

14. Olby, *Origins of Mendelism*, see n. 9, ch. 1; Herbert F. Roberts, *Plant Hybridization before Mendel* (Princeton: Princeton University Press, 1929); Conway Zirkle, *The Beginnings of Plant Hybridization* (Philadelphia: University of Pennsylvania Press, 1935); for a sharp contrast between Mendel the hybridizer and the twentieth-century view of Mendel the geneticist, see also Augustine Brannigan, "The Reification of Mendel," *Social Studies of Science* 9 (1979): 423–454.

15. L. A. Callender, "Gregor Mendel: An Opponent of Descent with Modification," *History of Science* 26 (1988): 41–75; Staffan Müller-Wille and Vítězslav Orel, "From Linnean Species to Mendelian Factors: Elements of Hybridism, 1751–1870," *Annals of Science* 64 (2007): 171–215; Pablo Lorenzano. "What Would Have Happened if Darwin Had Known Mendel (or Mendel's Work)?" *History and Philosophy of the Life Sciences* 33 (2011): 3–48. On Mendel's place in a broader European intellectual context: Margaret Campbell, "Mendel's Theory: Its Context and Plausibility," *Centaurus* 26 (1982): 38–69.

16. Gregor Mendel, *Versuche über Pflanzenhybriden*, facsimile of the original paper from *Verhandlungen des naturforschenden Vereines in Brünn* 4 (1866): 3–47 (Weinheim: H. R. Engelmann, 1960), on 3 (equal to section 1 of the Mendel-Web version). Translations are my own.

17. Ibid., 3 (section 1 on MendelWeb).

18. Floyd Monaghan and Alain Corcos, "Mendel, the Empiricist," *Journal of Heredity* 76, no. 1 (1985): 49–54.

19. Mendel, *Versuche über Pflanzenhybriden*, see n. 14, 3 (section 1 on MendelWeb).

20. Ibid., 5–7 (section 2 on MendelWeb).

21. Ibid., 10–12 (section 4 on MendelWeb).

22. Ibid., 12–18 (sections 5–7 on MendelWeb).

23. Ibid., 24–32 (section 9 on MendelWeb).

24. Ibid., 41–42 (section 11 on MendelWeb).

25. Olby rightly has cautioned against reading the modern gene concept back into Mendel's paper: Robert C. Olby, "Mendel no Mendelian?" *History of Science* 17 (1979): 53–72.

26. Thomas Hunt Morgan, "The Rise of Genetics," *Science* 76 (1932): 261–267, 285–288, 263.

27. L. C. Dunn, "Mendel, His Work, and His Place in History," in "Commemoration of the Publication of Gregor Mendel's Pioneer Experiments in Genetics," *Proceedings of the American Philosophical Society* 109, no. 4 (1965): 189–198, 191.

28. Ibid.

29. Mendel followed up the pea work with one more project on plant hybridization, and that one focuses on just such a counterexample, the hawkweeds: Gregor

Mendel, "Ueber einige aus künstlicher Befruchtung gewonnene Hieraciumbastarde" (1869), in *Versuche über Pflanzenhybriden: Zwei Abhandlungen (1865 und 1869)*, ed. Erich Tschermak (Leipzig: Wilhelm Engelmann, 1901). It is also clear from his surviving letters to the botanist Carl von Naegeli—his only known scientific correspondent—that there were questions about the generalizability of his laws: see Carl Correns, "Gregor Mendels Briefe an Carl Nägeli, 1866–1873: Ein Nachtrag zu den veröffentlichten Bastardierungsversuchen Mendels," in *Gesammelte Abhandlungen zur Vererbungswissenschaft aus periodischen Schriften, 1899–1924* (Berlin: Julius Springer, 1924).

LOUIS PASTEUR

THE CHEMIST IN THE CLINIC

INTRODUCTION

Born in 1822 to a modest tanner in the Jura, Louis Pasteur went on to become France's first and greatest popular scientific hero. His state funeral in 1895 brought thousands onto the streets of Paris to bid farewell to the "*bienfaiteur de l'humanité*."[1] In accordance with his wishes, Pasteur's family declined the honour of having his body placed in the Pantheon, and to this day it lies in a crypt under the internationally renowned research institute that bears his name. It is now impossible to abstract Pasteur from the context of his iconic status, particularly in France, and even those who have wanted to expose the reality behind the myth are obliged to take the myth into account as their starting point. One of the elements of this myth is Pasteur's place as an outsider: a scientist trained in chemistry who helped to transform the practice of medicine. This passage from chemistry to medicine was neither direct nor sudden, nor was it without its difficulties. After beginning his research in the field of crystallography in the 1840s, Pasteur went on to study fermentation; the question of spontaneous generation; spoiled milk, wine, and vinegar; and diseases affecting silkworms, chickens, sheep, cows and pigs, before developing a vaccine-based treatment for rabies in humans in the 1880s.

This remarkably productive scientific career, which spanned over forty years, generated a high public profile for Pasteur, with each new step bringing greater notoriety. In light of his trajectory, it is not surprising that Pasteur's disciplinary credentials should have been challenged along the way. While the problems posed for the medical establishment by his not being a doctor stand out as elements of his biography, his legitimacy had already been questioned during the debates over spontaneous generation. Indeed, Pasteur's entry into these debates in 1859 marks a plateau in his career, confirming his interest in the microscopic life responsible for fermentation and marking a definitive move away from chemistry into what we would today consider microbiology. He would never return to original research in chemistry but, after conducting several investigations into fields of industrially related microbiology, would move on to the domain of medicine. Starting with diseases in animals, an area where he developed and consolidated his germ theory, Pasteur built on his practical knowledge to put in place methods for preventing these diseases. His final area of research involved transferring these techniques to the treatment of human disease, notably rabies. Thus, Pasteur's scientific life is usually presented in four different stages: his work on isomerism and optical activity, his study of fermentation and spontaneous generation (generally treated together), his identification and treatment of animal disease, and finally his development of the rabies vaccine. Rather than looking at these four stages, I want here to consider successively the passage from one to another, as it was this progressive crossing of disciplinary boundaries that brought Pasteur from chemistry into medicine.

Before visiting these three borders, however, it is important to consider Pasteur's initial training in the sciences, as his education conditioned the approach he applied throughout his scientific life. It was this methodological approach that constituted the common factor running through all his scientific endeavours, which was characterized by a rigorous logic and experimental method combined with a theoretical intelligence that led him remarkably quickly to a number of fruitful hypotheses in diverse areas of research.

PASTEUR'S TRAINING AS A CHEMIST

Of modest provincial origins, Pasteur's intelligence, hard work at school, and good network of local relations (including the head of the local school) gave him access to a leading Parisian school where he could prepare his entry into one of the elite institutes of higher education, the *École normale supérieure* in Paris.[2] Set up to supply teachers for French secondary education, the *École normale* had mutated into a place for training the specialized scientists and humanists who increasingly taught at French universities. The rigorous demands of the preparation for France's teaching elite (he not only succeeded in obtaining the aggregation qualification through competitive examination, but also earned two doctorates in 1847, one in chemistry, and the other in physics) moulded Pasteur into a dedicated scientist, working long hours both in the library and at the laboratory bench. *Laboremus* (let us work) became his motto, a guiding principle that he imposed on his collaborators as well as himself throughout his life. Perhaps more than other scientists produced by this system of the "*grandes écoles*," however, Pasteur complemented his work ethic with considerable scientific flair and intuition. Thus, an enduring characteristic of Pasteur's scientific approach was his tendency to jump to bold general conclusions based on relatively little evidence and then to work in a thorough and persistent manner to demonstrate the correctness of his intuitions. Often his initial ideas failed to pan out, but he found no difficulty in refining them or simply changing his position in light of the accumulating evidence. When combined with practical applications, this approach evidently brought with it a fair degree of risk (intellectual, economic, and ultimately medical), but Pasteur's intuition and his pragmatic realism saved him from having to publicly retract any of his views.

Pasteur's formative years at the *Ecole normale* in the rue d'Ulm not only gave him a solid scientific training but also provided him with an important network of influential patrons and friends. Thus, Paris brought him into contact with the leading French chemists of the time, including Jean-Baptiste Dumas (1800–1884), Antoine Jérôme Balard (1802–1876), and Jean-Baptiste Biot (1774–1862). He also crossed the path of a younger but equally renowned chemist, Auguste Laurent (1807–1853). The contact with Laurent might have led him into "classic" organic chemistry, but instead Pasteur chose to work on the relationship between crystal structure and the power of solutions to rotate polarized light. His remarkable achievements in this area would in turn bring him the unfailing support of Biot, who, along with Dumas, would exercise a key influence on Pasteur's subsequent career.

The support of these influential scientists greatly facilitated Pasteur's passage across distinct domains, whether by underwriting his scientific competence or as the direct source of invitations (if not orders) to undertake research in areas that he would otherwise never have entered. Indeed, this patronage proved essential for Pasteur's career on a number of levels. Like other graduates of the *École normale*, Pasteur was a civil servant and ran the risk of being appointed as a teacher in a provincial secondary school. In 1848 he was sent to teach high school physics and chemistry in Dijon, removing him from the stimulation and exchange provided by the Parisian context and denying him the possibility of continuing his research. The intervention of his mentors and friends behind the scenes in Paris procured him the more promising position of a chemistry teacher at the *Faculté des sciences* in Strasbourg. From here, his growing renown earned him a nomination as the dean of a new *Faculté des sciences* founded in Lille in 1854. Finally, in 1857, Pasteur returned to Paris and the *École normale* as a teacher and administrator, where he would stay for the rest of his working life.

As we shall see, it was as much his connections as his scientific curiosity that initially led Pasteur across disciplinary boundaries, and it was this network of patrons that made such a journey possible. Nevertheless, after a certain point his institutional recognition was such that anything must have seemed possible; in 1882 he was even elected a member of the *Académie française*, traditionally reserved for religious and literary dignitaries. But his first such disciplinary transition was of quite a different order.

THE FIRST FRONTIER:
FROM CHEMISTRY TO FERMENTATION

Pasteur was already a promising member of the French chemical community when he came to prepare his theses in chemistry and physics. He might well have chosen to work on the theoretical problems in organic chemistry being addressed by Auguste Laurent. Indeed, thanks to a close reading of his laboratory notebooks, the historian of science Gerald Geison has shown the influence that Laurent exercised on Pasteur's thinking, with his initial steps in crystallographic research structured around an interpretation of Laurent's type theory.[3] Nevertheless, Pasteur's first published research was oriented toward the relationship between crystal structure (and projected molecular form) and optical activity (the capacity of solutions to rotate plane-polarized light). Crystallography was quite a distinctive area of chemistry where, starting in the eighteenth century, researchers were aiming to better understand the relationship between the crystalline form of matter and its composition. But it was also an area of chemistry where vitalism was more present than in others, as the production of characteristic forms always suggested an analogy between crystals and living organisms. Like Claude Bernard before him, Pasteur explicitly drew the parallel between crystal formation and healing in living creatures. In a project for an overview of the topic drawn up around 1871, the enigma of what might drive a chemical compound to always assume the same crystalline form was assimilated to the healing of a cut or other wound, where the damaged flesh unerringly regains its original form. In a note that he made to himself, Pasteur wrote:

> Nothing is more curious than to carry the comparison of living species with mineral species into the study of the wounds of one and the other, and of their healing by means of nutrition—coming from within in living beings, and from without through the medium of crystallization in the others.[4]

While this very general consideration is not without interest, we need to look deeper to understand the logic of Pasteur's initial move from crystallography to the study of fermentation. In 1847, Pasteur included a significant amount of discussion of crystal structure in his chemistry thesis on arsenic acid and arsenates, and his physics thesis of the same year was specifically on the rotation of polarized light by solutions. The publication of his discoveries on the chiral nature of the isomers composing racemic acid followed shortly afterward.[5] Through painstaking experimental work, Pasteur succeeded in showing that the optically inactive racemic (or paratartaric) acid was composed of two different forms of acid, one that in solution rotated polarized light to the right (a feature of naturally occurring tartaric acid) and another that rotated it to the left. Pasteur separated the two isomers (identical elementary composition with apparently identical chemical structure and properties) by crystallizing salts of racemic acid and separating the two different crystals. While these crystals exhibited similar (hemihedral) forms, they were non-superimposable mirror images of one another (like the left and right hand, hence the name of chirality given to this form of symmetry).

Both the racemic acid and the optically active tartaric acid were products of fermentation in brewing, and Pasteur had an early intuition concerning the interrelation of the phenomenon of optical activity and living processes. He was convinced that only these processes could produce optically active compounds, and thus it was logical for him to infer that fermentation must be a living process. Thus, as François Dagognet has incisively argued, Pasteur's exploration of fermentation was an extension of his interest in dissymmetry.[6] Indeed, Pasteur's publications from the mid-1850s on amyl alcohol (a combination of optically active and inactive components), and lactic acid (also optically active) were further investigations into optical activity. In this context, Pasteur used this property of optical activity to follow the process of fermentation.

At first, then, Pasteur was studying fermentation only secondarily to optical activity, but there was another sense in which the study of fermentation itself did not necessarily represent a move away from chemistry. To understand this, we need to consider the debates over the nature of fermentation in the nineteenth century. While Pasteur's idea that yeast was a living organism responsible for alcoholic fermentation was not original,[7] there were still many scientists, in particular the distinguished German chemist Justus von Liebig, who argued that fermentation was simply a chemical reaction. In Liebig's view, the yeast played the role of a catalyst, with its decomposition provoking the conversion of sugars into alcohol and carbon dioxide. Although its presence was necessary, the yeast did not actively participate in the reaction, and fermentation was understood as a purely chemical phenomenon. Thus, while Pasteur and his successors eventually succeeded in imposing the interpretation of fermentation as a biological activity, when he started his work it could still have been considered a domain of chemistry.

It was Pasteur's appointment as dean of the *Faculté des sciences* at Lille in 1853 that brought him to study fermentation for its own sake. One of the goals of this new faculty was to develop the teaching of science for those who would work in science-based industries in the region. In 1856, in line with this policy, Pasteur both taught a course entitled "The Beet Alcohol Industry" and studied the problems confronted by a local beet processor, M. Bigo. Following this research, Pasteur started to publish on the vital nature of fermentation rather than on issues of optical activity. He first wrote on the production of lactic acid from milk, arguing that it was a result of identifiable microscopic organisms.

Pasteur's detailed study of fermentation led him into the debate over spontaneous generation, where he vigorously attacked Félix-Archimède Pouchet's position that primitive forms of life could emerge spontaneously from organic matter. By a series of increasingly elaborate experiments, Pasteur demonstrated (to his satisfaction, at least) that a sterilized medium would not spontaneously produce life, and that the apparent spontaneous production of life was caused by invisible germs in the atmosphere. Rather than presenting this debate in detail, however, I simply want to point out that, much more than Pasteur's research on fermentation, it represented a move into another field.[8]

While spontaneous generation was more clearly a biological question than fermentation, it is not clear that Pasteur's contribution earned him a reputation as a biologist. When nominated by his mentor Biot for the botany section of the *Académie des sciences* in 1861, Alfred Moquin-Tandon (himself a prominent naturalist and member of the *Académie des sciences*) disqualified Pasteur's candidature in the following terms: "Let's go to M. Pasteur's home, and if we find a single volume of botany on his bookshelf, I will put him on the list."[9] Pasteur did not conform to the model of a naturalist, but was instead contributing to the foundation of what would become a new domain of biology: microbiology. However, neither is it clear that Pasteur was interested in integrating his approach into the existing schemas of the naturalist or physiological communities. Unlike Ferdinand Cohn or even Robert Koch, Pasteur's training did not push him toward any systematic project of identification and classification of microorganisms. Instead, Pasteur followed his own research path deep into the fledgling domain of microbiology, pursuing a growing interest in the microbe as pathogen that would guide the rest of his career.

What retrospectively appears to be a case of crossing frontiers (from crystallography to fermentation) was not necessarily one at the time. What should be noted is how profoundly Pasteur's own work transformed these different fields, particularly fermentation. Nevertheless, when Pasteur did venture into a territory—the origins of life—that was increasingly seen as being the domain of biologists, certain members of the community challenged the legitimacy of a chemist in this area. It was his rigorous experimental approach learned in the chemical laboratory that both set Pasteur apart from other naturalists and secured his public victory in the debate.

THE SECOND FRONTIER:
FROM FERMENTATION TO ANIMAL DISEASE

The debate over spontaneous generation confirmed Pasteur's belief that all microorganisms, including the germs responsible for disease that he would subsequently identify, have an external origin. In his germ theory of disease, just as yeast was needed for fermentation, a pathogenic germ had to be introduced for an animal or human to suffer from a disease. Even before he started to study infectious disease, therefore, he had a model for it. Beyond this, when considering milk, wine, and later animal disease, Pasteur and his colleagues thought in terms of a group of related concepts—spoiling, disease, fermentation, and putrefaction. When in 1882 Pasteur's foremost disciple Émile Duclaux explicitly exploited the parallel between ferments and disease, what he emphasized was the continuity of techniques that were applied in the two domains, the means of isolating and cultivating the ferment and the pathogen.[10]

Pasteur's interest in the phenomena of fermentation was, as we have already seen, intimately tied to their industrial applications, notably wine, vinegar, and later beer production. These economically important processes depended on what turned out to be complex biological phenomena. The successful brewer needed to find the means to favour certain microorganisms that would generate specific products, and to hamper the action of others. Thus, the technique of pasteurizing wine (heating the wine after the alcoholic fermentation was completed) was designed to eliminate undesirable microbes that might spoil the wine. The much more complex task, however, was the management of the functional microorganisms in a productive way. The idea that competition between microorganisms for a nutritive medium could determine a "healthy" or a defective fermentation process would play an important role in Pasteur's evolving vision of microbiology. The interplay between the culture medium (including the atmosphere, with Pasteur clarifying the distinction between aerobic and anaerobic fermentation processes) and these microorganisms was of crucial importance for the success or failure of such industrial processes. In turn, the possibility of acting on the function of a microorganism by altering the medium would find several practical applications in Pasteur's subsequent therapeutic research.

Although his study of fermentation laid much of the foundation for Pasteur's later approach to therapy in animal and human disease, it developed in continuity with his crystallographic research. The orientation provided by his new duties at Lille led Pasteur to consider fermentation increasingly as a biological phenomenon, but we can see as much continuity as reorientation in this work. If we are looking for discontinuity or rupture in Pasteur's scientific career, the move from fermentation to the disease of silkworms is a more promising candidate.

In 1865, Pasteur was asked to help with a major crisis in French silk production—a disease affecting silkworms; it was a problem that had nothing to do with chemistry. The issue was of particular concern to Jean-Baptiste Dumas, who after a year as minister of agriculture (1850–1851) was now a senator. As a native of Alès in the south of France, an important centre for silk production, Dumas was desperately trying to save the region from economic disaster, and this meant calling on his most brilliant student. While he would have liked to decline the offer, Pasteur felt under the obligation to accept. Knowing little or nothing about silkworms or silk production he set to work reading up on the subject and then investigating the process at a laboratory in Pont Gisquet. What differentiated Pasteur's approach from his predecessors was his insistent use of the microscope to try and identify the cause of the disease, an approach that paid off in both theoretical and practical terms. Nevertheless, Pasteur and his collaborators had a great deal of difficulty in resolving the problems of the silk farmers because of the presence of not one, but two diseases: pébrine and flachery.

Although this work on silkworms occupied Pasteur for five years, he considered it as an interlude in his research career, so I just want to point out how it confirmed several points concerning the complexity of disease. First, his success in identifying microorganisms responsible for pébrine, and thus being able to use the microscope as a diagnostic tool, confirmed Pasteur's commitment to the use of this instrument to reveal pathogens. Second, the work on silkworms gave Pasteur some first-hand experience of the complexity and variability of disease. Pébrine, for example, did not affect all the infected silkworms in the same manner, and the occurrence of flachery was favoured by certain conditions, such as the humidity of the hatcheries.

While Pasteur's work on silkworms took him fully into the area of animal disease, by removing him from his program of culturing and experimenting with bacterial cultures it nevertheless took him away from an important line of research in just this area. Other animal diseases, anthrax in particular, would bring him back to this fundamental bacterial research, this time with a new therapeutic perspective.

Thus, Pasteur's study of anthrax constitutes another important watershed in his research. The highpoint of this anthrax research was the demonstration of the vaccine's efficacy made at Pouilly-le-Fort in 1881. Bruno Latour has made much of this public trial as being exemplary of the dramatic translation of experimental results outside the laboratory.[11] Here, I want to consider another episode in the anthrax story, Colin's chickens, as illustrative of Pasteur's general approach to animal disease. This story has the advantage of illustrating Pasteur's combative streak, his desire to impress medical (and veterinary) doctors, and his inventiveness in generating productive experimental observations. In short, this episode serves as a good example of Pasteur's modus operandi as an outsider with respect to the medical community.

On March 19, 1878, Louis Pasteur attended a plenary meeting of the *Académie de médecine* carrying a dead chicken and a cage containing two live ones. These unusual exhibits were the outcome of an ongoing dispute with Gabriel Colin, a teacher at the veterinary school at Alfort. According to Pasteur's account, Colin refused to accept his claim that chickens were immune to anthrax. Pasteur demanded a chicken infected with anthrax, which Colin was unable to supply. Instead, it was Pasteur who came to the Academy with a chicken that had died of anthrax. He had worked from the hypothesis that chickens were immune because their elevated body temperature (approximately 42°C) killed the bacteria. Assisted by Émile Duclaux and Jules Joubert, he had used a cold bath to reduce the chicken's body temperature and then succeeded in administering a fatal dose of anthrax. The two live chickens were controls, one having been injected with twice as much of the same culture and the other having been simply kept in the cold bath.[12] It is easy to appreciate the incongruity of Pasteur's presentation before a public accustomed to detailed reports of clinical cases. What counted for doctors of the time were the symptoms and typical course of a disease, as well as successful treatments—information gleaned from the doctor's experience on the wards and in private specialist practice.[13] This was the kind of information that the members of the *Académie de médecine* were interested in rather than speculations about a general theory of disease. Nevertheless, Pasteur's display had the merit of confronting the doctors of the august company with the reality of experimental disease in animals, although it did not of course oblige them to accept its pertinence with respect to medical practice.

Apart from the theatrical value of Pasteur's chickens, we can also note some key ideas at play in this episode. First, the virulence or danger posed by a pathogen depended on the environment, in this case the chicken's body. Change the body temperature and a harmless bacterium becomes deadly, and vice versa! On the model of Jenner's cowpox vaccine, Pasteur's goal would be to transform the germ rather than the host, and to do this he would rely on in vitro rather than in vivo techniques. It was the discovery in 1880 that a neglected chicken cholera culture rendered chickens immune to the disease rather than killing them that set Pasteur on the path to producing his vaccines. This discovery in turn depended on new techniques for cultivating bacteria, notably a range of cultures appropriate for different germs. Air, oxygen, and acidity provided the favoured means for attenuating the microbes; indeed, the details of each vaccine were the result of more or less precise variations in conditions, with the confirmation of the effects in test animals. Pasteur had come a long way from his study of crystals, although the use of the microscope and an insistence on experimental proof still linked him to this past research.

THE FINAL FRONTIER:
FROM VETERINARY TO HUMAN MEDICINE

Although Pasteur was acclaimed and decorated for his contributions to wine production and sericulture, it was his vaccine-based treatment for rabies that was responsible not only for his reputation as a benefactor of humanity, but also for his posterity in the form of the Pasteur Institute.[14] While the development of this vaccine was a difficult and uncertain process, the general principles on which it was based had been worked out for the vaccines against anthrax and chicken cholera. In 1882, Pasteur already proudly proclaimed his capacity, in principle at least, to conquer infectious disease thanks to the production of vaccines through a universal technique of attenuation.

> To summarize, we are convinced that we possess a general method of attenuation, whose application only needs to be modified to suit the exigencies of the physiological properties of the diverse microbes.[15]

Thus, again in principle, to combat a disease like rabies, Pasteur needed to isolate the germ responsible, cultivate it in an appropriate medium, reduce its virulence, and then use the attenuated product as a vaccine. While in the most general terms one could argue that the principle of treatment by attenuation was respected, in all the details rabies turned out to be quite different from anthrax. For a start, Pasteur never succeeded in isolating the microbe responsible, and so had to rely on cultivating rabies in vivo using a number of different animals, in particular the dog, guinea pig, and rabbit. After finding a technique for attenuating the virus (exposure of the spinal cord of infected rabbits to air), the long incubation period of rabies permitted the use of the vaccine (building up from a highly attenuated to a highly active dose) before the appearance of the characteristic symptoms.[16] This postinfection prophylaxis was a new departure, and the experimental use of the treatment in humans was justified only by the disease's deadly profile; after the appearance of the first symptoms, there was simply nothing that could be done.

On the path to the failed isolation and the successful attenuation of the rabies virus, Pasteur and his colleagues put several new techniques into place, including the transfer of the infection by trepanation and the exposure of the spinal cord to air (figure 2.1). This whole approach was so far outside the normal practice of clinical or pharmaceutical medicine at the time that in itself it could serve to qualify Pasteur as an outsider, or at least as a radical innovator in medicine. Furthermore, while the move from animal to human medicine may have been a small step in Pasteur's research trajectory, the scepticism of the medical community probably presented the biggest hurdle that he would ever have to overcome.

The passage to rabies took place while Pasteur was gathering together the group of researchers who would staff the future Pasteur Institute. Among the three leading figures, Emile Duclaux, Charles Chamberland, and Emile Roux, only Roux was a medical doctor. Although he was an enthusiastic member of the *Académie de médecine* in Paris, Pasteur did not have a similar passion for his colleagues. It was in this confrontation with the medical profession as it existed at the time that Pasteur experienced his most evident clash as an outsider. Indeed, the techniques and reflection behind the rabies treatment, while a continuation of Pasteur's work on anthrax and chicken cholera, were completely foreign to the clinical medicine tradition. While diseases could be classified by characteristic symptoms, the art of the doctor was to understand the individual case using intuition as much as experience. The idea of a uniform microscopic causal agent responsible for each disease was antithetical to this way of thinking. This said, it would be a mistake to think that only a chemist could think of disease and therapy in these terms. There were others with a medical training, like Casimir Davaine and Pierre Victor Galtier, who were exploring the same terrain combating pathogenic germs; anthrax for Davaine and rabies for Galtier. Still, such approaches remained marginal in medicine until the introduction of serotherapy a decade later. To back up this observation, we only need to think about the status of the rabies vaccine, which was never treated as a medicine. According to the law in force at the time in France, pharmacists had a monopoly over the preparation and sale of medicines and were expected to conform to the pharmacopoeia for these medicaments. Vaccines, including Pasteur's rabies vaccines, were not to be found in the pharmacopoeia and were rarely produced or distributed by pharmacists. Pasteur's treatment for

Figure 2.1: Louis Pasteur in 1887. Pasteur is represented in his laboratory at the *École normale* on rue d'Ulm, preparing his treatment for rabies. Note the absence of animals, the presence of microscopes and glassware and—at the center of the portrait—the rabbit's spinal cord, used to attenuate the virus.
Painting by Albert Edelfelt (1854–1905).

rabies was so extraordinary as a medicament that it effectively functioned outside the legal framework of French medicine.

CONCLUSION:
PASTEUR AS OUTSIDER

There are two ways that we can think about Pasteur's place as an outsider. First, we can consider what his move across domains brought to the various fields in which he worked, and second, how this disciplinary trajectory contributed to the construction of Pasteur as a scientist.

The most obvious effect of Pasteur as an outsider was not the introduction of chemistry into biology or medicine, but rather the development and orientation of microbiology itself.[17] He did not privilege the naturalist's goal of classifying what he saw through the microscope but always had an eye to the utility of microorganisms, first in industrial production and then in treating disease. While microbiology represented a challenge to biology due to the extent of the vistas that it opened up at the end of the nineteenth century, the germ theory of disease with its accompanying vaccine programme was a much more radical challenge to medicine, promising to overthrow a centuries-old tradition of clinical practice. The microscope was reinforced as a symbol of innovation for science-oriented medical doctors who wanted to distance themselves from preceding generations, and microbiology became an important element in reforms that would radically change all aspects of health care in the twentieth century. Nevertheless, we should not exaggerate Pasteur's role in these wider movements. Although the French historiography tends to place Pasteur first above the schools of Koch and Cohn in Germany, the Germans preceded and went much further in diagnostic identification and taxonomy than the French. Furthermore, microbiology was only one of many factors contributing to the reconfiguration of modern medicine, with the Germans leading the way in histology, for example.

It is likely that not being a medical doctor helped Pasteur in his approach to developing a treatment for rabies, as he had no allegiance to the clinical tradition that emphasized the diversity of experience at the bedside rather than the uniformity aimed for in experimental medicine. Nor was he bound up in the broader concerns of medical physiology that determined the research of his colleague and friend Claude Bernard. The analogy between infectious disease and fermentation provided the basis of the theory and practice of attenuation behind his vaccinations, and Pasteur applied therapy in humans in continuity with that in animals, an unacceptable idea for the majority of French doctors.

Turning the question around, then, what did his passage through different fields bring to Pasteur? Most evidently, his move from chemistry through biology to medicine brought him a level of public recognition and wealth that he never could have hoped for had he studied crystal structure for the whole of his career. His "miracle" treatment for rabies formed part of a revenge narrative that was central to the defence of French national pride following the defeat by the Prussians in 1870. Pasteur fitted the role of the scientific genius who overcame the prejudices of the conservative medical community to introduce his new cure for the incurable disease of rabies, thereby saving the honour of France.

In conclusion, the category of outsider for Pasteur seems a lot more relevant in the case of medicine than biology, which had not yet constituted a comparable institutional, let alone professional, identity. Furthermore, microbiology was being constructed by the work of Pasteur and his contemporaries, and would only be fully integrated at a later stage into a broadened conception of biology.

FURTHER READING

Dagognet, François. *Méthodes et doctrine dans l'oeuvre de Pasteur*. Paris: Presses Universitaires de France, 1967.

Debru, Claude. "L'interdisciplinarité et la transdisciplinarité dans l'œuvre de Louis Pasteur," accessed December 4, 2012. http://www.snv.jussieu.fr/vie/dossiers/pasteur/index.htm

Geison, Gerald. *The Private Science of Louis Pasteur*. Princeton: Princeton University Press, 1995.

Latour, Bruno. *The Pasteurization of France*. Harvard: Harvard University Press, 1993.

Vallery-Radot, René. *La vie de Pasteur*. Paris: Flammarion, 1900. English translation, Mrs R. L. Devonshire, *The Life of Pasteur* (New York: Phillips, 1902).

NOTES

1. Pasteur was not the first French scientist to receive a state funeral; his friend Claude Bernard had that honour a few decades earlier. But while Claude Bernard was recognised as a great scientist, he never had the popular appeal achieved by Pasteur.

2. The French Republican educational system provided not only free access to state-funded schools based on competitive entrance exams, but also various scholarships that allowed promising children from modest backgrounds to live and study in Paris.

3. Gerald Geison, *The Private Science of Louis Pasteur* (Princeton: Princeton University Press, 1995), chapter 3 on crystallography, particularly 58–85.

4. Quoted in René Vallery-Radot, *The Life of Pasteur*, trans. Mrs R. L. Devonshire (New York: Doubleday, 1919), 199.

5. Louis Pasteur, "Recherches sur les relations qui peuvent exister entre la forme cristalline, la composition chimique et le sens de la polarisation rotatoire," *Annales de chimie et de physique* 24 (1848): 442–459.

6. François Dagognet, *Méthodes et doctrine dans l'oeuvre de Pasteur* (Paris: Presses Universitaires de France, 1967).

7. Theodor Schwann, for example, had argued that fermentation was the function of a living organism.

8. Much has been written on these debates over spontaneous generation, including Nils Roll-Hansen, "Experimental method and spontaneous generation: The controversy between Pasteur and Pouchet, 1859–64," *Journal of the History of Medicine and Allied Sciences* 34 (1979): 273–292, and chapter 5 of Geison, *Private Science*.

9. Vallery-Radot, *Life of Pasteur*, 113.

10. Emile Duclaux, *Ferments et maladies* (Paris: G. Masson, 1882).

11. Bruno Latour, "Give me a laboratory and I will raise the world," in *Science Observed: Perspectives on the Study of Science*, eds. Michael Joseph Mulkay and Karin Knorr-Cetina (London: Sage, 1983), 141–170.

12. Pasteur gleefully recounts the whole story himself. Louis Pasteur, *Oeuvres de Pasteur* (Paris: Masson, 1922), vol. 6, 205–211 (originally published in the *Bulletin de l'académie de médecine*).

13. For a history of medical specialization, see George Weisz, *Divide and Conquer: A Comparative History of Medical Specialization* (Oxford: Oxford University Press, 2006), particularly chapter 1, "The rise of specialities in early nineteenth-century Paris."

14. Founded in 1887 and inaugurated in 1888, rabies was the justification for the fund-raising that paid for the Pasteur Institute.

15. "L'atténuation des virus" speech made at the International Hygiene Conference in Geneva, 1882 (Pasteur, *Oeuvres*, vol. 6, 403).

16. Like Pasteur, I use the term "virus" quite loosely. At this time, virus, germ, and microbe were used interchangeably to denote the microscopic agent held to be responsible for the disease.

17. For an overview of the development of microbiology, see Olga Amsterdamska, "Microbiology," in *Cambridge History of Science*, vol. 6, *The Modern Biological and Earth Sciences*, eds. Peter Bowler and John Pickstone (Cambridge: Cambridge University Press, 2009), 316–341.

3

FELIX D'HERELLE
UNCOMPROMISING AUTODIDACT

INTRODUCTION

To his biographers, Félix d'Herelle (see figure 3.1), the discoverer of bacteriophage, is indeed an "international man of mystery." Not only is his place of birth uncertain, but the name he went by evolved over his lifetime. He was an autodidact who left school at age seventeen, worked alone most of his life far from the centers of academic science, and was in nearly constant conflict with the accepted ideas of his time. His complete lack of connection to the two major centers of the newly developing study of microbes forced him to rely on his own ideas, his own observations, and his own judgments; truly, he was an outsider looking in. His work ranged from fermentation chemistry to bacteriophage therapy, involved five continents, challenged Nobel Prize winners, included novel theories of life, and was rejected by the major textbook authors of his era. Yet his work led to the current use of bacterial toxins in pest control and to fundamental advances in molecular biology made possible by his discovery of bacteriophage. Clearly, d'Herelle is a case study of an outsider.

This case study, however, is not just a description of a quirky, sometimes irascible individual, but an exemplar of the evolution from the nineteenth-century amateur ideal to the professionalization of science in the twentieth century. Especially in Britain—and d'Herelle was certainly committed to the Baconian ideal of British science—science was the province of "amateurs," the "virtuosi" of the Royal Society, who explored nature for the love of the subject and, as some have argued, for the social status such knowledge conferred. It might be argued that without professional science, the category of "outsider" has little meaning, however, and indeed, in the nineteenth-century context, "outside" is more likely to mean individualistic or independent. As the historian Morris Berman noted: "The lack of scientific organization in the nineteenth century was paralleled by a series of discoveries almost pathologically individualistic."[1] D'Herelle started his work as a nineteenth-century amateur (in the traditional sense of the term) and would in the twentieth century become an ardent advocate of a new field

Figure 3.1: Felix d'Herelle.
Courtesy of the d'Herelle family.

that he called "protobiology," one characterized by all the intellectual commitments and organization that define boundaries between "inside" and "outside."

EARLY YEARS

Félix Hubert d'Herelle (the name he eventually adopted) was born on April 25, 1873, and died February 22, 1949. Even the place of d'Herelle's birth seems controversial: in his autobiography[2] as well as numerous surviving passports and government documents, he gives his place of birth as Montreal and his nationality as Canadian. However, records of both the Paris city archives (a birth certificate) and the church of Saint-Philippe-de-Roule de Paris (a baptismal record) indicate that he was born in Paris.[3] D'Herelle's earliest childhood memories were of school in Paris about age six. One is tempted to resolve this uncertainty by noting that his Parisian birth record indicates "father unknown" and mother as one Augustin Josephe Haerens (age twenty-four), living in Paris. Perhaps, d'Herelle was never even aware of his illegitimate origin in Paris and truly believed that Montreal was his native city.

The family name, too, is problematic. His birth certificate indicates his name as Hubert Augustin Félix Haerens. School records in Paris attest to a Félix Haerens as a student, and a military enrollment record also shows his name as Hubert Augustin Félix Haerens. His first publication, a newspaper article on millenarianism in January 1899 in a provincial Quebec weekly, was signed "F. Hoerens."[4] By August of the same year, however, Félix and his brother Daniel, using the surname "Haerens d'Herelle," bought land in Longueuil, a suburb of Montreal, on which to construct a chocolate factory. In early 1900 he was identified in legal papers as "Hubert (Félix) Haerens d'Herelle (chimiste)." By 1901 when he submitted a paper to *Le Naturaliste Canadien*, he had settled on what became his final name, "Félix d'Herelle."[5]

But the question remains: where did "Herelle" come from? In French, the particle *dit* ("called") has been appended to a surname to assert claims to nobility or ancient heritage. Thus, Félix and Daniel appear to have added "dit Herelle," or with the contraction, d'Herelle, to indicate a family connection, most likely to their French-Canadian father. D'Herelle's grandson related a family belief that the form dates back to the valiant acts of an ancestor at the siege of Metz in 1552.[6] Whether this account is true or not, the grandson has recently appended "d'Herelle" as a family name himself.

Young Félix appears to have attended two of the most prestigious schools in Paris, the Lycée-Louis-Le-Grand, and the École Monge (now the Lycée Condorcet). However, he left before completing the course of study at about age seventeen.[7] Beyond this schooling, it appears that d'Herelle was essentially self-taught. After a few years of leisurely travels, he signed on with the French military at age twenty for a term of four years. It seems, however, that military life did not suit him, for his military record abruptly ends almost thirteen months later, with the note that he deserted on November 25, 1894. This desertion may have been reason enough to adopt a new Canadian identity as well as explain the inconsistencies in his chronologies at this period of his life.

His early rigorous French academic education, coupled with the absence of advanced education under experienced mentors, set the stage for most of d'Herelle's mature life: a profound belief in a life of reason. Repeatedly he referred to the founders of the French Enlightenment as his guides: Malbranche and the Port Royal philosophers, with a generous balance of Baconian experimentalism. In later life he summed up his views in an unpublished, three-volume manuscript, *La Valeur de l'Experience*,[8] a remarkable attempt at a consistent philosophy of life synthesizing Watsonian behavioral psychology, Baconian empiricism, and neo-Lamarckian French thought. It is likely that d'Herelle's self-education and his own philosophical reflections provided him with an unusual sense of certainty and self-confidence, both qualities very much in evidence in all of his scientific work and the controversies they generated.

His life between ages twenty-one and twenty-four apparently was one of travel and self-indulgence, but in the spring of 1897 he noted: "I was 24 years old, it was time for me to make some choices: the conclusion was that it was wiser to return to the country where I was born [referring to Canada], and then I would see what would happen. I was, moreover, always thinking about bacteriology, so on my arrival I set up a laboratory and began to experiment, all alone because at this time there were only two French-Canadians interested in microbes, Dr. Bernier, who was later the first professor of this subject at the University of Montreal, and myself."[9]

As part of his self-education in bacteriology, he subscribed to key journals in the field, specifically naming *Annales de l'Institut Pasteur*, *Comptes Rendus de la Société de Biologie de Paris*, and *Centralblatt für Bakteriologie*. But he still was leading a "life of idleness" with pleasure travels, perfecting his technique in his little laboratory, and occasionally attending a summer course in Europe. A major change in his life occurred in 1897 when he was offered an opportunity to put his bacteriological knowledge to work by Sir Henri Joly de Lotbiniere, the Minister of Inland Revenue for Canada. Sir Henri was concerned over the falling prices for Canadian maple syrup in the U.S. market, and he reasoned that the excess crop might be fermented into a sort of whiskey for the Americans. Young Félix was given the commission to investigate the feasibility of this idea. Sir Henri was an old friend of d'Herelle's father, it seems, so this commission may have been a legitimate project or simply a helpful nudge to direct a young family friend into gainful work.

It is in this early work on fermentation that d'Herelle's scientific style began to emerge. Self-consciously following Pasteur, he realized that the nature of the yeast used in the fermentation would be crucial, so he meticulously cultured yeasts from many sources, both local and exotic, even writing to tropical fruit dealers in New York City for samples of sugar cane from which to isolate yeast strains. Unfortunately for this endeavor, the U.S. market quickly recovered and his fledgling distillery was doomed. Undaunted, however, Félix and his brother Daniel embarked on another venture: they built a chocolate factory on land in the Montreal suburb of Longueuil. The Herelle Chocolate Works made chocolates, some flavored with vanilla, sold for "health." This enterprise, too, soon failed, yet their impressive two-story brick structure remains today as the centerpiece of a small industrial park (Herelle Place).

His first formal scientific paper appeared from work in his small home laboratory and addressed no small problem: the global balance of carbon. This paper was entitled "De la Formation du Carbone par les Végétaux"[10] and was a direct challenge to the existing understanding of chemical theory. He reasoned that the absorption of carbon dioxide by the plant life on earth was insufficient to account for all the carbon in the plants, and therefore, they must obtain carbon from some other source. Since direct uptake of carbonates from the soil was not believed possible, he suggested that perhaps plants have some biosynthetic way of making carbon from simpler substances. This hypothesis, of course, assumes that carbon is not a chemical element, but rather a chemical compound, in the same way that ammonia is a compound of the elements nitrogen and hydrogen. To experimentally test his idea, d'Herelle carried out a complicated experiment to measure the carbon dioxide evolved from the germination of radish seeds in an environment that carefully eliminated any source of incoming carbon dioxide. When he observed that the seeds that sprouted into rather anemic little radish plants produced a small quantity of carbon dioxide, he concluded: "With all these facts, it means that the plants themselves make the carbon that they need, and therefore the conclusion of this experiment, a conclusion very important from the chemical point of view: carbon is not an element." He went on to boldly assert, "If carbon is not an element, the two substances with which it has the best analogies, boron and silicon, may not be [elements] either."[11] The next issue of the journal carried a rather derisive letter to the editor which noted that until the carbon content of the radish seeds and plants were known, the experiment showed nothing. This would be the first in a lifetime of bold and iconoclastic challenges that d'Herelle would send out into the world of science.

About this time, young Félix d'Herelle, self-confident and ambitious, answered an advertisement for a bacteriologist needed by the government of Guatemala. In 1901 he packed up his family, his small home laboratory, and headed to Central America, where the next phase of his scientific career would unfold.

THE START OF A CAREER

As one of two government scientists in Guatemala, d'Herelle had many assignments, but it was in his study of coffee blight that his flair for innovative investigation came to the fore. Since before 1900, a mysterious disease of the coffee plants had been spreading in some of Guatemala's major agricultural regions. The vascular tissue was infected with a black powdery fungus, which d'Herelle identified as a new species he named *Pthora vastatrix* after several laboratories in Europe to which he sent samples were unable to identify it. While he could transmit the disease between plants in natural growing conditions, when he cultured the fungus in the laboratory, using agar supplemented with extracts of coffee plants, he was unable to transmit the infection to healthy plants growing in the typical alkaline soil. He noted two important ecological features of the disease: the fungus grew well in acidic soil, while there were disease-free sections in the part of the country most heavily infected, and those healthy regions had recently been blanketed with the alkaline ash from a recent eruption of the nearby Santa Maria volcano. D'Herelle reasoned that soil pH was a primary determinant in the spread of the coffee blight, and he suggested immediately that a program of soil liming might be protective.

This investigation of the coffee blight in Guatemala is exemplary of what would become d'Herelle's basic approach to microbiology, an approach now characterized as "ecological." He saw infections as a complex interplay of microbes, hosts, and environmental factors that determined the observed outcome. Repeatedly, this mode of thought would be applied to each problem he faced in his long research career. It was novel at the time, and no doubt was one reason he was often seen as an "outsider," one whose conceptions of biological problems seemed unfamiliar, overly complicated, and sometimes fanciful.

Microbiology in the last decades of the nineteenth century was not yet a mature field; the Society of American Bacteriologists, for example, was founded only in December 1899. There were two competing schools of microbiology, one French, led by Pasteur, and the other German, led by Koch, which in the later decades of the nineteenth century and early decades of the twentieth century effectively defined the boundaries of this nascent discipline. To be sure, Joseph Lister in London, and to a lesser extent, scientists such as Daniel Salmon in the United States, were important members of this "invisible college" led by the Europeans. These schools established the theories, methods, and questions that defined this new field. Most scientists who called themselves bacteriologists (rarely microbiologists) traced their scientific "lineage" one way or another to either Louis Pasteur or Robert Koch. Félix d'Herelle, however, was not part of this scientific family. While he thought of himself as a disciple of Pasteur, it was an aspiration, not a fact. He had to improvise methods on his own, having no access to the "tacit knowledge" of the insiders. He had to follow research entirely from the journals and the few available textbooks and monographs. Without communication with the communities which set goals and adjudicated theories, he had to fill in gaps in his understanding based on his own reason and his own experience. It helped that his personality was suited to this independence and that his mind was powerful enough to meet these challenges of being outside existing disciplinary boundaries.

D'Herelle considered his time in Guatemala as his scientific coming-of-age period. In his later recollections, he wrote: "When I think of Guatemala, it is always with affection. It is where I carried out my apprenticeship in life, where I commenced my scientific career. Obliged to occupy myself with questions of a great variety of interests, microbiology, hygiene, medical examination, mycology, fermentation, chemistry, botany, agriculture, all these constituted a training which were a great aid to me all my life."[12] In spite of this affection, in 1907 he accepted a position with the Department of Agriculture in the Yucatan in Mexico based on his knowledge of fermentation; he was charged with developing a commercial alcoholic fermentation of the residue of the sisal plant (the fiber of which was used for rope). He would use this Mexican opportunity to make two world-shattering observations: that bacteria can be used as an effective insect control agent, and that there are submicroscopic agents that infect bacteria. The former discovery would provide him with fame and some fortune, and the honor of being the founder of biological pest control, while the latter would lead to the discovery of bacteriophage, first exploited as primitive antibacterial therapy, and later as the tool for unlocking the mysteries of the gene and the development of molecular biology. Both discoveries were the direct result of his keen ecological approach, his independence of thought, and his unshakable self-confidence.

After removal of the fibers from the sisal plant, a succulent relative of the more well-known agave plant used to make tequila, the residue is known as bagasse, normally discarded as waste. It seemed reasonable, by analogy with tequila production from agave, to recover material from bagasse by fermenting it into distillable alcohol. His work on fermentation carried out in the region around Merida in the Yucatan, carefully documented in reports to the Mexican government and in at least six patents, went well, and by June 1910 he had a fully operational ethanol production plant up and running in Merida, producing 1,200 liters a day. At this point d'Herelle saw no further scientific challenges, so made plans to move to Paris where he intended to join, finally, the center of microbiology, the famed Pasteur Institute. While waiting for the arrival of his replacement, always restless, he made frequent forays outside the laboratory to study the country, its peoples, and its natural resources. He later recalled, "A cloud of locusts descended on the little park near my home. My first thought was 'if I would be able to find a disease of locusts . . .' "[13] Again, thinking ecologically, he asked the farm manager to be on the lookout for any locusts that appeared to be sick or diseased. Shortly d'Herelle was presented with a container of three dead locusts. He autopsied many dead locusts and disputed the accepted diagnosis that the disease of locusts was caused by larvae of a parasitic fly of the genus *Muscidae*, and instead suggested that, since the intestines contained large amounts of a black material packed with a uniform coccobacillus, an intestinal infection with bacteria was a more likely cause. Unfortunately, because his locusts had all been preserved in alcohol and the locust plague had passed, his attempts to culture the organism were frustrated.

D'Herelle sent this work for publication in the *Journal d'Agriculture Tropicale,* where it was published in the August 1910 issue.[14] The rhetorical style that he used all his life was already well developed in this paper. And why not? It was straight out of Aristotle, a philosopher who certainly must have appealed to him. His paper opened with his statement of the problem, reviewing what he considered the relevant prior knowledge, followed by systematic, unfavorable critiques of these prior views, and then presentation of his own views and arguments followed by the conclusion that his views must prevail. While the *Journal* published his report, they did so with a commentary by the leading authority on insect pathology at the Museum of Natural History in Paris, Jules Künkle d'Herculais, who suggested, dismissively, that the bacteria that d'Herelle noted were no doubt postmortem saprophytic organisms and that the fly larvae had simply left the insect prior to death of the locust, thus accounting for d'Herelle's failure to observe them in his autopsies.

Soon, however, d'Herelle was able to study unpreserved specimens and isolated a pathogenic organism from the intestines of the locusts that he named *Coccobacillus sauterelle* (later renamed *C. acridiorum*), which he showed could infect and kill healthy locusts. He ended his first publication on this organism, in true Pasteurian fashion, with the bold proposal to employ cultures of this organism in the field to end locust plagues wherever they occurred.[15] Because this paper was published in the prestigious *Comptes Rendus de l'Académie des Sciences*, submitted by no less a personage than Emile Roux, the director of the Pasteur Institute, d'Herelle's reputation grew rapidly. He was now far from being an obscure, lone microbiologist on the periphery of the cultured European world, and was even invited to meet the Argentine minister to France, who gave him a commission to exterminate locusts in Argentina. Soon he and his family were on their way to Buenos Aires.

FIRST SUCCESSES

Field trials of d'Herelle's coccobacillus exposed the promise and problems with this approach to pest control. *C. sauterelle* was an organism that lost virulence upon prolonged passage in culture, thus necessitating repeated re-isolations from infected locusts. Further, since it was not a spore-forming organism, fresh cultures had to be prepared very near the sites of application. D'Herelle spent much effort establishing large-scale culture facilities at distant agricultural sites. While large field trials were not unknown—indeed, in Argentina it was standard to build vast fences of zinc sheeting set deep in the soil to create miles and miles of a "great wall" against the locusts—the biological-ecological approach was novel. Although one can see hints of ecological thinking in the work of the German school (for example, Max von Pettenkofer's ideas on cholera and ground water), the predominant mode of thought was reductionist. D'Herelle, however, valued experience in all its ramifications as the basis for theorizing, and it was this global approach that often made more sense to bureaucrats than it did to scientists intent on throwing off the legacy of nineteenth-century vagueness via their commitment to controlled laboratory studies. Both of these shortcomings would be eventually overcome by others following in d'Herelle's footsteps, the most successful approach being the introduction of *Bacillus thuringiensis*, also a wide-spectrum insect pathogen, but a spore-forming organism that thus overcame the two main deficiencies of d'Herelle's coccobacillus, loss of pathogenicity on passage and instability on storage.

After his initial trials of the coccobacillus in Argentina, d'Herelle's fame spread, and he carried out field trials in Columbia, Algeria, Tunisia, and Cyprus. These endeavors, however, were not without their detractors. In

Argentina, for example, his style and lack of political sensitivity soon ran afoul of the entrenched establishment in the Ministry of Agriculture, which disputed the efficacy of his approach to locust control and objected to the fact that he was a foreigner.[16] Indeed, the entire episode was later satirized by a famous Argentine author, Arturo Cancela in his short story "*El Coccobacilo de Herrlin*," a transparent roman à clef involving a nonexistent plague of rabbits routed by a pathogenic coccobacillus.[17]

By the 1930s and 1940s, authorities in the field of insect control and pathology were crediting d'Herelle with the first application of the principle of biological pest control. It was d'Herelle's Pasteurian colleagues Serge Metalnikov and Rudolph Wihelm Glaser in the United States who showed the value of the more practical organism, *B. thuringiensis*, as a biological insecticide in the late 1920s. This organism is now widely used, both in its intact form as Bt spores and by insertion of the Bt toxin gene into crops to provide genetically engineered protection against insect pests.

It was d'Herelle's chance observation of "cultural irregularities" in his coccobacillus cultures as well as his deepening belief in a distinction between "natural disease" and "artificial disease" that would eventually lead him to his most famous scientific work, the discovery and characterization of bacteriophage. Having obtained a post as an unpaid researcher at the Pasteur Institute (not uncommon in those days of "gentlemen scientists"), d'Herelle followed the Institute's interest in enteric diseases and their possible prevention by immunizations. In the summer of 1915 he investigated an outbreak of unusually severe hemorrhagic dysentery among a group of French soldiers in a garrison near Paris.[18] His interest in "natural" infections and his ecological views led him to consider the special nature of this particularly severe epidemic. He attributed his approach to his understanding of recent work on hog cholera by de Schweinitz and Dorset,[19] who had shown that this important agricultural disease was not caused by *Salmonella cholerasuis* (the hog cholera bacillus) but by a filterable virus (now known to be a small RNA-containing classical swine fever virus, or CSFV). The pathogenicity of this virus for other species, however, was variable, and these authors speculated that other co-infections, perhaps with *S. cholerasuis*, influenced the severity of the CSFV disease in some hosts. It was this belief in multiple pathogen interactions that led d'Herelle to his first observation of bacteriophage.

In experiments based on the hypothesis of de Schweinitz and Dorset, d'Herelle passed the cultures of the dysentery bacteria from the sick soldiers through a filter that was known to remove bacteria but that allowed passage of some other infectious agents, then described as "filter-passing virus" (known examples at the time included tobacco mosaic disease, foot and mouth disease, smallpox, vaccine, rabies, and poliomyelitis). He then added this bacteria-free filtrate *back* to cultures of the dysentery bacteria to see if there were two components in the material, one filter-passing and one not. When he spread this mixture on an agar culture medium, he observed what he christened (rather unimaginatively) "taches" or "spots." These spots later became "taches vierges" (clear spots) and later "plaques," the term still in use today. In later recollections, d'Herelle said he immediately connected these spots with the "cultural irregularities" observed several years earlier in some cultures of *C. sauterelle*. The presence of a second submicroscopic and filter-passing organism in his dysentery cultures was confirmation of his belief in the complex mechanisms involved in natural infections.

When he tried to culture bacteria from the spots, there were no live bacteria, but when he removed material from the clear spots and added it to fresh dysentery cultures, more spots appeared. He concluded that the spots represented "colonies" of some ultramicroscopic organism growing at the expense of the bacteria, just as bacteria grew as isolated colonies on a solid medium. He could count the ultramicroscopic organisms, which he named "bacteriophage," just as one could enumerate bacteria by dilutions and colony counting. With this new quantitative method, he showed by serial infections of fresh bacteria that the bacteriophage multiplied stepwise and indefinitely. While this discovery of a heretofore unknown submicroscopic agent might have been exciting enough, d'Herelle continued his investigations to determine what role, if any, such an organism might play in natural infections.

By carefully studying fecal samples during the course of the diarrhea in his population of sick soldiers, d'Herelle noted that as the symptoms and bacterial count subsided, the phage titers rose, and he made a direct leap to assert that recovery from natural infections resulted from the action of the endogenous phages killing off the invading pathogenic bacteria. In effect, the balance between phage and bacteria determined the course of a natural infectious disease. In his very first publication, in which he described the discovery of bacteriophage and his method of plaque counting for quantitation, he also claimed that this ultramicrobe is "the true microbe of natural immunity, an obligatory bacteriophage with strict parasitic specificity, acting on a specific species and at a given time, in turn, on various germs to which it is adapted."[20]

In true Pasteurian fashion, d'Herelle sought direct application of his laboratory discoveries to practical problems, anticipating by nearly a century the current mantra of "translational research." He reasoned that the "natural immunity" and recovery provided by bacteriophages could be augmented by therapeutic application of phage stocks already prepared in the laboratory, in what was again an ecological approach to the treatment of infections, this time for bacterial dysentery rather than locust infestations. After a bit of self-experimentation to show its safety and with his unbounded enthusiasm, d'Herelle convinced a noted Parisian pediatrician, Professor Victor Henri Hutinel, to conduct a clinical trial of phages on children with diarrhea. His field tests extended to chicken farms in France suffering from avian cholera epidemics as well as water buffalos in Vietnam experiencing bovine hemorrhagic septicemia. The results were promising enough for him to elaborate his full-blown theory of natural infections in a 1923 monograph, *Les Défenses de l'Organisme*,[21] which was later expanded into an English version, *Immunity in Natural Infectious Disease*.[22] Having already proclaimed the immunological work of the Nobelist Jules Bordet to be "an error,"[23] d'Herelle elaborated his own theory of immunity based on the role of the bacteriophage. Not content to base the entire field of immunity and resistance to natural infections on his discovery, he also developed a successor to the "cellular theory of life" with his own "colloid theory of life." He thought of bacteriophages as the basic units of all life processes and, indeed, proposed that they were the first step in the origin of life from inanimate matter. He coined the term "protobe" for bacteriophages, and when he took up a professorship at Yale in 1928, was appointed "professor of protobiology," probably the only scientist to ever hold such a title. Yet even at Yale he was still an outsider; he was irascible to his colleagues, spent half of each year in Paris, where he was commercializing phage therapy (at a time—no longer with us—when faculty businesses were frowned upon), and after several run-ins with the Yale administration was forced out.

Certainly d'Herelle's most famous dispute with the scientific mandarins was over the biological and physical nature of bacteriophage. While he maintained—correctly, as it turned out—that phages were tiny particulate entities with the capacity to multiply as obligate intrabacterial parasites, his opponents, who included Jules Bordet and nearly all the other famous microbiologists of his time, thought of phages as akin to an enzyme that required some sort of activation. Thus, they reasoned, cells harbored a natural precursor to bacteriophage that was activated upon treating that cell with external phage in some sort of autocatalytic or "vitiation" reaction. The historian Ton van Helvoort has argued that this dispute was based on a conflict between d'Herelle's "bacteriological style" and Bordet's "physiological style."[24] Again, d'Herelle as the autodidact relied on his own theoretical framework, his own reason, and his own self-confidence.

This dispute over the nature of bacteriophage became even more elaborate as Bordet and his colleagues challenged d'Herelle's priority claims to further attack him. In 1915, a British microbiologist, Frederick W. Twort, published observations on "glassy transformation" of bacterial cultures that was serially transmissible.[25] Twort was not able to follow up on his observation, nor did he advance any explanation for his phenomenon. For d'Herelle, the elegance and simplicity of the plaque assay and its quantitative behavior was evidence enough to prove the particulate nature of the phage as distinct from the hypothesis of its enzymatic nature. Still, it was not until more physically minded researchers entered the field in the late 1930s and the direct visualization of bacteriophage with the newly invented electron microscope that d'Herelle's conception became universally accepted.

Over the two decades from the mid-1920s until his death in 1949, d'Herelle's main ideas about microbiology went from being fringe science to becoming accepted dogma. Yet even as his ideas were accepted, he remained an outsider. In 1922, d'Herelle openly criticized the use of the BCG (*bacillus Calmette-Guérin*) vaccine in children, and soon he was forced out of the Pasteur Institute by Bordet's protégé and developer of the vaccine, Albert Calmette, who declared him *persona non grata*. It was only at the urging of a younger generation of Pasteurians that twenty-five years later he was invited to speak at the institute when it commemorated the thirtieth anniversary of his discovery of bacteriophages in 1947. This attempt at reconciliation, however, did not go well, as d'Herelle took the opportunity to launch new assaults on the hegemony of the small-minded thinkers in the audience. To the end, he remained the iconoclast, the free-spirit with his own, self-assured, reasoned views. It is instructive to consider his own statement of his philosophy: "I have invariably started by making a logical analysis, which consists of formulation for one's consideration all possible hypotheses, even those which at first glance appear dismal, absurd or opposed to accepted theories, then subject each of these diverse preliminary hypotheses to experimental verification to decide in the last resort which of these hypotheses conform to the facts, that is to say, to reality. 'A theory must take into account all of the facts, all of the explanations, and contradict none': such is the axiom which I have advocated in many writings—in short, just Logic."[26]

CONCLUSION

Félix d'Herelle was able to make major scientific contributions without the qualities usually assumed to be a prerequisite to world-class standing in science: he was an autodidact from about the age of seventeen; he lacked both patrons and mentors; until mid-career he had no stable institutional position in mainstream science; and his scientific and rhetorical styles were heterodox, to say the least. As an autodidact, d'Herelle developed his own philosophy of science without the guidance or moderating influence of experienced colleagues and mentors. Early in his career, he adopted an idealistic approach to science that he saw as an extension of classical logic. A belief in truth as revealed by empirical observation gave him the courage of his convictions, and he was impatient with those who did not recognize or accept his reasoned approaches to natural knowledge. His confidence in his ability to understand and to develop the empirical means to study phenomena were at odds with the developing nineteenth- and twentieth-century trends of specialization and knowledge collectives. His willingness to consider complex systems of nature in what is now considered an ecological manner was at odds with the growing reductionist trends in twentieth-century experimental science.

Perhaps his early peripatetic life and his rejection of any authority beyond logic and reason led to his neglect of diplomatic skills in scientific controversy. Not only was he blunt in his criticism of what he saw as error, he was equally confident in his own ability to reason out grand theoretical systems for understanding infectious disease as well as the nature and origin of life itself. He seemed to be a man out of his time, perhaps more at home in the period of the great European philosophes such as Voltaire, Descartes, Newton, Malbranche, and Rousseau.

FURTHER READING

Häusler, Thomas. *Viruses vs. Superbugs: A Solution to the Antibiotics Crisis*. Basingstoke: Macmillan, 2006.

Summers, William C. *Félix d'Herelle and the Origins of Molecular Biology*. New Haven: Yale University Press, 1999.

Van Helvoort, Ton. "Research Styles in Virus Studies in the Twentieth Century: Controversies and the Formation of Consensus." PhD diss., University of Limburg, 1993.

NOTES

1. Morris Berman, "'Hegemony' and the Amateur Tradition in British Science," *Journal of Social History* 8 (1975): 40.

2. Félix d'Herelle, "Les Périgrinations d'un Microbiologiste" (unpublished manuscript, ca. 1940–1947), Archives de l'Institut Pasteur, item FR AIP HER1. All translations by the author.

3. Alain Dublanchet, "La Vraie Vie de Félix d'Herelle avant la Découverte du Bactériophage," *Association des Anciens Elèves de l'Institut Pasteur* 175 (2004): 80–82.

4. F. Hoerens [Félix d'Herelle], "Comment le Monde Finira-t-il? Quatorze Fin du Monde," *La Patrie*, January 14 and 21, 1899.

5. Félix d'Herelle, "De la Formation du Carbone par les Végétaux," *Le Naturaliste Canadien* 28 (1901): 70–75.

6. William C. Summers, *Félix d'Herelle and the Origins of Molecular Biology* (New Haven: Yale University Press, 1999), 3.

7. Dublanchet, "La Vraie."

8. Félix d'Herelle, "La Valeur de l'Experience" (unpublished manuscript, ca. 1940–1947), Archives de l'Institut Pasteur, item FR AIP HER4-5.

9. D'Herelle, "Perigrinations," 38

10. D'Herelle, "De la Formation."

11. D'Herelle, "De la Formation."

12. D'Herelle, "Perigrinations," 143.

13. D'Herelle, "Perigrinations," 208.

14. Félix d'Herelle, "Note sur une Maladie des Sauterelles au Yucatan," *J. d'Agriculture Tropicale* 10 (1910): 237–238.

15. Félix d'Herelle, "Sur un Épizooie de Nature Bactérienne Sévissant sur les Sauterelles au Mexique," *Comptes Rendus Acad. Sci. Paris* 152 (1911): 1413–1415.

16. D'Herelle, "Perigrinations," 285–287.

17. Arturo Cancela, *Tres Relatos Porteños* (Madrid: Calpe, 1923).

18. Georges Bertillon, "Une Épidémique de Dysentérie Hemorragique dans un Escadron de Dragons," *Ann. de l'Institut Pasteur* 30 (1916): 141–144.

19. Emil A. de Schweinitz and Marion Dorset, "A Form of Hog Cholera Not Caused by the Hog Cholera Bacillus," *Bureau of Animal Industry: Circulars* 41 (1903): 1–4; 43 (1903): 1–3.

20. Félix d'Herelle, "Sur un Microbe Invisible Antagoniste des Bacillus Dysentérique," *Comptes Rendus Acad. Sci. Paris* 165 (1917): 373–375.

21. Félix d'Herelle, *Les Defenses de l'Organisme* (Paris: Flammarion, 1923).

22. Félix d'Herelle, *Immunity in Natural Infectious Disease* (Baltimore: Williams and Wilkins, 1924).

23. D'Herelle, *Immunity*, 162.

24. Ton van Helvoort, "Bacteriological and Physiological Research Styles in the Early Controversy on the Nature of the Bacteriophage Phenomenon," *Medical History* 36 (1992): 243–270.

25. Frederick W. Twort, "An Investigation on the Nature of Ultramicroscopic Viruses," *Lancet* 2 (1915): 1241–1243.

26. D'Herelle, "Perigrinations," 73–74.

THE PARADOX OF SAMUEL BUTLER
INSIDER OR OUTSIDER?

INTRODUCTION

Samuel Butler, the English writer, was born in 1835 and died in 1902 (see figure 4.1). He was therefore a complete Victorian (the queen was on the throne from 1837 to 1901). His grandfather and namesake, Dr. Samuel Butler, was headmaster of the English public school (in reality, a private school) Shrewsbury when Charles Darwin was a pupil. Dr. Butler was, expectedly, an ordained minister of the established Anglican church and later became bishop of Litchfield. His son Thomas (the father of our Samuel Butler) also became a clergyman, albeit with a far less distinguished career than his father. The intention had been that Samuel would in turn become a minister, but early on he started to lose his faith, and in 1859 he emigrated to New Zealand, where he took up as a sheep farmer. He returned to England in 1864 having turned a tidy profit on his farm and set up as a gentleman of leisure in London. He was not a good investor, however, and basically lived on family money for the rest of his life.[1]

Samuel Butler's claim to fame lies through the products of his pen, both fiction and nonfiction. He started writing in New Zealand, particularly letters and essays for the local paper. Much of this was used in Butler's first book, a satire on utopias, *Erewhon* ("nowhere" spelled backward more or less), published in 1872.[2] He then turned to writing about evolution, publishing four books: *Life and Habit*[3]; *Evolution, Old and New*[4]; *Unconscious Memory*[5]; and *Luck or Cunning*.[6] Butler wrote other works, including a memoir of his grandfather; a book claiming that the author of Homer's *Odyssey* was a woman; and another claiming that Shakespeare's sonnets reveal a homosexual affair. His greatest work is a novel published posthumously in 1903: *The Way of All Flesh*.[7] (It was written much earlier, around the time of the appearance of the evolutionary works, and was left in an unfinished state, appearing only after editing by a close friend.) The work (and the later-published, very heavily edited *Notebooks*[8]) brought Butler much attention and honor, and for the first part of the twentieth century he was considered one of the major thinkers of the later Victorian age. That respect

Figure 4.1: Samuel Butler.

and attention has faded, but there is still considerable scholarly interest in Butler and his work, and much effort has been put into the task (not always easy) of linking his fiction and his nonfiction.[9] This bears on the topic and theme of this essay, namely Butler's role and status with respect to the biological (specifically the evolutionary) community and, whatever the answer, the extent to which he can be said to have been innovative and made lasting contributions.

MACHINES

Butler's engagement with evolutionary ideas apparently began when he read the newly published *On the Origin of Species*[10] while in New Zealand. This inspired him to write a piece, "Darwin among the Machines," published in the local newspaper.

The piece was later incorporated pretty much wholesale into *Erewhon*, the story of an adventurer who leaves his sheep farm and ventures across mountains until he finds himself in a new land and society. A vehicle to poke fun at Victorian society, it is most notable for the way in which crime and illness are reversed in *Erewhon*. Someone who is sick is considered a bad person, condemned, and avoided. Someone who has committed a crime, like embezzlement, is supported and comforted, and makes use of a kind of doctor-equivalent. As with real medicine, the treatment can be very painful, although one suspects that few doctors even in Butler's time made the harsh beatings given to the criminals quite so central a feature of the cure. There are other reversals and bizarre juxtapositions, including an identification of Christianity with banking, along with suggestions that the moral and social value of religion was equivalent to that of brute commerce.

The chapters on machines fall very much into this overall pattern, because the traveler finds that the people of *Erewhon* do not like machines, even to the point of banning them. Through perusal of an ancient work, half history, half philosophy, the traveler learns that the worry apparently had been that machines, like humans, evolve, and that eventually they would gain a kind of consciousness and take over society. "Assume for the sake of argument that conscious beings have existed for some twenty million years: see what strides machines have made in the last thousand! May not the world last twenty million years longer? If so, what will they not in the end become? Is it not safer to nip the mischief in the bud and to forbid them further progress?"[11] Machines are like organisms. Future machines are like possible humans, and as such are gifted with the power of thought. Nothing loath, Butler engaged in a clever defense of this possibility, arguing that there truly are no significant, pertinent disanalogies. Take the eye, for example:

what is this but a machine for seeing? More than this, it can already be said that we exist for machines as much as machines exist for us. A man picks up a spade to dig; in this sense, the spade exists for our benefit. However, conversely, we also exist for the spade—it could not function were we not around, and we play a machine-like role for the spade. Our digestive organs exist for the spade no less than the energy-producing powers of a complex machine exist for us.

All of this implies that machines could evolve to consciousness and take over the world. Some readers saw this parody as an attack on Darwin and his theory of evolution. Why would they not, given that *Erewhon* as a whole was an attack on major features of Victorian society—medicine, law, the church, and more? Butler rushed to say that this had not been his intent at all. He was a firmly committed Darwinian and had great respect for the author of the *Origin*: "I regret that reviewers have in some cases been inclined to treat the chapters on Machines as an attempt to reduce Mr. Darwin's theory to an absurdity. Nothing could be further from my intention, and few things would be more distasteful to me than any attempt to laugh at Mr. Darwin." (This is from the preface to a later edition.) Butler was speaking truthfully here, for he did greatly admire Darwin. On his return to England, Butler made efforts to link up with the great scientist, visiting the Darwins at Down House at least twice, and making a good friend of Charles's son Frank (the botanist and, later, editor of his father's letters).

BUTLER'S EVOLUTIONARY THINKING

Thus far, let us say around 1875, it would be inappropriate to give Butler the label of "outsider." He was no biologist, true. He did not pretend to be one. He was a writer, a minor novelist, who had utilized some important biological themes to make not a biological point but a satirical point. Was he particularly innovative with respect to biology? Unsurprisingly, Butler's tongue-in-cheek *Erewhon* has received much critical attention, and increasingly so in recent years, given the advent of computers and the real possibility in many people's minds that machines will indeed start to think, with consequences not necessarily entirely beneficial for humankind. (Shades of *2001*!) So, in a sense, one might say that he was innovative about cultural change or evolution, although not much more. There was no theory. What we have are suggestive ideas and analogies.

Things were now about to change, for even from the beginning, one can see signs of the worm in the bud, so to speak. The essential message of the *Origin* (especially the earlier editions) is that evolutionary change comes about through natural selection working on random variations (what today we would call mutations). It is absolutely crucial to this vision that the variations have no direction. All apparent purpose in the evolutionary process, whether it be through the design-like nature of features or adaptations (like the hand and the eye) or through some general upward climb from the simple to the complex, comes through ongoing selection of undirected variations that can nevertheless be put to use. Clearly, Butler's machine evolution incorporates a form of intention that transcends this unguided process. You don't make a better watch just by chance, but rather by consciously thinking about the problem and then acting on one's conclusions. In other words, machine evolution involves real design. It is, in a sense, cultural evolution, not just biological evolution.

Whether directly from this or in parallel as it were, the fact is that, through the decade of the 1870s, Butler became more and more convinced of the irrelevance of natural selection as a process of change and of the importance of some kind of intention doing the heavy lifting. This was very much the theme of the first of the evolutionary works, *Life and Habit*, and it was repeated more or less without change in the later books. Butler argued that we (and here the "we" moves right down the evolutionary scale, from humans to the first organisms) do things, and that these get engrained in our memories—memories that increasingly become unconscious in some sense. Hence (and here he drew on then-popular ideas promoted by Ernst Haeckel about early stages of individual development representing early stages of evolution), organisms are motivated by what their ancestors did in the past, and as we grow we assume more and more layers of recent history: "It is admitted on all hands that there is more or less analogy between the embryological development of the individual, and the various phases or conditions of life through which his forefathers have passed."[12]

Consider a person who plays the piano well, and for whom this is a recent acquisition that still requires some conscious thought, Butler writes: "In playing, we have an action acquired long after birth, difficult of acquisition, and never thoroughly familiarized to the power of absolutely unconscious performance, except in the case of those who have either an exceptional genius for music, or who have devoted the greater part of their time to practising."[13] Before this we have writing: "In writing, we have an action generally acquired earlier, done for the most part with great unconsciousness of detail, fairly well within our control to stop at any moment; though not so completely as would be imagined by those who have not made the experiment of trying to stop in the middle of a given character when writing at fit speed."[14] Then before this reading, and even earlier walking, all adding up to a kind of recapitulation: "We may observe therefore in this ascending scale, imperfect as it is, that the older the habit the longer the practice, the longer the practice, the more knowledge—or, the less uncertainty; the less uncertainty the less power of conscious self-analysis and control."[15]

Butler was talking about thought and action, but he saw this as bound up with the physical. In order to perform these actions, one must have or develop features able to perform the actions. Note that he thought we can go back beyond conscious intention, to organisms that do things or respond to events without thought, but whose actions are then ingrained in us, in a kind of unconscious memory of the race. Nevertheless, and this comes through particularly in the second evolutionary work, *Evolution, Old and New*, Butler saw mind in some sense, real design, as being the key feature in ongoing organic change: "Can we or can we not see signs in the structure of animals and plants, of something which carries with it the idea of contrivance so strongly that it is impossible for us to think of the structure, without at the same time thinking of contrivance, or design, in connection with it?"[16] Butler thought we could see design, although note that, unlike Archdeacon William Paley (or, later, Darwin's American supporter, the botanist Asa Gray), Butler was not looking directly for God's design. It was rather a design put in place by organisms themselves.

> Not man, the individual of any given generation, but man in the entirety of his existence from the dawn of life onwards to the present moment. In like manner we say that the designer of all organisms is so incorporate with the organisms themselves—so lives, moves, and has its being in those organisms, and is so one with them—they in it, and it in them—that it is more consistent with reason and the common use of words to see the designer of each living form in the living form itself, than to look for its designer in some other place or person.[17]

This is not very Darwinian—or is it? At the very least, stuff like this coming from a man who claimed to be an admirer of Darwin and a good friend of the family was unexpected to the point of being bizarre. Butler himself admitted this in a letter to Frank Darwin. In part, this was happening because Butler was following a line that he thought right and compelling. He was not so very odd in his beliefs, since there were others who shared similar sorts of views. Most particularly, the ideas were not that far distant from the man who was (more than Darwin in some respects) becoming identified with the evolutionary story, Herbert Spencer. And in part, this was happening, or at least being made possible, because, at a personal level, Butler was starting to quarrel with Charles Darwin in particular and the Darwinian clique in general.

The spark that caused the rift was the English publication of a biography of Charles Darwin's evolutionist grandfather, Erasmus Darwin.[18] Charles wrote an introduction to the English version at which Butler took umbrage, because he thought that this introduction failed to note changes from the original (German) version and that—given the overlap between his own views and that of the older evolutionist—he himself was thereby slighted.

To be candid, Darwin and his party did not deal with this tiff quite as well as they might have, and essentially (although Darwin himself privately recognized that a word of soothing might be appropriate) followed a policy of silence. This was oil on the Butlerian fire, although one suspects that no amount of public groveling would have closed matters reasonably happily. By this point, around 1880, Butler was engaged in a full-blown attempt to suggest that while Darwin's ideas were good and new, that which was good was not very new and that which was new was not very good. In particular, Butler was arguing that his own evolutionary theory had its roots in earlier thinkers, most obviously and importantly the work of the French evolutionist Jean-Baptiste de Lamarck. It might seem strange that Butler would so readily give away credit for his own thinking, but it is in part because Butler stressed that he himself was no professional scientist, but an amateur dabbler. (He was also inclined to say that amateurs had an edge over professionals!) It was also really a consequence of his thinking. If everything is a matter of inherited memory, then this even applies to Butler on evolution, so now the question is where is it inherited from? The answer is Lamarck! Thus, in relinquishing claim to originality, Butler could also thereby pillory Charles Darwin. On the one hand, in Butler's opinion, Darwin got it all wrong about natural selection, since it is the inheritance of acquired character that counts. All Darwin did was publicize evolution, not create new ideas on the subject. On the other hand, likewise in Butler's opinion, Darwin is inconsistent and deceitful. In later editions of the *Origin*, Darwin himself was relying more and more on Lamarckian processes and so was passing off as his own what was truly the work of others. "I know no more pitiable figure in either literature or science," wrote Butler.[19]

Summing up, where do we put Butler now with respect to the outsider question? The answer is rather more complex than it appears at first sight. To see this, we must pull back for a moment and put things into historical context. In an almost Comtian fashion, we can divide the history of evolutionary thinking into three phases.[20] Before 1859, the year that Darwin published the *Origin of Species*, evolutionary thought basically had the status of a pseudoscience. It was rightly seen, and for that reason both praised and condemned, as a vehicle for thoughts of progress in biology, and by analogy in culture. After about 1930, with the coming of Mendelian genetics and its synthesis with Darwinian selection, evolutionary thought was included within the professional sciences—it was a mature (or potentially mature) branch of empirical inquiry. In the period in between, its status was ambiguous. There was professional morphological and embryological work, particularly in Germany. But judged as a whole, evolution was less than this, and more a popular science. Causal speculation, when it existed at all, tended to be wild and unfocused rather than laboratory based and empirically grounded. Its major home was in museums, making itself felt through displays to the public. In this it contrasted with fields like physiology, and then later cytology and genetics, areas in which people like Thomas Henry Huxley and his students and successors were striving hard to raise standards and inculcate professionalism.

All of this is obviously pertinent to assessing Butler and his status. He could no longer hide behind the veil of literary license—nor, to be fair, would he have wanted to. Although do note that, inasmuch as Butler was (at least at first) claiming to be an amateur, he was covering himself—he can perhaps be considered more of a science writer or journalist than a scientist as such.[21] However, even with respect to science, Butler was certainly not an outsider in the sense that no one, including no one in the biological community, shared any of his ideas. He was right that some of the ideas were in Darwin's own writings, and more so in later years. They are also to be found in some respects in other professional biological writers like Haeckel, not to mention Herbert Spencer, whose status in the biological community was considerably higher back then than it is today. And one can even perhaps say that Butler was innovative in the way that he put everything together, stressing memory—although I doubt too much should be claimed in this respect. Not that this deterred Butler. As the years of evolutionary writing progressed, he felt more and more confidence in his ideas and thinking and really was portraying himself as a qualified expert. And to be candid, given the then status of evolutionary biology, he was not entirely out of line in doing so.

This did not stop his critics, particularly those after Darwin's death (in 1882) determined to cherish and burnish the master's legacy and reputation. People like Thomas Henry Huxley and the younger George Romanes were savage in their criticisms of Butler, and at least part of the strategy was portraying him as an outsider, as someone who had strayed into areas in which he had no legitimate basis on which to write and pronounce. But to a certain extent one senses that these critics were hoist by their own petard. It is true that people like Huxley and Romanes were (irrespective of the standing of evolutionary theory) professional biologists in their own right. They could make the argument that, even though evolutionary theory was somewhat of a popular science, it is best left in the hands of those with independent, professional, biological qualifications—which Butler did not have—but ultimately they were the ones who did not push evolutionary studies as fully professional, and in respects did this deliberately. In Huxley's case particularly, he was using evolution as a kind of secular religion to combat what he saw as the anti-reforming, ideological opposition of traditional Christianity.[22] So when someone like Butler came along, who even went so far as to argue that not being a professional biologist gave him a fresher perspective than that of the professionals, they really only had themselves to blame. Butler was an outsider with respect to clubs and honors and the like, but not entirely in presuming to engage in this kind of discussion.

EVOLUTION AS AUTOBIOGRAPHY

If this were the entire story, then we would already have an interesting case study of insider-outsider status, even though it would hardly go down as one of those significant stories where a rank outsider really did make a difference to the science of the insiders, the professionals. He obviously crystallized worries that people had about Darwinian evolution—Could such a blind process really have such end effects?—but whether he himself made much significant difference is hard to assess. Already people like St. George Mivart had had their critical say; non-Darwinians like the American paleontologists (Alpheus Hyatt and Edward Drinker Cope) were simply going their own ways; and those few who did take selection seriously (like the father of mimicry theory, Henry Walter Bates) were somewhat fringe figures. Bates was secretary to the Royal Geographical Society and entirely consumed with his duties there.

But there is more. We have yet to reckon with *The Way of All Flesh*, the work on which Samuel Butler's reputation stands or falls. Through several generations, it tells the story of the Pontifex family. We start with John Pontifex, back in the eighteenth century. A warm and friendly man, he is a successful carpenter. His son George, taken in by richer relatives, becomes a publisher and is somewhat of a brute of a man. His son Theobald becomes a clergyman, marrying another clergyman's daughter (Christina), and he is particularly rough on his son Ernest, the hero of the story, beating him as a child, resenting him as an adult and more.

Ernest grows up as a rather timid child and then man, pushed into the Church and following his father. He goes to work in a poor part of London and starts to lose his faith. As the result of a combination of naivety and sexual desire, he mistakes an honest woman for a prostitute, is convicted of sexual assault and spends six months in jail (causing his father to cut him off dead). On release, Ernest becomes a tailor and seller of second-hand clothes—here, there could well be a conscious echo of Thomas Carlyle's *Sartor Resartus* with its theme of the philosophy of old clothes—and he marries a former family housemaid, Ellen. They have a couple of children, but the marriage breaks down because of Ellen's drinking. Fortunately it turns out that the marriage was a sham, she was married before, they break up, and the children are farmed out with a happy, working-class couple. At the age of twenty eight, Ernest inherits a large fortune from an aunt, settles down to the happy life of a bachelor and writer, and, on his mother's deathbed, makes up with his parents. There is significant satisfaction in the fact that Theobald the father is very much put out, both by the fact that Ernest is now rich and does not need his support and that the money came from a source that Theobald thinks should rightly have gone to him.

The Way of All Flesh is a very lightly fictionalized story about the Butler family.[23] Samuel Butler truly loathed his father and the feeling was more or less mutual. One does not have to be a Freudian to see that this is the key to Butler's life and much of his writing. The break with the Church was clearly in some sense a break with the father. The savage critique of the Church in *Erewhon* was a savage critique of the father and what he stood for. And the Darwinian episode is a reflection of this too. At first, Butler put Darwin in almost a parental position. No sooner had he done this, than he was rebelling—against the man himself and against everything he stood for. In a way, one might say that Butler could not help himself and his theory of evolution explains why. Significantly, Butler's father had had similar problems with his father, Dr. Samuel Butler.

But what about *The Way of All Flesh*? Does one see the Butler version of the theory of evolution played out here? The answer is that in some sense one has to. The novel was written at the very time that Butler was developing and writing on his theory of evolution and there are many heavy hints in the novel that the connection is there and is deep. However, as learned article after learned article shows full well, the precise nature of the connection is a lot more difficult to uncover. It is clear that we are supposed to see the accumulated memories of the past playing out in the present. Like his father, Ernest is somewhat repressed, and one of the reasons why he gives up his children is that he fears he will inadvertently be the brute his father was.

What about change? Lamarck's theory was unambiguously one of progression, from the blob to the human. Do we see progress in the family life of the Pontifexes? The passage quoted above about machines (from *Erewhon*) talks of progress, of things getting better. To be honest, although there is some hope, there is not a major amount of progress in the Pontifex story. George is worse than John, Theobald is about with George, Ernest is better than Theobald, and the children and their children are supposed better than Ernest, although whether they are better than John might be questioned. What is clear is that Butler saw not only the persistence of memory through the generations, but also of new variation—new directed variation—being significant in the overall tale. Ernest has the great shock of conviction and prison. This sets up all sorts of tensions—and opportunities in their way. No longer can he depend on his father for help; he must find his own way and use his own resources. This he does when he turns to tailoring and selling clothes, as well as when deliberately he marries a woman from a lower social class. As it happens, the marriage does not work out well and eventually Ernest is rescued (if that is the right word) by a legacy. But in the process of his response to challenges he has developed and become a better man—and this can be passed on to future generations. (Frankly, the wonder is that the children were not permanently afflicted by fetal alcohol syndrome.)

The tension and ambiguity of *The Way of All Flesh* comes from the fact that Butler was trying to do two things at once. Tell the tale of his own family and of his role within it, and tell the tale of humankind and of the ways in which it changes and responds to pressures. Is this cultural evolution—the family—or is this biological evolution—the human race? In the most important way, for Butler this was not really a meaningful question. For him, biological evolution is cultural. The point is that in the lives of a family, over a few generations, we see the playing out of the evolutionary scenario. Which is precisely what we would expect, given that the story of the individual recapitulates the story of the line.

AND THE WORLD THOUGHT?

Finally, we come to reception and back to our question about outsider status. Butler was known in his lifetime as the author of a clever satire, and also as a controversialist, particularly in the squabble with the Darwinians. But he was considered a minor figure. With the publication of *The Way of All Flesh*, his stock began to rise very rapidly. Although today it does not stand as high as it did once—to be candid, Butler is not a great writer, and for all of its virtues *The Way of All Flesh* is not a masterpiece of psychology as one finds in the best Victorian authors (George Eliot comes to mind)—there is still very considerable interest in his work. In part, obviously, his posthumous fame came because Butler's account of family dynamics—most especially of the conflict between father and son (or more between son and father)—hit a nerve. People were coming out of the Victorian era and starting to look back and reject and criticize. Butler's novel seemed to encapsulate just the general emotions that many had. Significantly he was much praised by the Bloomsbury Group and clearly had a major influence, for instance on the writing of Lytton Strachey's debunking *Eminent Victorians*—a hatchet job on Victorian worthies if ever there was one. Butler—and other writers like Edmund Gosse, who wrote about his father Philip in *Fathers and Sons*—was very much part of that whole reaction against the previous age. That so much was going on at an unconscious level didn't hurt either in an age of Freud.

In part, Butler's posthumous fame—and this is the matter that interests us—came because the beginning of the twentieth century was a time when many were reacting against what they saw as the materialism of the age. This was the time when people like Henri Bergson were promoting the philosophy of vitalism, arguing that there were spirit forces—*élans vitaux*—responsible for evolutionary change. This was a time when Lamarckism was still thought by many to be the key to evolutionary change. And certainly this was when many thought that natural selection was at most a minor blip on the biological scene. George Bernard Shaw took up Butler's ideas with enthusiasm. He absolutely loathed what he saw as the blind, uncaring materialism of Darwin and his acolytes like Thomas Henry Huxley. "What damns Darwinian Natural Selection as a creed is that it takes hope out of evolution, and substitutes a paralyzing fatalism which is utterly discouraging. As Butler put it, it banishes Mind from the universe."[24] Shaw set a tradition that persists to our day. Karl Popper,[25] for instance, was effusive in his praise of Butler and open in his discomfort with selection judged as an empirical process.

Although obviously—especially with respect to insider-outsider questions—this popularity is two-edged. It came at the cost of pushing Butler (qua science) increasingly out of the popular discourse and towards that of the pseudo. As evolutionary thinking moved toward the professional—fueled more than anything by getting rid of Lamarck and taking up Mendel—Butler's thinking on evolution was increasingly untenable scientifically and—a major mark of the pseudo—increasingly taken up by those with ideological objections to the philosophical underpinnings of professional evolutionary science. His was a gentler, more congenial view of world processes—a view that appealed to non-scientists like Shaw and Popper. So it seems that, for all of his posthumous success, Butler ends up as the paradigmatic outsider with no lasting legacy.

And yet! Leave on one side those fascinating and still-fresh speculations about the evolution of machines, and do the same for Butler's insights into the ways in which biology and culture intertwine to determine human fates.

In one of those paradoxes that would have delighted Butler himself, there is a delicious envoi to the whole story. William Bateson, in the first decades of the twentieth century was the leading British spokesman for Mendelian genetics. It was he of all people who was preparing the way for the coming of a professional evolutionary biology. It was he of all people who ought to have been scathing about Samuel Butler. Not a bit of it! He spoke of Butler as "the most brilliant, and by far the most interesting of Darwin's opponents."[26] And to rub it in, he made his comments in a volume celebrating the hundredth anniversary of Charles Darwin's birth and the fifty years since the *Origin* was published. A volume being edited by a close friend of Darwin's son Frank!

How could this have been so? What price "outsider" now? The answer I am afraid tells you more about the nastiness of brilliant scientists than about the true status of Butler. For some fifteen years, Bateson had been in a bitter battle with a group—the "biometricians"—who cherished the memory of Darwin and who put natural selection on a pedestal above all else.[27] For both sides, it was not a question of Mendel and Darwin (as we would now judge) but of Mendel or Darwin—usually a question of whether the chief building blocks of change are major mutations with selection having but a cleaning up action later (Mendelians), or whether the chief building blocks are micro mutations with selection much involved in the creative action (biometricians). Butler, the apotheosis of Darwinian opposition—of Darwinian hatred—was a perfect staff with which to beat the opposition. And so he was put to this use. But it was not a lasting victory. The overall biological community was moving away from anti-materialistic thinking and even those who rather liked vitalism (Julian Huxley was a good example) took care to distance themselves from it in their professional science. In his classic, *Evolution: The Modern Synthesis*, Huxley writes quite roughly of Shaw and Butler, stressing their non-standing as scientists. The views of such "literary men," apparently "are based not on scientific fact and method, but on wish-fulfilment."[28]

Butler would not have given a fig. Although in *Erewhon* it appears that life before birth is more significant than life after death, wherever he is Butler is hugging himself with delight. Paradox and reversal of patterns were his stock in trade. The difficulties in assessing his contributions make it all worthwhile.

FURTHER READING

Butler, S. *Erewhon, or Over the Range*. London: Trubner, 1872.
———. *The Way of All Flesh*. London: Grant Richards, 1903.
Festing, H. *The Notebooks of Samuel Butler, Author of Erewhon—Selections Arranged and Edited By Henry Festing Jones*. London: Fifield, 1919.

Paradis, J., editor. *Samuel Butler, Victorian Against the Grain*. Toronto: University of Toronto Press, 2007.

Raby, P. *Samuel Butler: A Biography*. Ames: University of Iowa Press, 1991.

NOTES

1. E. F. Jones, *Samuel Butler, Author of Erewhon (1835–1902)—A Memoir* (London: Macmillan, 1920).

2. Samuel Butler, *Erewhon* (London: Trubner, 1872).

3. Samuel Butler, *Life and Habit* (London: Trubner, 1878).

4. Samuel Butler, *Evolution, Old and New* (London: Hardwicke and Bogue, 1879).

5. Samuel Butler, *Unconscious Memory* (London: Bogue, 1880).

6. Samuel Butler, *Luck or Cunning* (London: Trubner, 1887).

7. Samuel Butler, *The Way of All Flesh* (London: Grant Richards, 1903).

8. H. Festing, *The Notebooks of Samuel Butler, Author of Erewhon—Selections Arranged and Edited by Henry Festing Jones* (London: Fifield, 1919).

9. See for instance, A. Federico, "Samuel Butler's *The Way of All Flesh*: Rewriting the family," *English Literature in Transition, 1880–1920*, 38 (1995): 466–82; D. Guest, "Acquired characters: Cultural vs. biological determinism in *The Way of All Flesh*," *English Literature in Transition, 1880–1920*, 34 (1991): 283–92; D. Nielson, "Samuel Butler's *Life and Habit* and *The Way of All Flesh*: Traumatic evolution," *English Literature in Transition, 1880–1920*, 54 (2011): 79–100.

10. Charles Darwin, *On the Origin of Species* (London: Murray, 1859).

11. Butler, *Erewhon*, 192.

12. Butler, *Life and Habit*, 126–27.

13. Butler, *Life and Habit*, 11.

14. Butler, *Life and Habit*, 11.

15. Butler, *Life and Habit*, 13.

16. Butler, *Evolution, Old and New*, 30

17. Butler, *Evolution, Old and New*, 30–31.

18. C. Darwin, "Preliminary notice," in *Erasmus Darwin*, E. Kraus (London: John Murray, 1879).

19. Butler, *Luck or Cunning, as the Main Means of Organic Modification?* (London: Fifield, 1887).

20. Michael Ruse, *Monad to Man: The Concept of Progress in Evolutionary Biology* (Cambridge, MA: Harvard University Press, 1996).

21. B. Lightman, "'A conspiracy of one': Butler, natural theology, and Victorian popularization," in *Samuel Butler, Victorian Against the Grain*, editor J. Paradis, 113–42 (Toronto: University of Toronto Press, 2007).

22. M. Ruse, *The Evolution-Creation Struggle* (Cambridge, MA: Harvard University Press, 2005).

23. P. Raby, *Samuel Butler: A Biography* (Ames: University of Iowa Press, 1991); S. Shuttleworth, "Evolutionary psychology and *The Way of All Flesh*," in *Samuel Butler,*

Victorian Against the Grain, editor J. Paradis, 143–69 (Toronto: University of Toronto Press, 2007).

24. G. B. Shaw, *Back to Methuselah: A Metabiological Pentateuch* (Harmondsworth: Penguin, 1988).

25. K. R. Popper, "Intellectual autobiography," in *The Philosophy of Karl Popper*, editor Paul A. Schilpp, 1:3–181 (LaSalle, IL: Open Court, 1974).

26. W. Bateson, "Heredity and variation in modern lights," in *Darwin and Modern Science*, editor A. C. Seward, 85–101 (Cambridge: Cambridge University Press, 1909).

27. W. B. Provine, *The Origins of Theoretical Population Genetics* (Chicago: University of Chicago Press, 1971).

28. J. S. Huxley, *Evolution: The Modern Synthesis* (London: Allen and Unwin, 1942), 458.

OUTSIDERS FROM THE PHYSICAL SCIENCES

SAHOTRA SARKAR

ERWIN SCHRÖDINGER'S EXCURSUS ON GENETICS

INTRODUCTION

In 1936, the theoretical physicist Erwin Schrödinger returned to Graz in Austria, close to his native Vienna, after fifteen years of living abroad.[1] It was a mistake. Schrödinger had been born in Vienna in 1887 and had entered the university there in 1906, shortly after Ludwig Boltzmann's suicide that summer. Nevertheless, he studied physics and mathematics as he had originally planned. After desultory military service during World War I, and after living briefly in Jena, Stuttgart, and Breslau, Schrödinger moved to Zurich in 1921 to assume the position once held by Albert Einstein. In 1926, he gained international renown for formulating wave mechanics using what has come to be called the Schrödinger equation. In 1927, he moved to Berlin to replace Max Planck in the prestigious chair for theoretical physics at the University of Berlin. The Nazi rise to power precipitated a move to Oxford in 1933, the year that he shared the Nobel Prize for physics with Paul A. M. Dirac. Schrödinger actively avoided politics throughout his life; he was nevertheless the only highly prominent German-speaking physicist with no Jewish heritage to have opposed the Nazis by any action.

The return to Austria was motivated in part by an uncertain academic future in Britain but mainly by a desire to return home. However, in accepting a chair at Graz, Schrödinger seems to have chosen to ignore the facts that the university was a center for the Styrian Nazi Party and that more than half its students were active Nazis who dominated campus life. As if this were not trouble enough, the *Anschluss* of 1938, greeted with overwhelming enthusiasm by the vast majority of Hitler's compatriot Austrians, made Schrödinger's position at Graz increasingly precarious. Schrödinger even wrote an infamous compromising letter in 1938 praising the new Germany in an effort to avoid reprisals for his original departure in 1933. It proved to be insufficient; he was shortly dismissed from all his positions.

The Irish Prime Minister, Eamon de Valera, originally trained as a mathematician, arranged an escape to Zurich in Fall, 1938. The idea was to provide an appointment for Schrödinger in the newly minted Dublin Institute for

93

Figure 5.1: Erwin Schrödinger.
Photograph by Francis Simon/AIP/Photo Researchers, Inc.

Advanced Study (DIAS), which would include a School for Mathematical Physics in addition to one for Celtic Studies. But the process took time, and Schrödinger spent most of 1939 at Oxford and in Belgium before moving to Dublin in October. The DIAS was formally established only in late 1940, with Schrödinger as the first Director of its School for Mathematical Physics.[2] He remained there until retirement in 1956, when he returned to his native Vienna (and died in 1961) (see figure 5.1).

The concern of this paper is not with the entirety of Schrödinger's career in Dublin, most of which was spent, in parallel with efforts by Einstein, in the pursuit of a unified field theory incorporating electromagnetism and, possibly, the nuclear forces, along with gravitation, into the geometry of space-time. Those efforts failed, and Schrödinger's contributions to physics during his sixteen Dublin years were rather limited compared to his pathbreaking earlier work. Much of his time was also spent espousing doctrines associated with the Hindu metaphysical school of Vedanta, which remained a lifelong fascination. Rather, the focus here is on a set of lectures on biology, and in particular on a model for the gene, that Schrödinger presented in 1943 and which were published as *What Is Life?* in 1944. In recent decades there has been controversy about the contribution of the book. While some notable commentators have found it wanting in what it says about chemistry and biology, others, to different extents, have held it to be of central importance to the rise of molecular biology.[3] The assessment, as we shall see, should be much more nuanced than either of these positions. The question that will occupy much of what follows is whether Schrödinger's "outsider" perspective on biology contributed significantly to whatever insight, if any, his excursus on genetics provided in 1943.

WHAT IS LIFE?

The terms under which the DIAS was founded required each of its two schools to deliver a set of public lectures in alternate years. Schrödinger willingly agreed to give them for 1943, and they were scheduled to be held at Trinity College in February. There appears to be no particular reason that he chose a biological topic, though, contrary to standard historical accounts,[4] Schrödinger was not entirely an outsider to biology. He was an internationally recognized authority on the physiology and biophysics of color vision.[5] Moreover, his familiarity with some parts of biology dated back to childhood, specifically to his father's amateur but competent forays into botany, and also to his friendship with Franz Frimmel during his university days in pre–World War I Vienna. In Berlin in the early 1930s, biology was a topic of discussion with Max Delbrück, who, under Niels Bohr's influence, had recently decided to abandon physics for the (unsuccessful) pursuit of paradox in biology.[6] Moreover, in a public lecture in 1933, "Why Are Atoms So Small?", Schrödinger had already broached many of the themes treated in greater detail in the Dublin lectures ten years later.[7]

The grandiose topic for the lectures was "What Is Life?", and most of Schrödinger's analyses concerned the physical properties of the gene. However, there is little evidence that Schrödinger was adequately familiar with contemporary genetics, though he was cognizant of some of the remarkable advances of the first few decades of the twentieth century. He was aware of the biophysical work of H. J. Muller, and he used J. B. S. Haldane's 1942 *New Paths in Genetics* as a source.[8] In Dublin, by September 1942, he had realized that, given the absolute intensity of *x*-rays and the mutation rate induced by them, he could calculate from the target area the effective size of a gene. The physicist Paul Ewald drew his attention to a 1935 paper by N. W. Timoféeff-Ressovsky, K. G. Zimmer, and Delbrück that had already essentially provided a sophisticated analysis along similar lines.[9] The theoretical contribution of this paper, the work of Delbrück, was a rudimentary physical model of the gene. By 1942, it was already out of date. Yet, in an apparent sign that Schrödinger had paid little attention to ongoing research in biology (even though he mentions recent results of C. D. Darlington—with whom he corresponded around 1941—and a few others, besides Haldane), Delbrück's model formed the basis for the 1943 lectures.

The first lecture was slated for the evening of Friday, February 5, 1943, with two more due for following Fridays. The number of people attempting to crowd into the Trinity College lecture hall was so large that each lecture had to be repeated on the subsequent Monday to accommodate all those who were interested. The initial audience was estimated to be about four hundred and did not diminish for the successive lectures. Besides de Valera, it included "other notables from Church and State, cabinet ministers, diplomats, socialites, and artists."[10] From the perspective of the DIAS, the lectures must have been an unmitigated success.

The contents of these lectures have been analyzed time and again; it will suffice here to develop the argument in seven main stages.[11] First, Schrödinger began with a loaded question followed by an equally revealing assessment: "How can events *in space and time* which take place within the spatial boundary of a living organism be accounted for by physics and chemistry? . . . The obvious inability of present-day physics and chemistry to account for such events is no reason at all for doubting that they can be accounted for by those sciences" (1–2). As we shall note later, this is the reductionist's dream:[12] once physics has been successfully completed, there will remain no mystery to life. It is probably not much of a surprise that such confidence would issue from a physicist who was equally confident that the apparent paradoxes of the interpretation of quantum mechanics would also eventually be dissipated. Meanwhile, the purpose of the lectures was to show that the reductionist assessment was epistemically warranted.

Second, as in the 1933 lecture, statistical thermodynamics occupied much of Schrödinger's attention. The central problems that baffled him—that is, the ones that he thought the then-current physics and chemistry could not solve—were the remarkable stability of the gene over hundreds of generations and the exactness of its rules of reproduction. Now, classical statistical thermodynamics (which was central to Schrödinger's thinking about physics) allowed such exactness or order to emerge from disorder only through the statistical averaging of the behavior of a very large number of particles, much higher than the number of atoms (a few million at most) in a typical gene that Delbrück had estimated (and which, as it turns out, was a serious overestimate). Thus, classical physics could not account for the stability of genes. The time had come to invoke quantum mechanics—biology was returned to Schrödinger's intellectual home. He interpreted the gene to be a quantum-mechanical system. The laws of quantum mechanics explained the gene's stability, just as they explained the stability of ordinary matter such as rocks and crystals.[13]

Third, quantum mechanics did more than just ensure the stability of genes. Genes, like any quantum-mechanical system, could exist in many different stable states, which Schrödinger called "isomers," borrowing terminology from organic chemistry. Further, quantum-mechanical systems could occasionally, at random, be transformed from one stable state to another through transition processes known as tunneling. Schrödinger argued that such transitions were responsible for mutations in genes. Thus, a mutation was a random event, and its result was as stable as the original. This model of the gene was originally due to Delbrück; all Schrödinger did was popularize it with somewhat astonishing confidence in its validity: "If the Delbrück picture should fail, we would have to give up further attempts" (57).

Fourth, while the Delbrück model could explain the stability of the gene, it was essentially a model for any solid, and solids, from a quantum-mechanical perspective, were crystals that were repeats of some basic pattern. The trouble with such a model was that genes are not only transmitted through reproduction, they specify the construction and functioning of organisms. Genes must, therefore, both serve as templates of some sort and also have immense variety. How was this variety possible? Schrödinger's answer was that the gene, though solid, was yet aperiodic.

Fifth, the idea of the gene being such an aperiodic crystal was elaborated in a remarkable passage. It introduced, fully and clearly, the idea of a genetic code:

> It has often been asked how this tiny speck of material, the nucleus of the fertilized egg, could contain an elaborate code-script involving all the future development of the organism? A well-ordered association of atoms, endowed with sufficient resistivity to keep its order permanently, appears to be the only conceivable material structure, that offers a variety of possible ("isomeric") arrangements, sufficiently large to embody a complicated system of "determinations" within a small spatial boundary. Indeed, the number of atoms in such a structure need not be very large to produce an almost unlimited number of possible arrangements. For illustration, think of the Morse code. The two different signs of dot and dash in well-ordered groups of not more than four allow of thirty different specifications. Now, if you allowed yourself the use of a third sign, in addition to dot and dash, and used groups of not more than ten, you could form 29,254 different "letters"; with five signs and groups up to 25, the number is 372,529,029,846,191,405.[14]

Even more possibilities opened up if the code were not restricted to being linear.[15]

Sixth, what forces would maintain the gene as an aperiodic solid? Ordinary quantum mechanics would require periodicity. Therefore, what was required, according to Schrödinger, were new "order-from-order" laws of physics. Schrödinger's assessment of such a future discovery was unabashedly lyrical: "From Delbrück's general picture of the hereditary substance it emerges that living matter, while not eluding the 'laws of physics' as established up to date, is likely to involve 'other laws of physics' hitherto unknown, which, however, once they have been revealed, will form just as integral a part of this science as the former."[16] The new laws that Schrödinger expected were not new forces but, rather, new principles such as the averaging techniques of statistical mechanics, whereby the bulk properties of bodies are obtained as averages of properties of the particles comprising them.

Seventh, Schrödinger extended his thermodynamic analysis to address the observation that living organisms seemed to generate order from their disordered environments. Thus they seemed to behave in a fashion that contradicted the second law of thermodynamics. According to Schrödinger, there was no contradiction, since living organisms "feed" on "negative entropy."[17] After an elaborate discussion of how such open systems could show the complex organization typical of living organisms, Schrödinger ended the lectures with the pious hope that the organism was "the finest masterpiece ever achieved along the lines of the Lord's quantum mechanics."[18]

IMMEDIATE RECEPTION AND INFLUENCE

It had always been Schrödinger's intention to publish the lectures, and he had made necessary arrangements with Cahill & Company, a respected Dublin publisher. However, for the published version, he prepared an epilogue, "On Determinism and Free Will," in which he "beg[ged] leave to add [his] own, necessarily subjective, view of [the] philosophical implications" of the scientific discussion.[19] He posed the problem of free will and argued—as he had been emphasizing for over a decade—that quantum indeterminacy played no role in its genesis. However, the presumed existence of free will seemed to contradict the claim that physics would ultimately explain all living phenomena because the laws of physics were at least statistically deterministic. One possibility was that mind was something different from matter, but, as Schrödinger noted, the mind only arose in living things composed of matter. Therefore, the mind must somehow be embodied in matter. Schrödinger resolved this quandary by endorsing the Vedanta metaphysical doctrine, which linked individual consciousness or "Athman" [sic] with the underlying universal reality of "Brahman" (88).[20] Individual consciousness was only an unfortunate illusion: the goal of spiritual exercise was to get rid of this illusion and for the self to merge into universal reality. Schrödinger endorsed the mystics' slogan (88): "I have become God."[21]

The spirited defense of Vedanta in the Epilogue included an implicit rejection of Christianity. In conservative Catholic Ireland, it was unacceptable. Even though the type had been set and corrected, Cahill & Company balked at publication and the type was dispersed. *What Is Life?* finally appeared in 1944, published in the United Kingdom by Cambridge University Press. It was an even greater sensation than the lectures, helped by Schrödinger's prominence as a physicist. At ninety-one pages, it is perhaps the shortest book to have had such a marked influence on the development of twentieth-century science. A second edition appeared in 1948 and, in English, the book has always remained in print. There were French, German, Japanese, and Spanish translations.[22] There were at least sixty-five reviews by 1948.[23] The more distinguished scientific reviewers included Darlington, Delbrück, Haldane, Muller, and Michael Polanyi.

By and large, the reviews were positive. However, to the extent that the scientific reviewers commented on the Epilogue at all, they expressed bemusement or, as in the case of Muller, outrage: "If the collaboration of the physicist in the attack on biological questions finally leads to his concluding that 'I am God Almighty,' and the ancient Hindus were on the right track after all, his help should become suspect."[24] Moreover, chemists, and especially physicists, especially in private, had strong criticism of Schrödinger's analysis of the thermodynamics of living systems, in particular that what was at stake was "negative entropy" and that living organisms "extract" entropy from their environment.[25] In particular, Franz Simon insistently pointed out that living organisms do not feed on extremely well-organized forms of matter such as crystals. Consequently, there was no valid sense in which they fed on "negative entropy." In later editions of the book Schrödinger appended a note that both acknowledged Simon's criticism and also admitted that a more technical discussion would have required referring to the (Gibbs) free energy of a system rather than its negative entropy.[26]

THE LEGACY

What Is Life? did not immediately end Schrödinger's interest in genetics. An exchange of letters with Haldane in 1945 explored the population genetics of hornless cattle (with a dominant "hornlessness" gene).[27] However, beyond this episode, on Schrödinger's part, there is no explicit evidence of any subsequent interest in genetics.

Meanwhile, *What Is Life?* had a role in drawing "outsiders" into biology, though the significance of this role remains contested. Those who reported a significant influence include Seymour Benzer, Francis Crick, George Gamow, Alfred Hershey, Salvador Luria, Gunther Stent, James Watson, and Maurice Wilkins.[28] Perhaps Crick (1965) expressed outsiders' perception most accurately when he noted its influence without specifying any definite intellectual contribution:

> On those who came into the subject after the 1939–1945 war, Schrödinger's little book, *What Is Life?* seems to have been peculiarly influential. Its main point—that biology needs the stability of chemical bonds and that only quantum mechanics can explain this—was one that only a physicist would feel it necessary to make, but the book was extremely well written and conveyed in an exciting way the idea that, in biology, molecular explanations would not only be extremely important but also that they were just around the corner. This had been said before, but Schrödinger's book was very timely and attracted people who might otherwise not have entered biology at all.[29]

In 1970, Francois Jacob noted a role the book may have played for physicists:

> After the war many young physicists were disgusted by the military use that had been made of atomic energy. Moreover, some of them had wearied of the turn experimental physics had taken[,] . . . of the complexity imposed by the use of big machines. They saw in it the end of a science and looked around for other activities. Some looked to biology with a mixture of diffidence and hope. Diffidence because they had about living beings only the vague notions of the zoology and botany they remembered from school. Hope, because the most famous of their elders had painted biology as full of promise. Niels Bohr saw it as the source of new laws of physics. So did Schrödinger, who foretold revival and exaltation to those entering biology, especially the domain of genetics. To hear one of the fathers of quantum mechanics ask himself "what is Life?" and to describe heredity in terms of molecular structure, of interatomic bonds, of thermodynamic stability, sufficed to draw towards biology the enthusiasm of young physicists and *to confer on them a certain legitimacy.*[30]

We will deal with such assessments in more detail in the next section.

Meanwhile, Schrödinger's birth centenary in 1987 provided an occasion for reassessment of his varied contributions.[31] *What Is Life?* did not emerge unscathed. Perhaps not noticing that such wisdom only comes with hindsight, virtually all commentators took Schrödinger to task for even suggesting new principles of physics (for which no evidence has ever emerged). Linus Pauling and Max Perutz were perhaps the most critical, pointing out both the inadequacy of Schrödinger's thermodynamic analysis and his unfamiliarity with recent developments in genetics that had turned out to be most significant.[32] Pauling argued that Schrödinger should have stuck to conventional thermodynamics; according to him, not only did *What Is Life?* make no positive contribution, but "by his discussion of 'negative entropy' in relation to life, [it] made a negative contribution."[33]

These assessments are unduly uncharitable. At the time that Schrödinger was composing his lectures, it was not clear that the ordinary principles of statistical mechanics would suffice even to explain all macroscopic properties of inanimate matter. In the early 1940s, Lars Onsager began to establish that fact. Given his professional background, Schrödinger must surely have been cognizant of the difficulties surrounding attempts such as Onsager's. Moreover, interpreting life in terms of negative entropy was part of the intellectual milieu of the time, going back at least to Leo Szilard in 1929 who was another physicist who was eventually to make seminal contributions to molecular biology.[34]

Strangely, what these centenary assessments ignored was the one indubitable contribution that the book had made: a clear and compelling formulation of the idea of a genetic code, which, as Carlson noted in 1966, was of signal importance to molecular biology, especially immediately after the 1953 formulation of the double helix model for DNA.[35] Even before the double helix model, in unabashedly speculative work, biochemists Kurt Stern (1947), P. C. Caldwell and C. N. Hinshelwood (1950), and A. L. Dounce (1952) attempted to provide chemical models for Schrödinger's abstract code.[36] However, the most direct influence on future developments was through the mediation of another physicist, George Gamow, whose popular 1953 work, *Mr. Tompkins Learns the Facts of Life,* was heavily influenced by Schrödinger,

> a celebrated Austrian physicist who once made a basic contribution to the Quantum Theory. Now he is all wound up about the fundamental problems of biology, and thinks it's just the time for physicists to *cut in.* In fact, this *maladia biologica*, as some people call it, seems to have spread far and wide among the physicists, both theoreticians and experimentalists. And, instead of following the latest views of Dirac about the existence of light-ether, or measuring the number of delayed fission neutrons, many of them devote all of their time to breeding bacteria or cutting open the tummies of white mice.[37]

Gamow's biology was no more current than Schrödinger's ten years earlier. Nevertheless, after the construction of the double helix model, Gamow developed a far-reaching research program to decipher the genetic code using formal methods—as it turns out, generating much enthusiasm but very little ultimate success.[38] Gamow's models for the code were based on abstract optimal informational properties that DNA was supposed to possess, none of which, except that the code was triplet, turned out to be true of living organisms.[39]

However, this is not to say that Schrödinger had no precedent at all.[40] Some biologists, such as D. Wrinch (1936) and M. Bergmann and C. Niemann (1937), had recognized that the sequence of units in apparently linear biological macromolecules could determine their conformation and function.[41] Moreover, two of the books Schrödinger used to prepare his lectures had highly suggestive remarks. In 1942, Haldane speculated on how the gene could act as a template for its own reproduction:

> [The] gene is within the range of size of protein molecules, and may be like a virus. If so, chemists will say, we must conceive reproduction as fol-

lows. The gene is spread out in a flat layer, and acts as a model, another gene forming on top of it from pre-existing material such as amino-acids. This is a process similar to crystallization or the growth of a cellulose wall.[42]

Moreover, C. S. Sherrington (1940) spoke of a genetic cipher in a book Schrödinger is known to have admired:

> Among the essential dynamic properties of the chromosomal gene are its catalytic property for self-reproduction and the automatic regulation of that capacity. One might think that the process would beggar the variants possible even in protein-kind. With even only thirty amino-acids to ring the changes, different proteins are possible to a number requiring twenty-three ciphers after the third figure.[43]

It remained for *What Is Life?* to meld these ideas together.

CONCLUSION

Any assessment of Schrödinger's role as an outsider in genetics must be tempered by a recognition that, by and large, he remained an outsider even in fields in which he claimed professional expertise. Along with Einstein, he was the only other prominent quantum theorist who remained unconvinced of the completeness of that theory and its standard (Copenhagen) interpretation. His fanciful excursions into Hindu metaphysics also made Schrödinger unique in the company of Western scientists.[44] Indeed, after claiming that "Atman = Brahman," dabbling in mere genetics seems rather mundane.

Moreover, among the prominent physicists who ventured into biology, however briefly, Schrödinger's position remains unique. That position had two components. First, Schrödinger assumed that the phenomena of biology would be reduced to the laws of physics. The contrast here is primarily with Bohr and Delbrück and their hope that molecular biology would generate paradoxes for the laws of physics.[45] Second, and despite their future promise, the present laws of physics were insufficient for such an explanation. As noted earlier, the reduction of biology would require new laws which, once discovered in animate matter, would also be found to be operative in inanimate matter. The contrast here is with the coterie of biophysicists and chemists—including, but not limited to, J. D. Bernal, Pauling, Perutz, and Crick—pursuing the physical explanation of biological phenomena with the currently known laws of physics (and chemistry). Schrödinger falls into neither camp.

Of much more interest is the question whether Schrödinger's outsider status led to any unique insight into the nature of biology. Here, there remains room for disagreement. Four issues are relevant. First, if the negative assessments of Pauling and Perutz are correct, there was little insight to be gleaned from *What Is Life?*, and Schrödinger's outsider status would then be at best of marginal historical interest in understanding the conceptual development of twentieth-century biology. Second, it is impossible to ascribe any definite positive impact to the thermodynamic analysis of life that Schrödinger offered. It is true that, during the last fifty years, such thermodynamic analyses have periodically come into high fashion,[46] but they have even more predictably fallen out of favor under scrutiny. Third, the idea of a genetic code was critical to the development of molecular biology and genetics in the 1950s and 1960s, even though its continued relevance has since been questioned.[47] It was noted earlier that Schrödinger had few antecedents. What *What Is Life?* brought to the table was a precise combinatorial model of the relation between genes and their (heterocatalytic) products; Sherrington's remark on ciphers aside, this is *not* how biologists (including geneticists) conceptualized their material. If an outsider perspective is supposed to have provided new insight into biological problems, this is where it did—but it had nothing to do with Schrödinger's unique competence as a quantum physicist.

Fourth, whether it was well deserved or not, as Crick, Gamow, and Jacob noted, *What Is Life?* exerted a singular influence on outsiders, especially physicists, who migrated into biology after World War II. What remains contentious is whether these physicists had any positive transformative effect on the conceptual foundations of the emerging discipline. A generation ago historians were glib about positing such an influence,[48] but there remains ample room for caution. Early molecular biology (roughly 1950–1970) involved the pursuit of both structural and informational explanations.[49] The former strategy, in particular, under the guise of "structure determines function," has a relatively continuous and straightforward history going back to the work of Pauling and Perutz, and even earlier to the work of biophysicists, which was independent of the post–World War II migration of physicists into biology. The latter strategy—and informational thinking in general—was hardly the province of physics, even though many physicists, including Crick, developed this pattern of reasoning *after* their migration into biology and during a period also marked by attempts to proselytize for informational reasoning within physics and the other sciences.[50] Given these disciplinary constraints, it would be hard to maintain, credibly, that the physicists brought informational reasoning into molecular biology.[51]

Rather, the emergence of the informational sciences during the 1950s was more likely a common source for informational metaphors that pervaded all disciplines.[52] However, informational analysis in the form of the combinatorics of coding was a major theme in *What Is Life?*. This may well be the most significant contribution of the outsider perspective of the book, but it had virtually nothing to do with Schrödinger's background as a physicist.

FURTHER READING

Kilmister, C. W., editor, *Schrödinger: Centenary Celebration of a Polymath*. Cambridge, UK: Cambridge University Press, 1987.

Moore, W., *Schrödinger: Life and Thought*. Cambridge, UK: Cambridge University Press, 1989.

Perutz, M. F., "Physics and the Riddle of Life," *Nature* 326 (1987): 555–558.

Sarkar, S., "*What Is Life?* Revisited," *BioScience* 41 (1991): 631–634.

Schrödinger, E., *What Is Life? The Physical Aspect of the Living Cell*. Cambridge, UK: Cambridge University Press, 1944.

NOTES

1. Unless explicitly otherwise indicated, biographical material is from W. Moore, *Schrödinger: Life and Thought* (Cambridge, UK: Cambridge University Press, 1989).

2. Details are from W. McCrea, "Eamon de Valera, Erwin Schrödinger and the Dublin Institute," in C. W. Kilmister, editor, *Schrödinger: Centenary Celebration of a Polymath* (Cambridge, UK: Cambridge University Press, 1987), 119–135.

3. For the negative assessments, see, e.g., L. Pauling, "Schrödinger's Contribution to Physics and Chemistry," in C. W. Kilmister, editor, *Schrödinger: Centenary Celebration of a Polymath* (Cambridge, UK: Cambridge University Press, 1987), 225–233; M. F. Perutz, "Erwin Schrödinger's *What Is Life?* and Molecular Biology," in C. W. Kilmister, editor, *Schrödinger: Centenary Celebration of a Polymath* (Cambridge, UK: Cambridge University Press, 1987), 234–251; and M. F. Perutz, "Physics and the Riddle of Life," *Nature* 326 (1987): 555–558. For more positive assessments, see D. Fleming, "Émigré Physicists and the Biological Revolution," *Perspectives in American History* 2 (1968): 152–189; R. Olby, "Schrödinger's Problem: What Is Life?" *Journal of the History of Biology* 4 (1971): 119–148; and E. J. Yoxen, "Where Does Schroedinger's 'What Is Life?' Belong in the History of Molecular Biology?" *History of Science* 17 (1979): 17–52. S. Sarkar, "*What Is Life?* Revisited," *BioScience* 41 (1991): 631–634, attempts a more balanced assessment.

4. See, for example, Fleming, "Émigré Physicists"; Olby, "Schrödinger's Problem"; and Yoxen, "Where Does Schroedinger's 'What Is Life?' Belong?"

5. See Moore, *Schrödinger*.

6. There are numerous histories and appraisals of these developments—see, in particular, R. Olby, *The Path to the Double Helix* (Seattle: University of Washington Press, 1974); H. F. Judson, *The Eighth Day of Creation* (New York: Simon and Schuster,

1979); L. Kay, "Conceptual Models and Analytic Tools: The Biology of the Physicist Max Delbrück," *Journal of the History of Biology* 18 (1985): 207–246; and L. Kay, "The Secret of Life: Niels Bohr's Influence on the Biology Program of Max Delbrück," *Revista di storia della scienza* 2 (1985): 487–510; E. P. Fischer and C. Lipson, *Thinking about Science: Max Delbrück and the Origins of Molecular Biology* (New York: Knopf, 1988); and S. Sarkar, "Reductionism and Molecular Biology: A Reappraisal," Ph.D. dissertation, Department of Philosophy, University of Chicago, 1989.

7. E. Schrödinger, "Warum sind die Atome so klein?" *Forschungen und fortschritte* 9 (1933): 125–126, as discussed in Moore, *Schrödinger*, 264. Section 4 of *What Is Life?* is also titled "Why Are Atoms So Small?" (4).

8. J. B. S. Haldane, *New Paths in Genetics* (London: Harper & Brothers, 1942). For discussion, see Yoxen, "Where Does Schroedinger's 'What Is Life?' Belong?", and Sarkar, "*What Is Life?* Revisited."

9. N. Timoféeff-Ressovsky, K. Zimmer, and M. Delbrück, "Über die natur der genmutation und der genstruktur," *Nachrichten der gelehrten Gesselschaften der Wissenschaften (Göttingen), math.-phy. kl., fachgruppe* 6 (1935): 190–245.

10. See Moore, *Schrödinger*, 395, who goes on to quote from a *Time* magazine story from April 5, 1943: "Schrödinger has a way with him. . . . His soft, cheerful speech, his whimsical smile are engaging. And Dubliners are proud to have a Nobel prizewinner living among them. But what especially appeals to the Irish is Schrödinger's study of Gaelic, Irish music and Celtic design, his hobby of making tiny doll-house furniture with textiles woven on a midget Irish loom—and, above all, his preference for a professorship at the Dublin Advanced Studies Institute for one at Oxford."

11. The discussion partly follows Sarkar, "*What Is Life?* Revisited."

12. H. Stein, "Some Philosophical Aspects of Natural Science," Ph.D. dissertation, Department of Philosophy, University of Chicago, 1958, noted this point explicitly in what may have been the earliest philosophical analysis of the disagreements between Bohr and Schrödinger on the nature of biology; see also Olby, "Schrödinger's Problem." For the sense of "reductionism" appropriate here, see Sarkar, "Reductionism and Molecular Biology," and S. Sarkar, *Genetics and Reductionism* (New York: Cambridge University Press, 1998).

13. E. Schrödinger, *What Is Life? The Physical Aspect of the Living Cell* (Cambridge, UK: Cambridge University Press, 1944). Indeed, part of the power of quantum mechanics lies in its being the first physical theory that could account for the stability of matter.

14. Schrödinger, *What Is Life?*, 61.

15. Schrödinger, *What Is Life?*, 62.

16. Schrödinger, *What Is Life?*, 68–69.

17. Schrödinger, *What Is Life?*, 74.

18. Schrödinger, *What Is Life?*, 86.

19. Schrödinger, *What Is Life?*, 87.

20. Schrödinger, *What Is Life?*, 88.

21. Schrödinger, *What Is Life?*, 88.

22. Unless otherwise noted, details in this paragraph are from Yoxen, "Where Does Schroedinger's 'What Is Life?' Belong?"

23. See Perutz, "Physics and the Riddle of Life."

24. H. J. Muller, "A Physicist Stands Amazed at Genetics," *Journal of Heredity* 37 (1946): 90–92, 92. Muller's outrage was particularly significant because he had long advocated collaboration of geneticists with physicists—see, e.g., H. J. Muller, "Physics in the Attack on the Fundamental Problems of Genetics," *Scientific Monthly* 44 (1936): 210–214.

25. Schrödinger, *What Is Life?*, 44, life "feeds upon negative entropy"; for an extended and highly critical discussion, see Olby, "Schrödinger's Problem," 125–128. L. Brillouin, "Life, Thermodynamics and Cybernetics," *American Scientist* 37 (1949): 559–568, was perhaps most notable because he found Schrödinger's extension of the concept of entropy useful.

26. E. Schrödinger, *What Is Life? and Other Scientific Essays* (New York: Doubleday, 1956), 83–86.

27. The problem remains unsolved today; for details of the correspondence, see J. Crow, "Erwin Schrödinger and the Hornless Cattle Problem," *Genetics* 130 (1992): 237–239.

28. See Yoxen, "Where Does Schroedinger's 'What Is Life?' Belong?", 18; Perutz, "Physics and the Riddle of Life" records that Benzer, Stent, and Wilkins claimed that the book was decisive in drawing them from physics to biology.

29. F. Crick, "Recent Research in Molecular Biology: Introduction," *British Medical Bulletin* 21 (1965): 183–186, 184.

30. See F. Jacob, *La logique de vivant* (Paris: Editions Gallimard, 1970) as quoted by Perutz, "Physics and the Riddle of Life," 555. Jacob's interpretation of Bohr (e.g., 1933) is idiosyncratic insofar as Bohr suggested that biology would contradict *all* laws of physics.

31. See C. W. Kilmister, editor, *Schrödinger: Centenary Celebration of a Polymath* (Cambridge, UK: Cambridge University Press, 1987).

32. See Pauling, "Schrödinger's Contribution," and Perutz, "Physics and the Riddle of Life."

33. See Pauling, "Schrödinger's Contribution," 229.

34. See L. Szilard, "Über die Entropieverminderung in einem thermodynamischen System bei Engriffen intelligenter Wesen," *Zeitschrift für physik* 53 (1929): 840–856; and the discussion in Yoxen, "Where Does Schroedinger's 'What Is Life?' Belong in the History of Molecular Biology?"; and H. S. Leff and A. F. Rex, editors, *Maxwell's Demon: Entropy, Information, Computing* (Princeton, NJ: Princeton University Press, 1990).

35. See E. A. Carlson, *The Gene: A Critical History* (Philadelphia: W. B. Saunders, 1966), 231–232; for the double helix, see J. D. Watson and Francis Crick, "Molecular Structure of Nucleic Acids—A Structure for Deoxyribose Nucleic Acid," *Nature* 171 (1953): 737–738; and J. D. Watson and Francis Crick, "Genetical Implications of the Structure of Deoxyribonucleic Acid," *Nature* 171 (1953): 964–967.

36. See K. G. Stern, "Nucleoproteins and Gene Structure," *Yale Journal of Biology and Medicine* 19 (1947): 937–939; P. C. Caldwell and C. N. Hinshelwood, "Some Considerations on Autosynthesis in Bacteria," *Journal of the Chemical Society* 4 (1950): 3156–3159; and A. L. Dounce, "Duplicating Mechanism for Peptide Chain and Nucleic Acid Synthesis," *Enzymologia* 15 (1952): 251–258.

37. G. Gamow, *Mr. Tompkins Learns the Facts of Life* (Cambridge, UK: Cambridge University Press, 1953), 66.

38. See G. Gamow, "Possible Relation Between Deoxyribonucleic Acid and Protein Structures," *Nature* 173 (1954): 318; and G. Gamow, "Possible Mathematical Relation Between Deoxyribonucleic Acid and Proteins," *Biologiske meddelelser udviket af det kongelige danske videnskabernes Selskab* 22 (1954): 1–11; G. Gamow and N. Metropolis, "Numerology of Polypeptide Chains," *Science* 120 (1954): 779–780, explicitly mention "coding." For relatively recent analyses of the history of coding in molecular biology, see S. Sarkar, "Biological Information: A Skeptical Look at Some Central Dogmas of Molecular Biology," in S. Sarkar, editor, *The Philosophy and History of Molecular Biology: New Perspectives* (Dordrecht: Kluwer, 1996), 187–231; S. Sarkar, "Decoding 'Coding': Information and DNA," *BioScience* 46 (1996): 857–863; and L. E. Kay, *Who Wrote the Book of Life? A History of the Genetic Code* (Stanford, CA: Stanford University Press, 2000). Earlier historical discussions include Olby, *The Path to the Double Helix* and Judson, *The Eighth Day of Creation*.

39. See Sarkar, "Reductionism and Molecular Biology," and Sarkar, "Biological Information," for an analysis of Gamow's work.

40. The possibility of a code within the hereditary material is mentioned in the letters of J. Miescher, the discoverer of DNA, and published in 1897 (see R. Olby and E. Posner, "An Early Reference to Genetic Coding," *Nature* 21 [1967]: 556). That had no impact on future developments.

41. See D. Wrinch, "On the Molecular Structure of Chromosomes," *Protoplasma* 25 (1936): 550–569; and M. Bergmann and C. Niemann, "Newer Biological Aspects of Protein Chemistry," *Science* 86 (1937): 187–190.

42. See Haldane, *New Paths in Genetics*, 43. The same idea is also found in an earlier piece by Haldane (J. B. S. Haldane, "Quantum Mechanics as a Basis for Philosophy," *Philosophy of Science* 1 [1934]: 78–98), which may have interested Schrödinger because it was on quantum mechanics.

43. See C. S. Sherrington, *Man on His Nature: The Gifford Lectures, 1937–8* (Cambridge, UK: Cambridge University Press, 1940), 163. For Schrödinger's admiration for this book and for an account of their correspondence, see Yoxen, "Where Does Schroedinger's 'What Is Life?' Belong?", 35, 51, n. 84.

44. However, Moore, in *Schrödinger*, has persuasively argued that Schrödinger's belief in Vedanta made no tangible difference to his practice as a physicist.

45. See N. Bohr, "Light and Life," *Nature* 131 (1933): 421–423, 457–459; G. S. Stent, "Waiting for the Paradox," in J. Cairns, G. S. Stent, and J. D. Watson, editors, *Phage and the Origins of Molecular Biology* (Plainview, NY: Cold Spring Harbor Laboratory Press, 1966), 3–8; and G. S. Stent, "That Was the Molecular Biology That Was," *Science*

160 (1968): 390–395; for philosophical analysis of these differences, see Stein, "Some Philosophical Aspects of Natural Science," and Sarkar, "Reductionism and Molecular Biology."

46. See, e. g., L. Brillouin, "Life, Thermodynamics, and Cybernetics," *The American Scientist* 37 (1949), 554–568, which was directly influenced by Schrödinger, as well as I. Prigogine, "Structure, Dissipation and Life," in M. Marois, editor, *Theoretical Physics and Biology* (Amsterdam: North Holland, 1969), 23–52.

47. See Sarkar, "Decoding 'Coding.'"

48. For nuanced discussions of disagreements, see Fleming, "Émigré Physicists"; Olby, "Schrödinger's Problem"; and Sarkar, "Reductionism and Molecular Biology."

49. For a short history of structural (or reductionist) explanation in molecular biology, see S. Sarkar, *Genetics and Reductionism* (New York: Cambridge University Press, 1998); for a history of informational ones, see Sarkar, "Decoding 'Coding'"; and L. E. Kay, *Who Wrote the Book of Life?*.

50. The work of Brillouin (e.g., L. Brillouin, *Science and Information Theory* [New York: Academic Press, 1956]) stands out.

51. Olby, *The Path to the Double Helix*, is the target of this criticism.

52. This was perhaps most strikingly seen in the emergence of Norbert Wiener's cybernetics (N. Wiener, *Cybernetics* [Cambridge, MA: MIT Press, 1948])—see S. Heims, *The Cybernetics Group* (Cambridge, MA: MIT Press, 1991), for a history.

LINUS PAULING
LEADING EXPORTER OF CHEMICAL
INSIGHTS INTO BIOLOGY

*Fred Stare [Professor at the Harvard School of Public
Health] said, "He's an outsider. He's never had a course in
nutrition. He'd probably flunk the course we'd give to our
first-year students."*[1] *—Linus Pauling 1992*

INTRODUCTION

At the Nobel ceremony in 1954, one of the laureates addressed
Swedish students with some words of wisdom. Linus Pauling, the laureate
for chemistry, said the following: "When an old and distinguished person
speaks to you, listen to him carefully and with respect—but do not believe
him. Never put your trust in anything but your own intellect. Your elder, no
matter whether he has gray hair or has lost his hair, no matter whether he is
a Nobel Laureate, may be wrong . . . so you must be skeptical—always think
for yourself."[2] In addition to being advice for others, this free-thinking attitude captures Pauling's own approach to experts in biology, medicine, and
politics, or so I will argue. As perhaps the leading chemist of the twentieth
century, Pauling made many forays into biology. His outsider status allowed
him to export ideas about chemical bonding and structure from chemistry
into biology, concepts that helped form modern molecular biology.

EARLY LIFE

Linus Carl Pauling was born February 28, 1901, in Portland, Oregon. His
father operated a drug store. As a young boy he showed signs of unusual
intelligence, which was fostered by his father until he died when Linus
was nine. Pauling began college at Oregon Agricultural College (OAC), now
called Oregon State University, in 1917. His achievements in his chemistry
classes in particular were spectacular, and he was offered a job as an in-
structor in quantitative chemistry as an eighteen-year-old sophomore. In
addition to his classes and teaching, he read chemistry journals and was impressed with landmark papers in theoretical chemistry by Irving Langmuir
and Gilbert N. Lewis.[3] Unlike the physicists' solar system model of the atom,

Lewis proposed a model of the atom with groups of eight electrons arranged at the corners of nested cubes. The focus on the structure of electrons suggested that there might be general explanations of then mostly descriptive chemical facts. Lewis's approach to science also provided a model for the young Pauling: Lewis was bold, theoretical, and focused on structure.[4] While teaching, Pauling met Ava Helen Miller, a very bright student in one of his classes, who would become his wife and a significant force in moving Pauling into political issues later in his life.

After graduating from OAC, Pauling turned down Harvard and enrolled at Caltech, a rapidly improving school that was being formed into a new sort of institution for California: small, focused on basic research, and with especially close contact between students and professors.[5] Pauling completed his doctoral research in x-ray crystallography in the laboratory of Roscoe Dickinson, solving the structure of molybdenite (MoS_2).[6] The focus on crystallography and molecular structure reinforced the idea for the young Pauling that properties of substances were based on their structures. With the aim of making chemistry more theoretical and less descriptive, he had begun to focus on an important bridge between physics and chemistry: the chemical bond. After being advised by the Caltech chemist Arthur Amos Noyes and Lewis that he should go to Europe to study with the major players in the quantum revolution, Pauling arranged a Guggenheim-funded trip to Germany in 1926.

PAULING'S EARLY CAREER: OUTSIDER TO PHYSICS

Many of the major physicists of the day had been through Arnold Sommerfeld's Institute in Munich, including Max von Laue, Wolfgang Pauli, Werner Heisenberg, Hans Bethe, Edward Teller, and Lawrence Bragg. It was an exciting time to be in Europe studying theoretical physics. Erwin Schrödinger had just published the idea that the electron could be seen as a wave, and Pauling attended a talk where Heisenberg, who had proposed the matrix form of quantum mechanics, publicly objected to Schrödinger's system. Pauling became more convinced that improvements in theoretical chemistry would involve applications of quantum mechanics. He wrote up his ideas and Sommerfeld submitted Pauling's paper, "The Theoretical Prediction of the Physical Properties of Many-Electron Atoms and Ions," to *Proceedings of the Royal Society.*[7] He adopted what might be called a "semiempirical method,"[8] calculating various values, comparing them with experimental values and then correcting the theory to more closely agree with what was measured. Although Pauling's work in Europe was on the border of physics and chemistry, Pauling thought himself an outsider to physics. He put it this way: "Most people seem to think that work such as mine dealing with properties of atoms and molecules should be classed with physics, but I (as I have said before) feel that the study of chemical substances remain chemistry even though it reaches the state in which it requires the use of considerable mathematics. The question is more than an academic one, for the answer really determines my classification as a physicist or chemist."[9] Pauling met with other scientists with similar interests, including Walter Heitler and Fritz London, who had used ideas of rapid electron exchange (resonance) to explain the energy of the bond between two hydrogen atoms. This approach would be adopted and extended by Pauling as he returned to the United States in late 1927.

Arguably Pauling's most important chemical work came in 1931, with a series of papers on the chemical bond.[10] Extending some of John Slater's insights into how to simplify the Schrödinger wave equation, he developed what became known as valence-bond theory. Among other things, he was able to predict that carbon should form four hybrid orbitals and provide an explanation of the tetrahedral bonding properties of carbon. The power of his approach could be seen with his analysis of benzene in 1932 and 1933.[11] Pauling and his student G. W. Wheland proposed that benzene was in resonance among five canonical structures. Measured properties of benzene could be seen as an average of the five structures. The benzene work drew Pauling into organic chemistry and closer to biology.

THE MOVE INTO BIOLOGY

Pauling's first significant move into biology was partly driven by economics. The Great Depression adversely affected Caltech's finances. Faculty members, including Pauling, were forced to take pay cuts. Noyes hoped that major grants from the Rockefeller Foundation could help fix Caltech's operating deficit. He encouraged Pauling to think more about organic chemistry, and in 1932 Pauling applied for a $15,000 per year grant. He proposed expanding his research success with benzene to other molecules including proteins and hemoglobin. Pauling's timing was fortuitous. As Lily Kay has described, Warren Weaver, a former member of the Caltech faculty, became director of the natural sciences division within the Rockefeller Foundation and was interested in promoting Pauling's approach.[12] Pauling was funded at $10,000 per year for the next several years. Weaver envisioned a "Science of Man" and, while visiting Caltech in October of 1933, pushed Pauling to concentrate his research on biological problems.

With the visit of Alfred Mirsky, a protein chemist from the Rockefeller Institute, Pauling was able to realize some of Weaver's aspirations for his group. In 1936 they published a paper titled "On the Structure of Native, Denatured and Coagulated Proteins," that suggested hydrogen bonds were important in stabilizing the native forms of proteins. This idea formed a cornerstone of the structural biology of proteins. They wrote, "[The native protein] molecule consists of one polypeptide chain which continues without interruption throughout the molecule (or in certain cases two or more such chains); this chain is folded into a uniquely defined configuration, in which it is held by hydrogen bonds."[13] Unlike the laboratories of J. D. Bernal and William Ashbury in the United Kingdom, which hoped to solve protein structure using x-ray analysis more directly, Pauling looked for shortcuts based on an understanding of physical and structural chemistry, an application of the approach Pauling called "the stochastic method." His approach produced results: understanding hydrogen bonding already had allowed him to formulate a useful way of looking at protein structure. Ahead, Pauling imagined, lay the fruits of being an outsider from chemistry.

The death of Noyes, one of Pauling's biggest supporters at Caltech, in 1936 allowed Pauling to assume the chairmanship of the Division of Chemistry and Chemical Engineering, despite some colleagues' worries that he was too self-centered and too young for the position. Before he died, Noyes had submitted a large grant to the Rockefeller Foundation for "bio-organic" research that he wanted Pauling to direct, and Pauling would successfully realize Noyes's vision.

Pauling was also successful in his choice of collaborators, and, given his growing reputation, important collaborators were coming to him. Robert Corey, a protein crystallographer, successfully petitioned Pauling for a position at Caltech in 1937. Together they decided to attack the protein structure problem indirectly: first, determine the structure of amino acids such as glycine, and then use this information to build models of protein chains. Earlier, Pauling had successfully used a similar approach to solve various silicate structures. Corey and Pauling were opposites: Pauling was extroverted, speculative, and good at the big picture; Corey was quiet, thorough, and detail oriented. Together they formed a formidable team. Neither was a biologist in the traditional sense.

IMMUNOLOGY AND COMPLEMENTARITY

In 1936, another important influence also sought out Pauling. Karl Landsteiner won the Nobel Prize for the discovery of the ABO blood groups and was a giant in the field of immunology. He wanted to talk to Pauling about the different specificities of antibodies and discuss how it was possible for a given protein antibody to be highly specific for one antigen, but not for another apparently similar antigen. After giving it some thought, Pauling hypothesized that if an antibody could denature and then renature next to an antigen, it might renature in a way that maximized complementarity, which for Pauling meant it maximized the number of van der Waals bonds and electrostatic attraction between the antibody and antigen. Through his correspondence with the biologist-physician Landsteiner, Pauling was able to understand the differences in their approaches to science: "Landsteiner would ask, 'What do these experimental observations force us to believe about the world?' and I would ask, 'What is the most simple, general, and intellectually satisfying picture of the world that encompasses these observations and is not incompatible with them?'"[14] To paint with broad strokes, this methodological difference also characterized the contrast between the outsider Pauling and biology in general. The physical chemist aimed for simple, elegant, general theories. Biologists were wearier of letting abstract, theoretical considerations oversimplify the complexity of life, preferring to let the data shape any conclusions drawn. Although he found them intellectually satisfying, Pauling was in no rush to publish his speculative ideas about immunology, and it would take others in the field starting to think along similar lines before he published his paper in 1940.[15] He also thought that if he was right about antibody formation it should be possible to make artificial antibodies by denaturing protein and finding the right conditions to allow it to renature in the presence of the antigen.

During World War II, much of Caltech's research effort was refocused towards war research. Pauling's work on immunology continued because it had clear relevance for the war effort. In 1941, Pauling's collaborator, Dan Campbell, thought that he might have created artificial antibodies using a procedure suggested by Pauling's theory. Pauling used the preliminary findings to write an optimistic press release about the synthesis of antibodies in vitro. The story was picked up by the press and reported widely. Pauling applied for a patent and was given an additional $20,000 grant for artificial antibody development by the Rockefeller Foundation. Unfortunately, it turned out to be a case of Pauling jumping on the bandwagon, running with limited experimental support, and having too much faith in the power of his general framework. Further experimental work did not live up to the hype, and the "artificial antibodies" did not offer protection to animal test subjects. Although immunologists knew that Pauling had made a serious mistake, few were willing to say so in print, presumably because of his stature as an eminent chemist, and Pauling's reputation was only slightly tarnished by the episode.

In the mid-1940s, Pauling made another pitch to fund "chemical biology" on a large scale to the Rockefeller Foundation. He envisioned a new biology based on molecular structure and chemistry. To help him achieve this new biology, Pauling helped convince George Beadle to join Caltech as the chairman of the Division of Biology, a division that had been without a strong leader since the retirement of Thomas Hunt Morgan. Beadle and Pauling believed that biology was poised for a period of significant progress, similar to that which chemistry and physics had experienced in the two or three preceding decades. They asked for $400,000 per annum for fifteen years and, in 1948, were given $100,000 per annum for seven years.[16] This money would allow for the creation of one of the world centers for molecular biology.

Perhaps the central concept that structured Pauling's thinking about molecular biology was complementarity. This concept had been fine-tuned by Pauling's work in immunology and his thinking about molecular biological structure from a chemical point of view. The key to the function of large biological molecules was not ionic and covalent bonds, as some thought; rather, it was complementary shapes and a multitude of weak bonds that held structures together. Contrary to Max Delbruck's hope, no new biological laws were needed; the key to many life processes was complementary structures. In 1948, in the Boot Lecture at the University of Nottingham, Pauling applied the concept of complementarity with prophetic effect: "In general, the use of a gene as a template would lead to the formation of a molecule not with identical structure, but with complementary structure. . . . If the structure that serves as a template consists of, say, two parts which are themselves complementary in structure, then each of these parts can serve as the mold for the production of a replica of the other part, and the complex of two complementary parts thus can serve as the mold for the production of duplicates of itself."[17] Unfortunately, Pauling's remarks did not spark further insights into the nature of the gene, but would be seen as deeply insightful once Watson and Crick's structure for DNA was published several years later. A chemical understanding of biological complementarity is arguably the most important general concept exported into biology by Pauling.

THE PAYOFF FOR BIOLOGY:
THE ALPHA HELIX AND SICKLE CELL ANEMIA

Pauling and his students continued to work on the structure of proteins. He visited England in 1948, and while resting with a cold played around with paper drawings of peptide chains. Using his knowledge of the planarity of the peptide bond, Pauling discovered that it was possible to make a single helix in which hydrogen bonds are formed between adjacent turns of the helix (see figure 6.1). However, the model Pauling created could not account for a reflection in the diffraction pattern that indicated a 5.1 Å repeat. At least for now, what would be called the alpha helix was merely a pretty idea that needed more experimental support.

Starting in 1946, Pauling had his PhD student Harvey Itano work on trying to find the difference between normal hemoglobin and the hemoglobin of those who suffer from sickle cell anemia. By the late 1940s, Itano had worked out that the two hemoglobins were remarkably similar. It was the new technique of electrophoresis that was the key to finding the difference. The sickle cell hemoglobin differed by a measurable electrical charge. As Bruno Strasser has discussed, the importance of the sickle cell work was that it showed, for the first time, that a disease could be caused by a single heritable change in molecular structure.[18] Pauling's molecular approach to life could now be applied to medical problems.

Figure 6.1: Linus Pauling in his Crellin Laboratory office at Caltech, September 15, 1955. A large space-filling model of the alpha-helix is in the background.
Image from the Ava Helen and Linus Pauling Papers, Special Collections, Oregon State University.

Molecular disease provided a bridge between Pauling's scientific and political lives. With the dramatic display of atomic power at Hiroshima and Nagasaki, Pauling was becoming increasing alarmed at the prospect of mutational damage due to atomic explosions. Radiation could cause heritable mutations much like the one that causes sickle cell anemia. Over the next decade and beyond, Pauling would devote more and more time to left-wing political causes and come to the view that scientists have a moral obligation to get involved in politics.

Despite his increasing political activism, Pauling and his collaborators remained remarkably fecund, producing many original papers. Prompted by Lawrence Bragg, John Kendrew, and Max Perutz's attempt to describe configurations of polypeptide chains in a 1950 volume of *Proceedings of the Royal Society*,[19] Pauling felt a sense of urgency to publish his ideas on the alpha helix that he and Herman Branson had been refining. The British team made some obvious mistakes: they did not assume a planar peptide bond and assumed that a helix had to have an exact number of amino residues per repeat of the helix. After an exhaustive search of all possibilities, Branson was confident that there were only two possible helices: one that had 3.7 residues (and 5.44 Å) in each turn of the helix, and one that had 5.1 residues (and 5.03 Å) per turn. While Pauling was confident in the structures, there was one piece of evidence that the models still could not explain: the 5.1 Å reflection in the diffraction pattern. Nonetheless, Pauling published a series of papers in *Proceedings of the National Academy of Sciences* (*PNAS*) in 1951 with faith that the reflection could be accounted for in some other way.[20] Bragg, England's x-ray crystallography pioneer, had been beaten. Max Perutz went about trying to confirm the helix. Eventually, the solution to the puzzling 5.1 Å reflection was found: Francis Crick and Pauling somewhat independently realized that if alpha helices were coiled around each other, it would account for the data.

The *PNAS* papers made a big splash. Pauling was invited to present his findings at the Royal Society in London. To the dismay of the British scientists, Pauling was unable to attend, as the State Department would not issue him with a passport. McCarthyism was gaining strength and, although there was no clear evidence and Pauling swore to the contrary, he was labeled a communist or communist sympathizer. Corey presented in his place to the often-skeptical British scientists. Pauling fought back by publicizing the restrictive passport policy and had a partial victory against McCarthyism when he was granted a passport to visit France and England in 1952.

PAULING OVERREACHES:
THE DNA TRIPLE HELIX

In late 1952, Pauling started thinking about possible structures for DNA. Electron microscope data from Robley Williams at Berkeley suggested that DNA was cylindrical, and Pauling inferred that it was probably helical. Based on some density calculations, Pauling suggested a three-chain model with the phosphates in the middle of the helix. However, there were problems with the model. It could not account for the formation of sodium salts of DNA since the negative phosphates were packed into the center. Even so, his gamble publishing the alpha helix model that also could not account for all the data had paid off, and, likewise, Pauling felt that his DNA structure should be published notwithstanding the apparent problems. As before, he had faith that his chemistry-informed approach was the correct one. When Watson and Crick saw Pauling's model, they were elated, since it was clear to them that his three-stranded model, similar to a model they earlier had proposed and rejected, would not be stable at physiological pH. Ironically, they consulted Pauling's *General Chemistry* to confirm their initial reaction. Sure enough, Pauling had made a major mistake. As is well known, Pauling's attempt at a solution convinced Lawrence Bragg that Watson and Crick should resume looking for a structure for DNA at Cambridge even though the relevant data were being generated by Rosalind Franklin at Kings College in London. Bragg presumably did not want to be beaten to the DNA structure by Pauling, as he had been with silicate structure and the alpha helix.

The DNA episode dulled Pauling's interest in solving large biological molecules somewhat, though as chairman he did help build a new building at Caltech, the Church Laboratory, which joined Beadle's biology building with the chemistry building. The connecting building serves as an apt metaphor for Pauling and Beadle's vision for a new chemical biology articulated in the mid-1940s.

In early 1954, Pauling learned that he had won the Nobel Prize for chemistry. Although the State Department had repeatedly denied Pauling's most recent requests for foreign travel, they decided to avert a potential public relations disaster by allowing Pauling to travel to Stockholm to collect his prize. After visiting Sweden, Pauling traveled around the globe giving talks. In Japan, he learned from Japanese scientists about the nature of the radioactive fallout from the Bikini bomb tests (see figure 6.2). The Nobel Prize allowed Pauling to speak more freely about political issues such as nuclear testing, a freedom that he fully exercised over the next decade.

Building on the success of the sickle cell anemia work, Pauling was successful in obtaining a $450,000 Ford Foundation grant to look at molecular

Figure 6.2: Linus Pauling lecturing on hemoglobin and sickle cell anemia at Osaka University in Japan in 1955.
Image from the Ava Helen and Linus Pauling Papers, Special Collections, Oregon State University.

diseases, such as what became known as phenylketonuria (PKU). Additionally, Pauling speculated that many mental diseases, such as schizophrenia, were due to a molecular imbalance and not rooted in experience as many psychiatrists believed.

Over the course of the 1950s, Pauling devoted more and more time to his peace work. The conservative administration of Caltech was not supportive and, after a combative interview on *Meet the Press* where Pauling was accused of having communist sympathies, the President of Caltech, Lee DuBridge, asked Pauling to resign as chairman of the Division of Chemistry and Chemical Engineering.[21] Pauling was upset, but the change did free up more time to devote to promoting a nuclear testing ban.

THE BEGINNINGS OF MOLECULAR EVOLUTION

In 1959, a young French post-doc came to work under Pauling. Emile Zuckerkandl was interested in hemocyanin, but Pauling pushed him into an evolutionary analysis of hemoglobin. Pauling saw connections between molecular disease and molecular evolution and hoped that knowing more about molecular evolution would enlighten the nuclear testing debate. He wanted to be able to better understand the mutational damage of an atomic explosion to human life. Zuckerkandl used protein "finger printing" (paper chromatography and electrophoresis) to compare the hemoglobin of various species. Hemoglobins from various primates were remarkably similar. Zuckerkandl also considered the amino acid composition of the alpha chain of human and gorilla hemoglobin.

Pauling used his invitations to contribute to Festschrifts to publish speculative papers on molecular evolution with Zuckerkandl. In a Festschrift to Albert Szent-Györgyi, the Nobel laureate who nominated Pauling for the Nobel Prize, Zuckerkandl and Pauling proposed that molecular evolution could be thought of as a stochastic clock. One works out the mutation rate— 11–18 million years per amino acid substitution in the case of hemoglobin— and then uses that rate to make estimates of the timing of speciation events. They later called this idea the molecular evolutionary clock, and it came to be one of the central concepts of the new field of molecular evolution.

Insider biologists Ernst Mayr and G. G. Simpson disliked the molecular clock and the use of molecular characters more generally. Simpson went so far as to say that traditional morphological characters allow for a better understanding of evolutionary history and that molecular characters lie.[22] Zuckerkandl, with Pauling's support, championed the new approach at conferences and in a series of papers. He would remedy one of the weaknesses of the new field by becoming the first editor of the first journal: *The Journal of Molecular Evolution*. Zuckerkandl and Pauling would also suggest that anthropology, as well as evolutionary biology, should be molecularized, coining the term "molecular anthropology." While the clock hypothesis has come to be associated with the neutral theory of evolution, as originally presented, Zuckerkandl and Pauling thought the molecular clock was compatible with natural selection operating at the molecular level. In fact, that molecular changes can make a significant difference to the survival and reproduction of organisms was Pauling's motivational principle in this area.

Pauling won the 1962 Nobel Peace Prize for his work in mobilizing public opinion in support of a nuclear test ban. While the Biology Division held a small party for him, the reaction from the Chemistry Division was muted. Many of the chemists were politically more conservative than Pauling and had mixed feelings about his political crusade. The contrast with the jubilant celebration for his Nobel in chemistry was stark. Pauling was hurt that his fellow chemists and the Caltech administration, especially President Lee DuBridge, were not more congratulatory. The lack of recognition reinforced Pauling's decision to leave Caltech after forty-one years. Although the disconnect with the Caltech chemists was largely rooted in differences in political views, Pauling's success with sickle cell anemia and the alpha helix had moved him closer to the cutting edge of biology and distanced him from pure chemistry. While not quite an insider in biology, Pauling had become one of the fathers of molecular biology.

Institutional movement would characterize Pauling in the next decade or so. Pauling took a position at a The Center for Democratic Institutions, a think tank in Santa Barbara run by Robert Hutchins, the former president of the University of Chicago, but found its lack of laboratory facilities too limiting and left three years later for UC San Diego, and then Stanford University. When Stanford would not provide him more space, he and his former student Art Robinson founded a private institute first called The Institute of Orthomolecular Medicine and later renamed The Linus Pauling Institute of Science and Medicine. With his own institute, Pauling would finally be free of university constraints and politics. He could play the free-thinking outsider role in science and politics with little risk to his career.

VITAMIN C

In the mid-1960s, Pauling began a new research endeavor, one that perhaps made him most famous with the public at large. Influenced by the work of the biochemist Irwin Stone, Pauling began to think that the inability of humans to synthesize vitamin C was a molecular disease and consequently that most humans suffer from "chronic subclinical scurvy."[23] The solution was to take a daily supplement of pure vitamin C at many times higher dosage than the recommended daily amount. Pauling himself took megadoses of vitamin C for three years before he began to publicize the practice. He observed that his colds were much milder and less frequent than they were in the past. The idea that many molecular diseases could be treated by the correct amount of a naturally occurring molecule formed the basis of what Pauling called "orthomolecular medicine." Pauling wove these ideas into speeches, and the popular press picked them up in late 1969.

The reaction of the established medical community was almost uniformly negative. As was becoming a recurring theme, Pauling's detractors, in this case nutrition experts, accused Pauling of promoting an idea that did not have sufficient experimental backing. After reviewing the literature on vitamin C, Pauling thought that the medical establishment mistakenly treated vitamin C like a drug rather than a nutrient. Most published studies did not give the participants more than a few hundred milligrams of the vitamin. Pauling advocated several grams of vitamin C. To promote his ideas, he published *Vitamin C and the Common Cold* in 1971, which sold well with the public but elicited negative reviews in the medical literature. The reviewer in the *Journal of the American Medical Association* put it especially harshly: "Here are found, not the guarded statements of a philosopher or scientist seeking truths, but the clear, incisive sentences of an advertiser with something to sell."[24] Pauling was also accused of short-circuiting the peer review process by publishing popular books instead of peer-reviewed articles in medical journals. The negative feelings were mutual, and Pauling thought many insider physicians were arrogant rule-followers who could not think outside the box.

The success of the book led many people, from cranks to eminent scientists, to write to Pauling. One letter that would broaden Pauling's thinking about vitamin C came from a Scottish doctor named Ewan Cameron, who had been using vitamin C to treat patients with terminal cancer. Cameron claimed that patients taking 10+ grams of vitamin C per day lived longer and more comfortably than historical controls. Cameron and Pauling corresponded and wrote a scholarly paper together. Here was a chance for Pauling to counter his critics and publish clinical data in support of the therapeutic value of vitamin C. Pauling submitted the paper to *Science*, but it was rejected. In a highly unusual move, the paper was also twice rejected by *PNAS* even though Pauling was a member of the National Academy. Among other things, the reviewers rightly worried about the use of "historical controls," the intentional matching of the current case with what appear to be similar cases in the past. They preferred studies that randomly assign patients to a control wing or placebo wing of a study.

Pauling saw the need for basic animal studies and eventually a large double-blind clinical trial. He lobbied the National Cancer Institute (NCI) to fund such studies, but was repeatedly turned down for funding. By the mid-1970s, finding funding for his private institute was proving difficult. Pauling and the institute decided to bypass traditional funding agencies and raise money from the public directly. Given the penetration of the idea of the alleged therapeutic benefits of vitamin C, the campaign was successful and kept the Linus Pauling Institute out of financial hardship for several years. Bypassing the medical community's traditional forms of funding proved a successful way to fund the "outsider" institute.

In the meantime, Ava Helen Pauling was diagnosed with stomach cancer. After an operation to remove the tumor, Pauling convinced his wife to skip the usual chemotherapy and radiation and use megadoses of vitamin C instead.

Eventually in 1977, the NCI agreed to fund a large clinical trial to test the efficacy of vitamin C on cancer. It was to be run by Charles Moertel at the Mayo Clinic. When the study was published two years later it was devastating for Pauling. The title, "Failure of High-Dose Vitamin C (Ascorbic Acid) Therapy to Benefit Patients with Advanced Cancer—A Controlled Study" said it all.[25] Undeterred, Pauling looked for flaws in the study. Many patients had prior chemotherapy, and Pauling argued that their compromised immune systems would not be helped by vitamin C. As the popular press picked up Moertel's study, Pauling continued to play defense, but it was a losing battle with the medical establishment. As Thomas Hager put it, "Pauling was now an outsider in an established field, the intruder whose ideas were dismissed by those in power."[26] Eventually the NCI would fund a second study excluding patients with prior chemotherapy from participation. Unfortunately for Pauling, the second trial would also not find evidence that vitamin C is an effective treatment for terminal cancer.

Pauling had other problems, too. He had fallen out with his partner Art Robinson over some skin cancer trials using mice and vitamin C. Eventually Pauling fired Robinson from the Linus Pauling Institute, and, in return, Robinson filed suit against Pauling and the Institute for $25.5 million.[27] In addition, Ava Helen's cancer had returned. Pauling upped her dose of vitamin C, but to no avail. She died December 7, 1981. The publicity over the lawsuits (which were eventually settled) and the Mayo Clinic studies diminished the fund-raising ability of the Linus Pauling Institute. Pauling himself did not relinquish the idea that vitamin C had therapeutic uses, and in the last decade of his life still pushed the idea that vitamin C could help prevent heart disease, among other benefits. Pauling would not abandon his orthomolecular theoretical framework, in which vitamin C had to have therapeutic value. Of course, many of Pauling's earlier successes were due in part to holding fast to an elegant theoretical framework, so his intransigence in the face of conflicting data does not necessarily indicate irrationality. Additionally, changing his mind also would be to acknowledge that his advocacy of vitamin C over chemotherapy might have hastened his wife's death.

CONCLUSION

What does the life of Linus Pauling show us about outsiders in biology? Pauling was able to bring a focus on chemical structure to biology and shape the emerging field of molecular biology. His methodological focus on heuristics informed by chemical considerations and model building allowed him to solve the alpha helix and other structures faster than more rigorous but slower methods. Watson and Crick used his approach to discover the structure for DNA and were able to beat him at his own game.

Pauling was unafraid to cross disciplinary boundaries, as can be seen by his early attempts to apply quantum mechanics to chemical structures. In both cases, whether into physics or into biology, Pauling was driven by a belief in what I call methodological structural reductionism—that advances in a discipline often occur by linking it with structures posited by a "deeper" discipline. Given his outsider status, he saw himself as the person to make the links. This belief was vindicated by the success of the chemical concepts in the new biology and by the success of twentieth-century chemistry based on quantum mechanics. Another aspect of his success was his ability to pick excellent collaborators. Corey allowed him to work on protein crystallography, Zuckerkandl allowed him to make inroads into molecular evolution, and Cameron allowed him to publish medical articles.

It is interesting to contrast Pauling's warm reception in the new field of molecular biology, which internalized most of his insights about hydrogen bonding and complementarity, with his cool reception in medicine, which largely rejected his advocacy of megavitamin therapy. The difference is partly evidential and partly sociological. Pauling could not marshal convincing clinical evidence, and physicians did not share his commitment to the ortho-molecular vision of life. Pauling's success with molecular biology was reinforced by the fact that the biologists were scientists like Pauling, but there is no denying that in molecular biology he was able to ground his speculation in less controversial data. Given his training in chemistry, he was able to leverage his quantitative skills as molecular biology became more quantitative and driven by physical instrumentation. Furthermore, to move from chemistry to molecular biology is arguably much easier than vice versa. Chemical concepts advanced twentieth-century molecular biology, and Pauling's expertise as a chemist allowed him to make conceptual advances that simply were unattainable for biologists with only training in biology. As Pauling put it in a retrospective interview:

> Even fifty years ago I was recommending to students in the California Institute of Technology who came to me for advice, to do graduate work in chemistry rather than in biology, even if they were interested in biology. They could take some classes in biology, but they could do reading by themselves and learn most of biology.[28]

This somewhat chauvinistic approach and a personality that was not intimidated by experts in other fields marked Pauling's career. He was not afraid to publicize ideas on topics outside his central expertise that were short on evidential support, but made theoretical sense given his overall framework. Such an optimistic approach based on prolific disciplinary outreach allowed both for Pauling's greatest discoveries and also his greatest failures.

FURTHER READING

Hager, Thomas. *Force of Nature*. New York: Simon and Schuster, 1995.

Kay, Lily E. *The Molecular Vision of Life: Caltech, The Rockefeller Foundation, and the Rise of the New Biology*. New York: Oxford University Press, 1993.

Mead, Clifford and Thomas Hager, editors. *Linus Pauling: Scientist and Peacemaker*. Corvallis: Oregon State University Press, 2001.

Rich, Alexander and Norman Davidson, editors. *Structural Chemistry and Molecular Biology*. San Francisco: W. H. Freeman, 1968.

NOTES

I thank Thomas Hager, whose biography of Pauling has influenced my thinking about Pauling's overall scientific career. I thank Jonathan Fitzgerald and David Silverstein for stylistic suggestions, and Oren Harman for comments on an earlier draft. John Horgan generously shared his 1992 interview with Pauling. Trevor Sandgathe helped with selecting images.

1. Pauling, interview with the science writer John Horgan, September 14, 1992.

2. Cited in Thomas Hager, *Force of Nature* (New York: Simon and Schuster, 1995), 457.

3. Irving Langmuir, "The Arrangement of Electrons in Atoms and Molecules," *Journal of the American Chemical Society* 41 (1919): 686–734; Gilbert N. Lewis, "The Atom and the Molecule," *Journal of the American Chemical Society* 38 (1916): 762–85.

4. Hager, 64.

5. Judith Goodstein, *Millikan's School: A History of the California Institute of Technology* (New York: W. W. Norton, 1991).

6. R. Dickinson and Linus Pauling, "The Crystal Structure of Molybdenite," *Journal of the American Chemical Society* 45 (1923): 1466–71.

7. Linus Pauling, "The Theoretical Prediction of the Physical Properties of Many-Electron Atoms and Ions. Mole Refraction, Diamagnetic Susceptibility, and Extension in Space," *Proceedings of the Royal Society A* 114 (1927): 181–211.

8. Hager, 129.

9. Pauling to A. A. Noyes, 1926, quoted in Clifford Mead and Thomas Hager, Editors, *Linus Pauling: Scientist and Peacemaker* (Corvallis: Oregon State University Press, 2001), 249.

10. Linus Pauling, "The Nature of the Chemical Bond. Application of Results Obtained from the Quantum Mechanics and from a Theory of Paramagnetic Susceptibility to the Structure of Molecules," *Journal of the American Chemical Society* 53 (1931): 1367–400; Linus Pauling, "The Nature of the Chemical Bond. II. The One-Electron Bond and the Three-Electron Bond," *Journal of the American Chemical Society* 53 (1931): 3225–37; Linus Pauling, "The Nature of the Chemical Bond. III. The

Transition from One Extreme Bond Type to Another," *Journal of the American Chemical Society* 54 (1932): 988–1003.

11. Linus Pauling and G. W. Wheland, "The Nature of the Chemical Bond. V. The Quantum Mechanical Calculation of the Resonance Energy of Benzene and Naphthalene and the Hydrocarbon Free Radicals," *Journal of Chemical Physics* 1 (1933): 362–74. Before 1931, Pauling could be seen as an important chemist, but to some extent an outsider more interested in theoretical issues and less in experimental questions than the typical chemist, who did not have the mathematical and physical training to fully appreciate what Pauling was doing. After 1931, chemistry itself started to change, making Pauling's theoretical approach based on quantum mechanics more acceptable and valued. For example, that year Pauling won the Langmuir Prize from the American Chemical Society.

12. Lily E. Kay, *The Molecular Vision of Life: Caltech, The Rockefeller Foundation, and the Rise of the New Biology* (New York: Oxford University Press, 1993).

13. A. E. Mirsky and Linus Pauling, "On the Structure of Native, Denatured, and Coagulated Proteins," *Proceedings of the National Academy of Science* 22 (1936): 439–47.

14. Linus Pauling, "Fifty Years of Progress in Structural Chemistry and Molecular Biology," *Daedalus* 99 (1970): 988–1014, 1005.

15. Linus Pauling, "A Theory of the Structure and Process of Formation of Antibodies," *Journal of the American Chemical Society* 62 (1940): 2643–57.

16. Hager, 278–79.

17. Linus Pauling, "Molecular Architecture and the Processes of Life," Sir Jesse Boot Foundation, Twenty-First Sir Jesse Boot Foundation Lecture, 1–13, May 28, 1948, Nottingham, England.

18. Bruno Strasser, "Sickle-Cell Anemia," in Mead and Hager 2001. Linus Pauling, Harvey A. Itano, S. J. Singer, Ibert C. Wells, "Sickle Cell Anemia, a Molecular Disease," *Science* 110 (1949): 543–48.

19. Lawrence Bragg, J. C. Kendrew, and M. F. Perutz, "Polypeptide Chain Configurations in Crystalline Proteins" *Proceedings of the Royal Society London A* 203, no. 1074 (1950): 321–57.

20. Linus Pauling, Robert B. Corey, H. R. Branson, "The Structure of Proteins: Two Hydrogen-Bonded Helical Configurations of the Polypeptide Chain," *Proceedings of the National Academy of Science* 37 (1951): 205–21.

21. Hager, 493.

22. Gregory J. Morgan, "Emile Zuckerkandl, Linus Pauling, and the Molecular Evolutionary Clock, 1959–1965," *Journal of the History of Biology* 31 (1998): 155–78.

23. Hager, 566; Irwin Stone, "Hypoascorbemia, The Genetic Disease Causing the Human Requirement for Exogenous Ascorbic Acid," *Perspectives in Biology and Medicine* 10 (1966): 133–34.

24. Franklin C. Bing, "Vitamin C and the Common Cold," *JAMA* 215 (1971): 1506.

25. Edward T. Creagan, Charles G. Moertel, Judith R. O'Fallon, Allan J. Schutt, Michael J. O'Connell, Joseph Rubin, and Stephen Frytak, "Failure of High-Dose

Vitamin C (Ascorbic Acid) Therapy to Benefit Patients with Advanced Cancer—A Controlled Trial," *New England Journal of Medicine* 301 (1979): 687–90.

26. Hager, 609.

27. Hager, 604.

28. Pauling 1990, quoted in Mead and Hager, 50.

HALLAM STEVENS

FROM BOMB TO BANK
WALTER GOAD AND THE INTRODUCTION
OF COMPUTERS INTO BIOLOGY

INTRODUCTION

GenBank—the preeminent online repository for DNA sequence data—is a symbol of biology's recent transformation into an information science.[1] In 2013, it contains roughly 150 billion base pairs of sequence from about 300,000 different organisms and processes and adds millions more base pairs to its collection every day.[2] This information is used routinely by biologists all over the world. Beginning in 1979 as a pilot database containing only about 100,000 base pairs of sequence, GenBank gradually became an internationally recognized tool for biological work. It assumed a vital role in the planning and execution of the Human Genome Project. Until his retirement in 1988, Walter Goad (1925–2000) played the central role in the creation and development of the database (see figure 7.1). As a leading scientist in Los Alamos National Laboratory's (LANL) Theoretical Biology and Biophysics Division (T-10), Goad was instrumental in convincing the National Institutes of Health (NIH) to fund the project; from 1982 to 1988 he oversaw the construction and management of the database.[3]

However, Goad began his scientific career not as a biologist, but as a theoretical physicist. Coming to Los Alamos as a graduate student in 1950, Goad made important contributions to the ongoing work to design a thermonuclear device. This raises several questions. First, how can we account for Goad's transition from physics to biology? What motivated his transformation from a researcher pursuing mostly classified weapons work to his involvement in a highly public and community-engaged project? Second, what influence did Goad's training and early research career in physics have on his later work in biology? Since Goad's work with GenBank came to exert such an important influence on biology, is his background of any significance in understanding why and how computers came to be used in the life sciences from the 1980s onward?

In answering this set of questions, I wish to show here that there is a clear continuity between some of Goad's earlier (physics) and later (biological) work: both used numerical and statistical methods to solve data-intensive

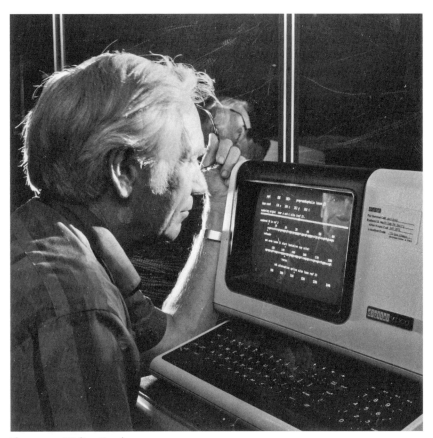

Figure 7.1: Walter Goad.
Image from the Walter Goad Papers, American
Philosophical Society Library, Philadelphia, PA.

problems. Not surprisingly, digital electronic computers were Goad's most important tool. As a consequence of this, Goad's work imported specific ways of doing and thinking into physics from biology. Especially, it brought ways of using computers as data-management machines. Goad's position as a senior scientist in one of the United States's most prestigious scientific research institutions imparted a special prestige to these modes of practice. In other words, Goad's outsider status with respect to biology allowed him to import new practices that were conjoined and contained in the tools of his work.[4] Ultimately, the physics-born computing Goad introduced played a crucial role in redefining the types of problems that biologists addressed; the reorganization of biology that has accompanied the genomic era can be understood as, in part, a consequence of the modes of thinking and doing that the computer carried from Los Alamos.

PHYSICS

Born into the rural South (Marlowe, Georgia) during the Depression, Goad longed to escape what he saw as the "stultifying and hypocritical"[5] surroundings of his upbringing. Goad's mother was from a farming family, his father a lumberman from the hill country of Kentucky; their tumultuous family life was marked by frequent displacement as Goad's father searched for work. At twelve, Goad began working with a local radio repairman and started hanging around a local radio station. When the radio station fell short of an engineer, Goad volunteered to take the Federal Communications Commission exam and fill in. This was his ticket out of the South. At age sixteen, with America's involvement in World War II just beginning, Goad landed a job at a new station in Schenectady, New York. Seeing potential in him, the station's owner suggested he attend nearby Union College, and he commuted by bus from the room under the transmitter where he lived. At Union he studied physics but also enrolled in the Navy's V-12 officer training program. By the time he graduated in the spring of 1945, the war was all but over. Goad was assigned to a small ship in Manila and spent less than a year in service.[6]

Discharged in San Francisco in June 1946, Goad "somehow, already had it in his head that the Department of Physics at Berkeley was his goal."[7] Berkeley proved a disappointment—although it was an exciting place to be doing physics, the faculty was overwhelmed with close to three hundred graduate students, and J. Robert Oppenheimer soon left for Princeton. After a year, Goad reconsidered and moved to Duke University. The physics faculty there included Fritz London and Lothar Nordheim. Although Goad intended to work with London, it was Nordheim who suggested an appealing problem in cosmic ray physics. In 1950, Nordheim accepted an invitation to work at Los Alamos for a year; Goad was invited to accompany him and finish his thesis there. For a young physicist, this was a dream come true; Los Alamos was famous due to its wartime work, and for Goad, "there in the shadowy southwest, [it] had a Shangri-La aura . . . alive with science and determination."[8]

When Goad arrived at Los Alamos, work on the hydrogen bomb was in full swing under President Truman's crash program. Although Goad was there ostensibly to finish his thesis, it was hard not be caught up in this urgent and interesting work being led by Hans Bethe, Enrico Fermi, George Gamow, and Edward Teller. Goad began working on problems of neutron transport—that is, how neutrons moved around inside a supercritical mass of uranium. This had close parallels with his thesis work on how cosmic rays move through the atmosphere. Goad proved an able contributor:

> Sometime in the spring of 1951, I was given a crucial opportunity. Carson Mark described a set of questions that needed resolution, all involving the number of neutrons required to initiate the chain reaction in a supercritical system. . . . He asked me if I'd be interested in working on the problem. A great fog seemed to me to cloak the matter and without a great deal of confidence I tackled it. . . . In a week or so I understood the situation thoroughly. In another week I had a definitive treatment that answered not only the questions Carson had raised, but a number of others that one could only see as the fog lifted. In retrospect I can see that I was established in Carson's and others' eyes as a principal member of the team, and that changed everything for me. . . . I began to share a key satisfaction of the work at Los Alamos: the physics was of a high level, and unlike much academic work, its consequences—tangible, real—were, for better or worse, an immediate part of one's life.[9]

We can construct an idea of the kind of problems that Goad was tackling by examining both some of Goad's later published work and his thesis on cosmic ray scattering.[10] This work has three crucial features. First, it depended on modeling systems (like neutrons) as fluids using differential or difference equations. Second, such systems involved many particles, so their properties could only be treated statistically. Third, insight was gained from the models by using numerical or statistical methods, often with the help of a digital electronic computer. During the 1950s, Los Alamos scientists pioneered new ways of problem-solving using these machines.

Electronic computers were not available at the time Goad first came to Los Alamos in 1950 (although Los Alamos had had access to computers elsewhere since the war). However, by 1952 the laboratory had the MANIAC (mathematical analyzer, numerical integrator, and computer), built under the direction of Nicholas Metropolis. Between 1952 and 1954, Metropolis worked with Fermi, Stanislaw Ulam, Gamow, and others on refining numerical methods for use on the new machine, applying them to problems in phase-shift analysis, nonlinear coupled oscillators, two-dimensional hydrodynamics, and nuclear cascades.[11] Of particular importance were stochastic methods such as "Monte Carlo," in which systems were simulated by using the computer to repeatedly generate random numbers.[12] Los Alamos also played a crucial role in convincing IBM to turn its efforts to manufacturing digital computers in the early 1950s. Los Alamos was the first institution to receive IBM's "Defence Calculator," the IBM 701, in March 1953.[13]

When attempting to understand the motion of neutrons inside a hydrogen bomb, it is not possible to write down (let alone solve) the equations of motion for all the neutrons (there are far too many). Instead, it is necessary to find ways of summarizing the vast amount of data contained in the system. Goad did this in some cases by treating the moving neutrons like the flow of a fluid. Fluid flow can be described by well-known differential equations that can be solved by "numerical methods"—that is, by finding approximate solutions through intensive calculation.[14] In other cases, Goad worked by using Monte Carlo methods—that is, by simulating the motion of neutrons as a series of random moves.[15] In this kind of work, Goad used electronic computers to perform these calculations: the computer acted to keep track and manage the vast amount of data involved. The important result was not the motion of any given neutron, but the overall pattern of motion, as determined from the statistical properties of the system.

When Goad returned to his thesis at the end of 1952, his work on cosmic rays proceeded similarly. He was attempting to produce a model of how cosmic rays would propagate through the atmosphere. Since a shower of cosmic rays involved many particles, once again it was not possible to track all of them individually—instead, Goad attempted to develop a set of equations that would yield the statistical distribution of particles in the shower in space and time. These equations were solved numerically based on theoretical predictions about the production of mesons in the upper atmosphere.[16] In both his work on the hydrogen bomb and in his thesis, Goad's theoretical contributions centered on using numerical methods to understand the statistics of transport and flow using electronic computers.

BIOLOGY

Despite the success of his work in physics, from the early 1960s, Goad became increasingly interested in biological problems, and especially issues in molecular biology. Although he did not stop working on physics or weapons problems, from the mid-1960s onward, Goad devoted an increasing amount of his professional attention to biology. Goad's interest in biological problems was piqued by what he saw as the fundamental, elegant, and logical structure of molecular biology. The work that Goad pursued in biology had important similarities to his work in physics: not only did it center on the electronic computer as an important tool, but like his work on neutrons, depended heavily on numerical and statistical methods. Goad's turn to biology should not be understood as a radical disjuncture: instead, Goad put the same tools and the same methods to work on a new set of problems.

In doing so, he imported the computer into biology as an instrument with a small set of specific uses in solving data-intensive problems.

Goad's transformation from a successful nuclear physicist in the 1950s to a biologist in the 1970s depended in part on the unique institutional context in which it occurred. First, Los Alamos's wartime successes and its ongoing importance for national security afforded work there both prestige and a degree of protection from outside scrutiny. For its scientists, this translated into a great deal of latitude to pursue curiosity-driven basic research. Moreover, the exigencies of wartime work promoted and valued a style of highly collaborative, interdisciplinary work.[17] Second, the life sciences, and especially molecular genetics, were of immediate relevance and importance to the laboratory's mission. Los Alamos had always devoted resources to biology and medicine. From the beginning of the Manhattan Project, it was well understood that radioactive materials—including the materials to be used in a bomb—would be hazardous, and as such the project included a medical division with a mandate to conduct research on the effects of radiation on the body.[18] By 1947, the Atomic Energy Commission had begun to establish a vigorous research program in biology and medicine, both at its own laboratories and at universities around the nation.[19]

At Los Alamos, as elsewhere, such work initially focused on industrial health and safety, but after the war Los Alamos established a health division (known as "H") that did basic research in biomedicine, especially in the new field of health physics. By the early 1950s the H Division employed 125 researchers; in 1952, a $1.5 million lab was constructed across the canyon from the main laboratories.[20] As scientists and the public became more concerned with the longer-term effects of radiation, biomedical work increasingly focused on the impact of radiation on heredity. The new field of molecular biology was of special relevance since the mutation of genes due to radiation exposure was a molecular process.

By the late 1950s, the basic physics and engineering problems of both fission and fusion weapons were well understood. Although, of course, research into making more efficient and more deliverable thermonuclear weapons would continue, the principles of weapons design remained largely static. Goad's weapons work increasingly moved away from the problems in bomb design toward problems concerning the effects of nuclear weapons, which involved conducting research on weapon yield and weapon vulnerability. He used his expertise to serve on the Air Force's Foreign Weapons Evaluation Group and Scientific Advisory Committee.[21] Listening to Oppenheimer's lectures at Berkeley in 1946, Goad developed "a predisposition to always look for a connection to the most basic science of everything I thought about."[22] But Goad's work on weapons must have been increasingly unfulfilling in this respect—the fundamental problems had been solved, and he was spending most of his time on second-order issues.

If work on nuclear weapons now seemed increasingly routine, molecular biology was quite the opposite: here the fundamental problems were just beginning to emerge. Designing a hydrogen bomb had required understanding the logic and mathematics of the interactions of fundamental physical particles; molecular biology seemed to require similar insight into the fundamental logical and mathematical structure of living things. In the periods immediately before and after World War II, a number of physicists began to turn their attention to biological problems, exerting a strong influence on the formation of molecular biology. Inspired by Ernst Schrödinger's "What Is Life?" (1944), physicists including Gunther Stent, Maurice Wilkins, Francis Crick, and Seymour Benzer began to frame biological problems in terms of the interactions between simple, tractable, identifiable, controllable elements (genes, master molecules, code scripts).[23] In other words, the physicists remade biology as the fundamental science of life. It was exactly this that attracted Goad and his colleagues. Moreover, Goad realized that such problems might be profitably attacked using resources he had ready in hand. His skills in statistical, numerical, and computational methods were extremely unusual among biologists (even among those who had come from physics), and as such Goad quickly found a range of problems with which he could productively engage.

The event that triggered interest in biology in the T Division at Los Alamos was a visit by the molecular biologist Leonard Lerman from the University of Colorado. Invited as part of a collaboration with H Division, Lerman told Goad, George Bell, Ulam, and a few others of the "revolution" occurring in biology. This small group began a journal club.[24] Bell recalled that one of the first papers he reported on was Jacob and Monod's model of the *lac* operon; another was on a model of the immune system by Leo Szilard.[25] What interested the group was such precise, physical, and mathematical models of biological systems. For Goad, Lerman's accounts of new experiments and models in molecular biology showed the "clarity and concreteness with which the mechanisms of life would emerge from such analysis . . . many of us were galvanized."[26] In 1964–65, Goad spent a sabbatical at the University of Colorado Medical Center in Denver. This marked his first attempt to tackle biological problems. This work that Goad began to pursue in the 1960s—first in Colorado and then back in Los Alamos—has three important features. First, it made heavy use of electronic computers at Los Alamos.

Second, a large fraction of the work involved problems of transport and flow. Finally, it focused on analyzing biological systems using statistical and numerical methods.

For instance, in Colorado, Goad collaborated extensively with the physical chemist John Cann, examining transport processes in biological systems. First with electrophoresis gels, and then extending their work to ultracentrifugation, chromatography, and gel filtration, Goad and Cann developed models for understanding how biological molecules moved through complex environments.[27] The general approach to such problems was to write down a set of differential or difference equations that could then be solved using numerical methods on a computer. This work was done on the IBM-704 and IBM-7094 machines at Los Alamos. These kinds of transport problems are remarkably similar to the physics that Goad had contributed to the hydrogen bomb: instead of neutrons moving through a supercritical plasma, the equations now had to represent macromolecules moving through a space filled with other molecules.[28]

Here, too, it was not the motion of any particular molecule that was of interest, but the statistical or average motion of an ensemble of molecules. Such work often proceeded by treating the motion of the molecule as a random walk and then simulating the overall motion computationally using Monte Carlo methods.[29] Goad himself saw some clear continuities between his work in physics and his work in biology. Reporting his professional interests in 1974, he wrote: "Statistics and statistical mechanics, transport processes, and fluid mechanics, especially as applied to biological and chemical phenomena."[30] By 1974 Goad was mostly devoting his time to biological problems, but "statistical mechanics, transport processes, and fluid mechanics" well described his work in theoretical physics too. Likewise, in a Los Alamos memo from 1972, Goad argued that the work in biology should not be split up from the Theoretical Division's other activities: "nearly all of the problems that engage LASL have a common core: . . . the focus is on the behavior of macroelements of the system, the behavior of microelements being averaged over—as in an equation of state—or otherwise statistically characterized."[31]

Los Alamos provided a uniquely suitable context for this work. The laboratory's long-standing interest in biology and medicine—and particularly in molecular genetics—provided some context for Goad's forays. Few biologists were trained in the quantitative, statistical, and numerical methods that Goad could deploy; even fewer had access to expensive, powerful computers. Mathematical biology remained an extremely isolated subdiscipline.[32] Those few who used computers for biology were marginalized as theoreticians in an experiment-dominated discipline.[33] Margaret Dayhoff at the National

Biomedical Research Foundation (NBRF), for instance, struggled to gain acceptance among the wider biological community.[34] Goad's position at Los Alamos was such that he did not require the plaudits of biologists—the prestige of the laboratory itself, as well as its open-ended mission, allowed him the freedom to pursue a novel kind of cross-disciplinary work.

Goad's work from 1965 onwards may have dealt with proteins instead of nucleons, but his modes of thinking and working were very similar. His biology drew on familiar tools, particularly the computer, to solve problems by deducing the statistical properties of complex systems. The computer was the vital tool here, because it could keep track of and summarize the vast amounts of data present in these models.

GENBANK

In 1974, the Theoretical Division's commitment to the life sciences was formalized by the formation of T-10, Theoretical Biology and Biophysics. The group was formally headed by Bell, but by this time Goad too was devoting almost all his time to biological problems. The group worked on problems in immunology, radiation damage to nucleotides, transport of macromolecules, and human genetics. The senior scientists saw their role as complementary to that of experimenters, building and analyzing mathematical models of biological systems that could then be tested.[35]

It was around this time that Goad and the small group of physicists working with him began to devote more attention to nucleic acid sequences. For biologists, both protein and nucleotide sequences were the keys to understanding evolution. Just as morphologists compared the shapes of bones or limbs between different species, comparing the sequence for dog hemoglobin to cow hemoglobin allowed inferences about the relatedness and evolutionary trajectories of dogs and cows. The more sequence that was available, and the more sensitively it could be compared, the greater insight into evolution could be gained. In other words, sequence comparison allowed biologists to study evolutionary dynamics very precisely at the molecular level.[36] Sequences were an appealing subject of research for physicists for several reasons. First, they were understood to be the fundamental building blocks of biology—studying their structure and function was equivalent in some sense to studying electrons and quarks in physics. Second, their discrete code seemed susceptible to the quantitative and computational tools that physicists had at their disposal. Computers were useful for processing the large quantities of numerical data from physics experiments and simulations; the growth of nucleotide sequence data offered similar possibilities for deploying the computer in biology.[37] The T-10 group immediately attempted to formulate sequence analysis as a set of mathematical problems. Ulam realized quickly that the problem of comparing sequences to one another was really a problem of finding a "metric space of sequences."[38]

Under the supervision of Goad, a group of young physicists—including Temple Smith, Mike Waterman, Myron Stein, William Beyer, and Minoru Kanehisa– began to work on these problems of sequence comparison and analysis, making important advances both mathematically and in software.[39] T-10 fostered a culture of intense intellectual activity; its members realized that they were pursuing a unique approach to biology with skills and resources available to few others.[40] Within the group, sequence analysis was considered a problem of pattern matching and detection: within the confusing blur of As, Ts, Cs, and Gs in a DNA sequence lay hidden patterns that coded for genes or acted as protein-specific binding sites. Even the relatively short (by contemporary standards) nucleotide sequences available in the mid-1970s contained hundreds of base pairs—far more than could be made sense of by eye. As a tool for dealing with large amounts of data and for performing statistical analysis, the computer was ideal for sequence analysis.[41] Goad's earlier work in physics and biology had used computers to search for statistical patterns in the motion of neutrons or macromolecules; here, also by keeping track of large amounts of data, the computerized stochastic techniques (e.g., Monte Carlo methods) could be used for finding statistical patterns hidden in the sequences. As the Los Alamos News Bulletin reported on Goad's work on DNA in 1982, "Pattern-recognition research and the preparation of computer systems and codes to simplify the process are part of a long-standing effort at Los Alamos—in part the progeny of the weapons development program here."[42] Goad's work used many of the same tools and techniques that had been developed at the laboratory since its beginnings, applying them now to biology instead of bombs.

Los Alamos's Sequence Database—and eventually GenBank—evolved from these computational efforts. For Goad, the collection of nucleotide sequences went hand in hand with their analysis: a collection was necessary in order to have the richest possible resource for analytical work; but without continuously evolving analytical tools, a collection would just be a useless jumble of base pairs. In 1979, Goad began a pilot project with the aim of collecting, storing, analyzing, and distributing nucleic acid sequences. This databasing effort was almost co-extensive with the analytical work of the T-10 group: both involved using the computer for organizing large sets of data. "These activities have in common," Bell wrote, "enhancing our understanding of the burgeoning data of molecular genetics both by relatively straightforward organization and analysis of the data and by the development of new tools for recognizing important features of the data."[43] Databasing meant knowing how to use a computer for organizing and keeping track of large volumes of data. In other words, data management—the organization of sequence data into a bank—depended deeply on the kinds of computer-based approaches that Goad had been using for decades in both physics and biology. Goad's experience with computers led him (and the T-10 group) to understand and frame biological problems in terms of pattern matching and data management—these were problems that they possessed the tools to solve. In so doing, these physicists brought not only new tools to biology, but new kinds of problems and practices.

In 1979, Goad submitted an unsolicited proposal to the NIH in the hope that he might receive funding to expand his database. After some hesitation, a competitive request for proposals was issued in 1981. Goad was not the only person attempting to collect sequences and organize them using comput-ers. Elvin Kabat had begun a collection of sequences of immunoglobulins at the NIH, while Kurt Stüber in Germany, Richard Grantham in France, and Douglas Brutlag at Stanford also had their own sequence collections. Day-hoff's collection of (mostly protein) sequences at the NBRF utilized computer analysis to compile the *Atlas of Protein Sequence and Structure*. However, this kind of collection and analysis was not considered high-prestige work by biologists, and Dayhoff struggled to find funding for her work.[44] Goad's po-sition as a physicist at a prestigious laboratory afforded him independence from such concerns: he could pursue sequence collection and comparison just because he thought it was valuable scientific work.

Ultimately, the $2 million, five-year contract for the publicly funded se-quence database was awarded to Goad's group in June 1982. Both the origins and the subsequent success of GenBank have been detailed elsewhere.[45] Goad's scientific biography, however, suggests how GenBank was partly a product of his background in physics, as he imported a statistical and data-management style of science into biology via the computer.

CONCLUSION:
COMPUTERS AND THE HUMAN GENOME PROJECT

Goad's work did not have an immediate effect on biology. By the mid-1980s, only a handful of biologists were routinely using computers for their re-search. In 1984, for instance, GenBank distributed the database to about 120 laboratories on magnetic tape and about five users accessed the database online each day.[46] This number was growing steadily, but it represented only a very small percentage of biologists. Indeed, computational work in biology continued to be largely isolated from the mainstream experimental community. Experimentalists did not see how computers could make important contributions to biological knowledge, and as such, they were mostly ignored.

This changed dramatically with the Human Genome Project (HGP), which was, in essence, a large-scale effort in sequence collection that required the kinds of tools and techniques in information processing and data management that Goad had pioneered at Los Alamos. At a celebration of Goad's retirement, Elke Jordan, one of the database's managers at the NIH, wrote to congratulate him: "I think the roots of the Human Genome Project are in GenBank, myself, because GenBank allowed everyone to see the benefit of central, accessible data resources and that is what the final outcome of the HGP will be."[47] GenBank was critical to the HGP because the project had to rely on the accumulation, standardization, and organization of vast amounts of sequence data from laboratories around the world. But, as Jordan implies, GenBank also formed a model for the kind of large-scale, collaborative work that the HGP would involve. Just before his death in 2000, Goad remembered that one of the most important things he learned from the Theoretical Division at Los Alamos was "how really effective collaboration is done."[48] Los Alamos thrived on a practical, interdisciplinary, large-scale teamwork that included not only physicists but also engineers, biologists, chemists, managers, and computer scientists. GenBank brought some of this spirit into biology: it was a way of generating collaborative, interdisciplinary modes of working organized in and through the computer.

In the longer term, GenBank has had very important effects on biological work. Most obviously, it has provided an international, centralized resource for sharing, communicating, organizing, and distributing DNA sequence data. In the 1990s, the statistical, numerical, and data management techniques developed by Goad were increasingly incorporated into textbooks and undergraduate and graduate courses. More and more computer scientists and mathematicians began to turn their attention toward developing algorithms and software for biology that elaborated, extended, and refined the kinds of work that had initially been done at T-10 in the 1970s. Journals, courses, and jobs in the new subdisciplines of "bioinformatics" and "computational biology" began to appear in the mid-1990s. These fields are dedicated to solving the problems posed by the large-scale collection and comparison of sequence information. Many of the statistical and database techniques of bioinformatics have origins in the kinds of numerical and computational physics pursued by Goad.

It is no exaggeration to say that Goad's work was ultimately responsible for the introduction of computers into the mainstream of biology. In this respect, Goad succeeded where many others had failed—before GenBank, computers had remained esoteric to biologists.[49] Goad achieved this not by adapting computers to biological problems, but rather by adapting biological problems to computers; statistical and numerical methods first applied in Goad's work on neutron transport in the hydrogen bomb were repurposed first to understand macromolecular transport, and later for pattern recognition and data management of nucleic acid sequences. It was not the case that computers suddenly became useful for solving old biological problems, but rather that the computer came to redefine the kinds of problems that were considered relevant to biology and the kinds of solutions that one could expect.

Goad's position as a physicist at a world-renowned laboratory allowed him to import ways of working into biology from his own discipline. Goad's techniques and tools—particularly the computer—carried their prestige from his work in physics and had an almost automatic credibility because of his status as an outsider to biology. This allowed not only the introduction of a new tool (the computer), but specific ways of thinking centered on statistics, pattern recognition, and data management. Moreover, the introduction of computers went hand in hand with a dramatic shift in the organization of biological work: biologists began to work in interdisciplinary teams (which included engineers, managers, computer scientists, mathematics, and others), international collaborations, and large, centralized laboratories. The computer played a key role in facilitating, organizing, and justifying these modes of work. Although some biologists have continued to resist the centralization and scaling-up of biological work that has accompanied the HGP and genomics, the computer—and the ways of thinking and doing it enables—is an increasingly dominant tool. Goad's role as an outsider meant that the computer came to biology not as a machine for solving biological problems, but rather as a technology that imported ready-made ways of thinking, doing, and organizing from physics.

FURTHER READING

Hunner, Jon. *Inventing Los Alamos: The Growth of an Atomic Community.* Norman: University of Oklahoma Press, 2004.

McElheny, Victor K. *Drawing the Map of Life: Inside the Human Genome Project.* New York: Basic Books, 2010.

November, Joseph A. *Biomedical Computing: Digitizing Life in the United States.* Baltimore, MD: Johns Hopkins University Press, 2012.

Smith, Temple F. "The history of genetic sequence databases," *Genomics* 6: 701–707.

Stevens, Hallam. "Coding sequences: A history of sequence comparison algorithms as a scientific instrument," *Perspectives on Science* 19, issue 3 (2011): 263–299.

1. On biology becoming an information science see Timothy Lenoir, "Shaping biomedicine as an information science," in *Proceedings of the 1998 Conference on the History and Heritage of Science Information Systems*, Mary Ellen Bowden et al., editors (Medford, NJ: Information Today, 1999), 27–45.

2. The most recent data are reported at http://www.ncbi.nlm.nih.gov/genbank/.

3. For a history of GenBank's founding see Bruno J. Strasser, "The experimenter's museum: GenBank, natural history, and the moral economies of biomedicine," *Isis* 102 (2011): 60–96.

4. This might be described as "legitimacy exchange"; see Geof Bowker, "How to be universal: Some cybernetic strategies, 1943–70," *Social Studies of Science* 23 (1993): 107–127.

5. Walter B. Goad, "Memoir of Walter Goad" (organized by Baruch Blumberg during summer of 2000), Box 11, Walter B. Goad papers, MS Coll. 114, American Philosophical Society. Sources from this collection will hereafter be cited as WBG papers.

6. "Memoir of Walter Goad," WBG papers.

7. "Memoir of Walter Goad," WBG papers.

8. "Memoir of Walter Goad," WBG papers.

9. "Memoir of Walter Goad," WBG papers.

10. It is not clear exactly which problem Goad is referring to in the extended quotation; however, it is likely that it refers to work that was written up as W. B. Goad, G. I. Bell, and H. Bethe, "Time dependence of the slowing down of neutrons with constant scattering mean free path," LA-1445 (1952). This is one of the earliest reports co-authored by Goad at Los Alamos. Box 2, Publications 1 (1952–1965), WBG papers.

11. On computers at Los Alamos see N. Metropolis, "The Los Alamos experience, 1943–45," in *A History of Scientific Computing*, Stephen G. Nash, editor (New York: ACM Press, 1990), 237–250; Herbert L. Anderson, "Metropolis, Monte Carlo, and the MANIAC," *Los Alamos Science* (Fall 1986): 96–107; N. Metropolis and C. Nelson, "Early computing at Los Alamos," *Annals of the History of Computing* 4, issue 4 (1982): 348–357.

12. For example, imagine a complicated shape drawn within a square. Suppose you wanted to estimate the area of that shape. Rather than performing any calculations, imagine dropping grains of sand at random into the square. The ratio of the number of grains that fall inside the shape to those that fall outside gives an approximation of the area of the shape. The more grains, the better the estimate. A computer could be used to generate pseudo-random numbers to simulate the random dropping of the sand grains. This is a Monte Carlo calculation. Since the method is based on playing the odds, Ulam and von Neumann named it after the Monte Carlo Casino in Monaco.

13. Donald Mackenzie, "The influence of the Los Alamos and Livermore National Laboratories on the development of supercomputing," *Annals of the History of Computing* 13 (1991): 179–201, 189.

14. See, for instance, W. B. Goad, "Wat: A numerical method for two-dimensional unsteady fluid flow," LASL Report, LAMS-2365 (1960). Box 2, Publications 1 (1952–1965), WBG papers.

15. See, for instance, W. B. Goad and R. Johnson, "A Monte Carlo method for criticality problems," *Nuclear Science and Engineering* 5 (1959): 371–375. On the use of Monte Carlo methods in physics, see Peter Galison, *Image and Logic: A Material Culture of Microphysics* (Chicago: University of Chicago Press, 1997), chapter 8.

16. Walter B. Goad, "A theoretical study of extensive cosmic ray air showers" (PhD diss., Duke University, Department of Physics 1953). See Box 2, WBG papers.

17. On the organization of the wartime laboratory, see Lillian Hoddeson et al., *Critical Assembly: A Technical History of Los Alamos during the Oppenheimer Years, 1943–1945* (Cambridge, UK: Cambridge University Press, 1993).

18. Timothy Lenoir and Marguerite Hays, "Manhattan Project for biomedicine," in *Controlling our Destinies: Historical, Philosophical, Ethical, and Theological Perspectives on the Human Genome Project*, Phillip R. Sloan, editor (Indiana: University of Notre Dame Press, 2000), 29–62, 33.

19. For an account of the biomedical programs under the AEC, see Richard G. Hewlett and Francis Duncan, *Atomic Shield: A History of the United States Atomic Energy Commission 1947/52*, vol. 2 (US Atomic Energy Commission, 1972), 112–113, 251–255.

20. Jon Hunner, *Inventing Los Alamos: The Growth of an Atomic Community* (Norman: University of Oklahoma Press, 2004), 169–170.

21. Walter B. Goad, "Vita," January 30, 1974, Box 1, Los Alamos Reports 1974, WBG papers; Walter Goad, list of publications, December 1977, Box 1, Los Alamos Reports 1977, WBG papers.

22. "Memoir of Walter Goad," WBG papers.

23. Evelyn Fox Keller, "Physics and the emergence of molecular biology: A history of cognitive and political synergy," *Journal of the History of Biology* 23 (1990): 389–409. It is unclear whether Goad himself was aware of either Schrödinger's work or the role of physicists in molecular biology.

24. Walter B. Goad, "Sequence analysis: Contributions by Ulam to molecular genetics," in *From Cardinals to Chaos: Reflections on the Life and Legacy of Stanislaw Ulam*, Necia Grant Cooper et al., editors (Cambridge, UK: Cambridge University Press, 1989), 288–293.

25. George Bell, "George Bell on George Bell," Theoretical and computational biology workshop, September 12–13, 1994, LANL, Box 11, Articles about Goad—from Los Alamos Bulletin (1982–1987), WBG papers. Bell would go on to pursue work on the immune system, developing his own mathematical models.

26. Goad, "Sequence analysis."

27. Walter B. Goad and John R. Cann, "Theory of moving boundary electrophoresis of reversibly interacting systems," *Journal of Biological Chemistry* 240 (1965): 148–155; Walter B. Goad and John R. Cann, "Theory of zone electrophoresis of reversibly interacting systems," *Journal of Biological Chemistry* 240 (1965): 1162–1164; Walter B. Goad and John R. Cann, "The theory of transport of interacting systems of biological macromolecules," *Advances in Enzymology* 30 (1968): 139–175.

28. This is not a complete account of Goad's work during this period. He also contributed to studies on the distribution of disease, the digestion of polyoxynucleo-

tides, and protein-nucleic acid association rates, among others. Some of this work involved other kinds of transport phenomena, and all of it involved statistical methods and computers.

29. A random walk is a method of simulating motion (usually of atoms of molecules) using random numbers. A starting direction is picked at random (imagine rolling an eight-sided die marked with north, northeast, east, southeast, etc.). The object moves one unit in that direction. Then another direction is picked at random and the object moves one unit in that direction. The process is repeated many times, giving a random path, or "walk." The result often looks like a complex scribble or the flight path of a fly. Random walk simulations are used in economics (stock market prices), ecology (animals movement), population genetics (genetic drift), neuroscience (neuron firing), computer science, and psychology, as well as physics.

30. Goad, "Vita."

31. Walter Goad, "T-division," office memorandum to T-division leader search committee, November 29, 1972, Box 1, Los Alamos Reports 1972, WBG papers.

32. Historians have noted that mathematical biology has had difficulty establishing itself within the mainstream of biological work. See Evelyn Fox Keller, "Untimely births of a mathematical biology," in *Making Sense of Life* (Cambridge, MA: Harvard University Press, 2002), 79–112.

33. For an account of the use of computers in biology and medicine in this period, see Joseph A. November, "Digitizing life: The introduction of computers to biology and medicine" (PhD diss., Princeton University, Department of History, 2006).

34. On Dayhoff, Bruno J. Strasser, "Collecting, comparing, and computing sequences: The making of Margaret O. Dayhoff's *Atlas of Protein Sequence and Structure*," *Journal of the History of Biology* (2010), accessed June 29, 2011, doi: 10.1007/s10739-009-9221-0.

35. "T-10 theoretical biology and biophysics," Box 1, Los Alamos Reports 1977, WBG papers.

36. See Michael R. Dietrich, "Paradox and persuasion: Negotiating the place of molecular evolution within evolutionary biology," *Journal of the History of Biology* 31 (1998): 85–111; Joel B. Hagen, "Naturalists, molecular biologists, and the challenge of molecular evolution," *Journal of the History of Biology* 32 (1999): 321–341.

37. Undated letter from Goad to Japan[?], Box 12, Laboratory notes, n.d., WBG papers.

38. S. M. Ulam, "Some ideas and prospects in biomathematics," *Annual Review of Biophysics and Biomathematics* 1 (1972): 277–292; Goad, "Sequence analysis."

39. For a history of sequence comparison algorithms see Hallam Stevens "Coding sequences: A history of sequence comparison algorithms as a scientific instrument," *Perspectives on Science* 19, issue 3 (2011): 263–299. Some seminal papers from this group: William A. Beyer, Myron L. Stein, Temple F. Smith, Stanislaw Ulam, "A molecular sequence metric and evolutionary trees," *Mathematical Biosciences* 19 (1974): 9–25; Michael S. Waterman, Temple F. Smith, William A. Beyer, "Some biological sequence metrics," *Advances in mathematics* 20 (1976): 367–387; Temple F. Smith and

Michael S. Waterman, "Identification of common molecular subsequences," *Journal of Molecular Biology* 147 (1981): 195–197.

40. On T-10 culture see Michael Waterman, "Skiing the sun: New Mexico essays," accessed June 29, 2011, available at: http://www.cmb.usc.edu/people/msw/newmex .pdf.

41. "Proposal to establish a national center for collection, and computer storage and analysis of nucleic acid sequences," Theoretical biology and biophysics group, University of California, LASL, December 17, 1979, Box 2, Publications 3 (1978–1981), WBG papers.

42. "$2 million earmarked for genetic data bank at LANL," LANL News Bulletin, 1982, Box 11, Articles about Goad—from Los Alamos Bulletin (1982–1987), WBG papers.

43. George Bell, letter to George Cahill, April 22, 1986, Box 7, Human Genome Project 1 (1986), WBG papers.

44. For more, see Strasser, "Collecting, comparing, and computing."

45. Bruno J. Strasser, "GenBank: Natural history in the 21st century?," *Science* 322 (2008): 537–538; Strasser, "The experimenter's museum."

46. Christine Carrico, memorandum to director, NIGMS, Report of Second Advisory Meeting, February 17, 1984, Box 4, Genbank Advisory Group Minutes and Notes 1, WBG papers.

47. Elke Jordan, note card to Goad, November 12, 1991, Box 4, "Genbank: The first 15 years," WBG papers.

48. "Memoir of Walter Goad," WBG papers.

49. See November, "Digitizing life."

III OUTSIDERS FROM MATHEMATICS

MICHAEL R. DIETRICH & ROBERT A. SKIPPER, JR.

R. A. FISHER AND THE FOUNDATIONS OF STATISTICAL BIOLOGY

INTRODUCTION

In July of 1951, Sir R. A. Fisher (1890–1962) used the occasion of his Bateson Lecture to reflect on statistical methods in genetics (see figure 8.1). Having made foundational contributions to both statistics and genetics beginning arguably in 1918, Fisher saw them as quintessential twentieth-century disciplines. While nineteenth-century antecedents could be easily found for both, statistics and genetics came to maturity as "distinct points of view" in the twentieth century.[1] Fisher played an important role in articulating the point of view of both modern genetics and modern statistics, but more importantly, he successfully managed their integration. In doing so, Fisher did more than bring his training in mathematics to bear on biological topics. He used his mathematical abilities to reconceive statistics, experimentation, and evolutionary biology.

Fisher came to biology as an outsider in the sense that his formal training as a student emphasized mathematics. His interests in genetics and eugenics, however, began early as well. Even though there were other well-known synthesizers of mathematics and biology at the time, Fisher's interests and training were not typical of the great majority of biologists of his day. Fisher's innovative work in statistical biology met both with indifference from many biologists, as a result, and often opposition from more established statistical biologists.

The opposition that Fisher's innovations faced led him to comment, "A new subject for investigation will find itself opposed by indifference, by inertia, and usually by ridicule. A new point of view, however, affecting thought on a wide range of topics may expect a much fiercer antagonism."[2] The statistical, genetic, and experimental "points of view" that Fisher championed developed significantly over the first decades of the twentieth century. The antagonism they faced only pushed Fisher to refine them further. A more passive advocate may have been pushed to the margins, but Fisher's persistence, personality, and patrons allowed him to redefine how mathematics could be brought into the core of biological thought and practice.

Figure 8.1: R. A. Fisher.
Photograph by SPL/Photo Researchers, Inc.

ARTICULATING NEW POINTS OF VIEW

In retrospect, Fisher claimed that genetics was a natural source for statistical thinking because of its heavy use of frequencies and the natural randomization of genotypes that make genetic experimentation easy. However natural the synergy between genetics and statistics may be, Fisher was initially drawn to their intersection through eugenics and biometry. As a student at Cambridge from 1909 to 1913, Fisher studied mathematics but developed broad interests. As he himself commented, he entered Cambridge on "the centenary of Darwin's birth and the jubilee of the publication of *The Origin of Species*."[3] William Bateson, Mendel's English champion, had been given a professorship the year before, and in 1912 it would be endowed as the Arthur Balfour Chair of Genetics.[4] Darwinism, Mendelism, and the debates over their differences would have been unavoidable for Fisher at Cambridge. Eugenics made them irresistible.

Sir Francis Galton inaugurated the English eugenics movement in 1869 with his book *Hereditary Genius*.[5] Galton's eugenics rested on the scientific management of human heredity by encouraging reproduction among those supposed to be "fit" and discouraging reproduction of the supposedly "unfit." In doing so, eugenicists hoped to direct the course of human evolution.[6]

Once introduced to the rising eugenics movement, Fisher's youthful enthusiasm led him to help form the Cambridge University Eugenics Society in 1911. Eugenics for Fisher was not a passing fad, however. His intellectual interest in the topic never faltered, and Fisher sought to live by the eugenic principles that he advanced by having eight children.[7]

Eugenics formed the natural bridge between Fisher's interest in mathematics and the new fields of genetics and statistics. In turn-of-the-century England, Francis Galton and Karl Pearson occupied this intersection, and it was to their work that Fisher turned as a college student. In order to understand the resemblance between generations, Galton had developed techniques of correlation and regression to represent heredity from a statistical point of view. Like his cousin, Charles Darwin, Galton believed that hereditary traits were continuous gradations of form best described by a normal distribution. An individual's heredity then was represented in terms of a law of ancestral heredity, in which an individual's ancestors each made a diminishing contribution.[8] Inspired by Galton's approach, Karl Pearson developed biometrics, a statistical approach to biology, which supported a Darwinian view of the gradual evolution of continuous traits under the direction of natural selection.

In 1911, Fisher addressed the Cambridge Eugenics Society on the intersection of Mendelism and biometry. At the time, Mendelism was cast by William Bateson in diametric opposition to the biometrical approach advocated by Karl Pearson. For Bateson, Mendelism supported his view of discrete hereditary characters and saltational evolution. However, as a dispute, the arguments between Mendelians and biometricians also raised questions about the appropriate role of probability and statistics. Fisher was taken by exactly these questions, and in his overview of both positions notes the role of probability in each and praises the power of the biometrical use of statistics to analyze biological observations without relying on theory or abstraction. Moreover, from a eugenic point of view, the biometrical approach convinced him that it was possible to create "a slow and sure improvement in the mental and physical status of a population" without the complications of the "experimental breeding" that Mendelism would require.[9]

While still an undergraduate, Fisher made his first foray into statistics with his 1912 paper, "An Absolute Criterion for Fitting Frequency Curves."[10] Inspired by work on the theory of errors in astronomy and mathematics, Fisher criticized the least squares method and the method of moments. John Aldrich claims that Fisher's true target here was Karl Pearson's 1902 essay, "On the Systematic Fitting of Curves to Observations and Measurements."[11] Pearson favored both of these methods, but Fisher found their justification to be arbitrary, and so their agreement with each other was problematic rather than reassuring. In their place, he championed an approach to error he had learned from the astronomers based on Gauss's least squares method.[12]

The dominant school of thought at the time was Bayesian and employed what was called the method of inverse probability. Named for Thomas Bayes, the Bayes theorem for any two events A and B claims that the conditional probability of event A given event B is:

$$P(A \mid B) = \frac{P(B \mid A)P(A)}{P(B)}.$$

In this equation, the probability of event A, P(A), is called the prior probability, meaning that it is the probability of A before event B. The probability of A given B or after B has occurred is the posterior probability. The Bayes theorem tells us how to adjust our prior probability of A in light of a new event B; or the prior probability of a hypothesis, H, in light of evidence, E. In more direct terms, Bayes theorem addresses the problem of how new evidence can guide the revision of our previous beliefs—it addresses the problem of induction.

At the time Fisher was writing, it was common practice to recognize that if the prior probability was unknown, then one could assume that there was, in Pearson's words, an "equal distribution of ignorance" so all probability values of H are the same, or all beliefs about the probability of H are equivalent. This means that the probability of H given E does not depend on the prior probability of H, but on the remaining term, P(E/H)/P(E). Fisher was critically examining the claim that the curve that best fits the data was the one that maximized the posterior probability. What he was proposing was that maximizing P(H/E) and maximizing P(E/H)/P(E) were not equivalent. In fact, they weren't even both probabilities.

Fisher was a frequentist. He believed that probabilities described relative frequencies of events in a certain number of trials or experiments. The probability of a kind of occurrence was estimated by the ratio of the number of observed occurrences of an event for a given total number of trials. Moreover, given an infinite number of trials, the ratio of the number of events to the number of trials will converge on the true probability value. Fisher understood the Bayesian approach as trying to assign a probability value to something that was unique—not subject to repeated trials, and not subject to sampling as a result. For Fisher, it was legitimate to ask about the probability of observing an experimental outcome, but it did not make sense to speak of the probability of a hypothesis. To differentiate these approaches, Fisher distinguished the probability of a hypothesis from its likelihood. The likelihood of a hypothesis given some evidence is equivalent to the probability of the evidence given the assumption of the hypothesis. For Fisher, maximizing likelihoods was the more statistically sound method of estimation and curve fitting since it could be grounded in observed frequencies. In making his case for the distinction between probability and likelihood, Fisher would transform the foundations of statistics. Translating the resulting authority from statistics to biology and demonstrating the deep relevance of statistics to biology were crucial in Fisher's transformation from a mathematician to a biologist.

Fisher was drawn to the theory of errors by a paper by "Student," really W. S. Gosset who was not allowed by his employer, Guiness Brewing, to publish under his own name since the statistical test was considered a trade secret. Gossett addressed the problem of estimating error when using a small sample size. This led Fisher to consider how to estimate error when calculating a correlation coefficient for small samples. Moreover, Fisher's maximum likelihood method gave importantly different results from Gossett's. The fine points of Fisher's essay were not fully grasped by Gossett or Pearson, who read it and discussed it in correspondence, but they did appreciate that Fisher was a young talent in statistics.

After graduation Fisher spent some time working on a farm in Canada, presumably to rest his notoriously poor eyesight. However, Fisher seemed to genuinely love farm life and would settle on a farm when he married in 1917. By then Fisher had returned to England and, determined to do his part for the war effort, taught mathematics and physics at public schools since he was not eligible for military service.

In 1914, Fisher published a proof of Gosset's solution using multidimensional geometry. While other statisticians did not share Fisher's fondness for geometrical reasoning about distributions, Pearson did publish it in his journal *Biometrika*.[13] Pearson's group was also occupied during the war with distributions of correlation coefficients, and in 1916, his group published a paper critical of Fisher's approach, claiming that it depended on the Bayes theorem. The injustice of this claim stung Fisher and motivated him clarify his stance on likelihood as an alternative to Bayes. At the same time it signaled a declining relationship with Pearson, who declined to publish Fisher's 1916 paper on correlation with regard to Mendelian traits. Pearson probably thought of himself as the senior professor helping to sort out the younger Fisher, and sent Fisher extensive comments on another 1916 note regarding error and estimation. Fisher responded by seeking to put statistics on a firmer mathematical foundation that directly took aim at the substantial work of Pearson.[14] When Fisher published his own criticisms of Pearson, their enmity became mutual. However, as historian Stephen Stigler has convincingly argued, the antagonism that Fisher felt toward Pearson certainly spurred him to develop his method of maximum likelihood and articulate the grounds for the justification of statistical methods.[15]

Throughout this period Fisher had maintained his interest in eugenics and sought an active role in the Eugenics Education Society. This interest brought him into contact with Major Leonard Darwin, Charles' fourth son and an avid eugenicist himself.[16] Darwin became Fisher's mentor and advocate around 1914, when Fisher began publishing book reviews for *Eugenics Review*. While Fisher scraped by financially as a farmer and teacher, Darwin arranged for a stipend from the Eugenics Society. This undoubtedly helped fuel the fire that led Fisher to write over 200 reviews on eugenics over twenty years.

But the relationship was something more. Fisher revered Darwin and sought his approval. When he discovered that Darwin and Pearson had had a disagreement on the effects of natural selection on the correlation of hereditary traits, Fisher sought to vindicate Darwin, but Darwin counseled restraint.[17] Nevertheless, Darwin encouraged Fisher to continue his work on the intersection of statistics and heredity. Indeed, when Fisher's paper on the correlation of relatives was rejected by Pearson for *Biometrika*, Darwin sponsored its publication in the *Transactions of the Royal Society of Edinburgh*. This paper helped Fisher land two job offers the next year: one in the Galton Laboratory and one at the Rothamsted Experimental Station. He chose Rothamsted.

EVOLUTION FROM A STATISTICAL GENETICS POINT OF VIEW

When Fisher was at Rothamsted Experimental Station from 1919 to 1933, he engaged with a mass of biological data that allowed him to revolutionize statistics, and he become convinced of the value of engaging statistical analysis with real biological data. Researchers at Rothamsted had accumulated years of data on crop growth and yield under a wide range of different conditions. Fisher's challenge was to find something biologically interesting in that data. What he found and published in a series of papers was important both biologically and statistically. Biologically, Fisher was able to disentangle the effects of various fertilizer treatments from soil, weather, and cultivation conditions. At the same time, Fisher developed methods for statistical experimentation based on randomization and the analysis of variance that would radically change the way in which any researcher could conduct a statistical experiment in the future. These methods were published in 1925 as *Statistical Methods for Research Workers*.[18] Initial reviews were critical as primarily English statisticians struggled to make sense of this new empirical approach. Reflecting on this later, Fisher offered that recognition takes time when the revision of cherished beliefs bruises the feelings of their holders. Fisher's recognition came first from abroad, where experimental agriculture was established and appreciated in institutions such as the USDA and many land grant universities in the United States.[19]

Fisher's work on the analysis of variance as applied to biology had begun much earlier in his work on correlation in evolution. In his 1918 paper, "On the Correlation of Relatives on the Supposition of Mendelian Inheritance," Fisher considered the statistical consequences of dominance, epistatic gene interaction, assortative mating, multiple alleles, and linkage on the correlations between relatives. Fisher argued that the effects of dominance and gene interaction would confuse the actual genetic similarity between relatives. He also knew that the environment could confuse such similarity. Fisher here formally introduced the concepts of variance and the analysis of variance. He wrote:

> When there are two independent causes of variability capable of producing in an otherwise uniform population distributions with standard deviations σ_1 and σ_2, it is found that the distribution, when both causes act together, has a standard deviation $\sqrt{\sigma_1^2 + \sigma_2^2}$. It is therefore desirable in analyzing the causes of variability to deal with the square of the standard deviation as the measure of variability. We shall term this quantity the Variance of the normal population to which it refers, and we may now ascribe to the constituent causes fractions or percentages of the total variance which they together produce.[20]

Fisher used this new tool to partition the total variance into its component parts. He labeled that portion of the total variance that accurately described the correlation between relatives the "additive" genetic component of variance. The "nonadditive" genetic component included dominance, gene interaction, and linkage. Environmental effects, such as random changes in environment, comprised a third component of the total variance. In 1922, on the basis of his 1918 work, Fisher argued that the additive component of variance was the most important for evolution by natural selection. Indeed, he argued that, particularly in large populations, nonadditive and environmental components of the total variance are negligible. Selection would act most strongly on any factor with a large additive contribution to the total genetic variance, usually by eliminating them from the population.[21] Most of the time, evolution, and especially adaptation, proceed very slowly, with low levels of selection acting on mutations of small effect and in large populations holding considerable genetic variation. Where Fisher's 1918 paper defended the principles of Mendelian heredity against the criticisms of the biometricians, his 1922 paper defended Darwinism *using* the principles of Mendelian heredity. Fisher's aim was to respond to a set of criticisms that Darwinian natural selection cannot be the correct mechanism of evolution because the genetics of populations are such that there is not enough variation available for selection to act upon. During the course of the paper, Fisher eliminated from consideration what he took to be insignificant evolutionary factors, such as epistatic gene interaction and genetic drift, and placed his confidence in natural selection.

Fisher's synthesis of Mendelism and Darwinism within a mathematical framework culminated in his 1930 book, *The Genetical Theory of Natural Selection*, which became one of the principal texts, along with those of J. B. S. Haldane and Sewall Wright, establishing the field of theoretical population genetics.[22] Fisher begins his book with his case for the mutual compatibility of Darwin's mechanism of natural selection and Mendelian genetics. He ends it by exploring the eugenic consequences of this statistical and genetic understanding of the evolutionary process. Fisher considered the first two chapters, on the nature of inheritance and the "fundamental theorem of natural selection," the most important of the book. Indeed, these two chapters accomplish the key piece of the reconciliation by continuing the general argument strategy he had used in 1918 and 1922 of defending Mendelian particulate inheritance and then demonstrating how Darwinian natural selection may plausibly be the principal cause of evolution in Mendelian populations.

Fisher's mathematical approach is most fully developed in his second chapter of *The Genetical Theory of Natural Selection*. The arguments here are drawn largely from Fisher's 1922 paper "On the Dominance Ratio" and his 1930 paper "The Distribution of Gene Ratios for Rare Mutations," which was a response to Sewall Wright's correction of Fisher's 1922 paper. Three key elements may be distilled from Fisher's "heavy" mathematics in the second chapter of *The Genetical Theory*. The first is a measure of average population fitness, Fisher's "Malthusian parameter" (i.e., the reproductive value of all genotypes at all stages of their life histories). The second is a measure of variation in fitness, which Fisher partitions into genetic and environmental components (based on his distinctions from 1918 and 1922). The third is a measure of the rate of increase in fitness (i.e., the change in fitness due to natural selection). For Fisher, *"the rate of increase of fitness of any organism at any time is equal to its genetic variance in fitness at that time"* (emphasis in original).[23] This last element is Fisher's "fundamental theorem of natural selection," and it is the centerpiece of his theory of natural selection.

Interestingly, inasmuch as Fisher considered his fundamental theorem the centerpiece of his evolutionary theory, it happens that the theorem is also the most obscure element of it. The theorem was often misunderstood until 1989, when Warren Ewens rediscovered George Price's 1972 clarification and proof of it.[24] Fisher's original statement of the theorem in 1930 suggests that mean fitness can never decrease because variances cannot be negative. Price showed that in fact the theorem does not describe the total rate of change in fitness but, rather, only one component of it. That part is the portion of the rate of increase that can be ascribed to natural selection. And, actually, in Fisher's ensuing discussion of the theorem, he makes this clear. The total rate of change in mean fitness is due to a variety of forces including natural selection, environmental changes, epistatic gene interaction, dominance, and so forth. The theorem isolates the changes due to natural selection from the rest, a move suggested in Fisher's 1922 paper. The relative importance of the additive component of genetic variance was increasingly appreciated in the genetics community after the Second World War. Price and Ewens recognized this and clarified the statement of the theorem by substituting "*additive* genetic variance" for "genetic variance" (since genetic variance includes both an additive and nonadditive part). With the theorem clarified, however, Price and later Ewens argue that it is not so fundamental. Given that it is a statement about only a portion of the rate of increase in fitness, it is incomplete. The Price-Ewens interpretation of the theorem is now the standard one.

Fisher compared both his 1922 and 1930 exploration of the balance of evolutionary factors and the "laws" that describe them to the theory of gases and the second law of thermodynamics, respectively. Of the 1922 investigation, Fisher says, the investigation of natural selection may be compared to the analytic treatment of the Theory of Gases, in which it is possible to make the most varied assumptions as to the accidental circumstances, and even the essential nature of the individual molecules, and yet to develop the natural laws as to the behavior of gases, leaving but a few fundamental constants to be determined by experiment.[25]

He continues the analogy in 1930, adding that

the fundamental theorem . . . bears some remarkable resemblances to the second law of thermodynamics. Both are properties of populations, or aggregates, true irrespective of the nature of the units which compose them; both are statistical laws; each requires the constant increase in a measurable quantity, in the one case the entropy of the physical system and in the other the fitness . . . of a biological population. . . . Professor Eddington has recently remarked that "The law that entropy always increases—the second law of thermodynamics—holds, I think, the supreme position among the laws of nature." It is not a little instructive that so similar a law should hold the supreme position among the biological sciences.[26]

The received view of these comparisons is that Fisher's interests in physics and mathematics led him to look for biological analogs.[27] No doubt this is part of the story. However, a more plausible interpretation of the comparison comes from treating Fisher's 1918, 1922, and 1930 works as one long argument. If we do so, we find that Fisher's strategy in synthesizing Darwinian natural selection with the principles of Mendelian heredity was to defend, against its critics, selection as an evolutionary cause under Mendelian principles. Following this argument strategy, Fisher built his genetic theory of natural selection piecemeal, or from the bottom up. That is, Fisher worked to justify the claim of his fundamental theorem by constructing plausible arguments about the precise balance of evolutionary factors. Thus, his piecemeal consideration of the interaction between dominance, gene interaction, genetic drift, mutation, selection, etc. led to his theorem. It was not, at least not primarily, the search for biological analogues to physical models and laws that underwrites the theorem.

No one has thought that Fisher's contribution to evolutionary genetics was less than groundbreaking. Rather, precisely what Fisher established, its nature and scope, and exactly how he did so, has been less than clear. With Fisher's work on variance in 1918, his work on the balance of factors in evolution in 1922, and his fundamental theorem of natural selection in 1930, we have a unified argument setting aside pervasive anti-Darwinism, originating a new mathematical approach to the evolution of populations, and establishing the very essence of natural selection.

The Genetical Theory sealed Fisher's reputation as a biologist. In 1933, he succeeded Karl Pearson as the Galton professor at University College London. His work on mathematics had earned him a place in the Royal Society in 1928. His 1925 book, *Statistical Methods for Research Workers*, was changing the way experimental biology could be conceived.

CONCLUSION

R. A. Fisher made extraordinary contributions to the mathematical foundations of statistics, statistical methods for experimentation, and the creation of population and evolutionary genetics. When combined with his strong belief in the social value of eugenics, the range of Fisher's interests pose a serious challenge to historians who would like to make sense of how one person could simultaneously pursue such disparate topics and make such important contributions to each. Fisher's daughter and biographer, Joan Fisher Box, offered separate chapters tracing simultaneous trajectories through his mathematics, statistics, genetics, and eugenics.[28] Other historians treat Fisher's interests in isolation from each other, and in doing so cast themselves in sharp contrast to historians who see Fisher's interests as mutually informed.[29] Having posed the question of how nonbiological training can foster innovation in the life sciences, we find ourselves seeking points of intersection in Fisher's work. These intersections are plentiful, but what were the historical conditions that allowed Fisher to find success in these intersections?

We claim that patronage and persistence played crucial roles in allowing Fisher to successfully bring his mathematical and statistical perspectives into biology. The patronage of Leonard Darwin early in his career facilitated the publication of Fisher's seminal 1918 paper on the correlation of relatives, and supported his work at the intersection of biology, eugenics, and statistics. At the same time, Fisher's personality allowed him to stubbornly persist in the face of Pearson's criticism and later opposition. Fisher's willingness to engage in a dispute and to oppose Pearson and the entrenched views on inverse probability were crucial in the history of his development of the foundations for theoretical statistics, on the one hand, and the methods of estimation and experimentation crucial to biology and evolutionary genetics, more specifically.

We do not wish to claim that the success of Fisher's contributions are solely the result of patronage or Fisher's personal advocacy. Their value and utility were recognized by many. In the case of experimental design and statistical inference, Fisher's work found an eager audience among agricultural experts around the world following the path of Mendelism in an age of faith in scientific progress.[30] However, intellectual worth alone did not overcome the barriers set by Pearson and other statisticians. Fisher's results were innovative, but the intersection of statistics, genetics, eugenics, and evolution was not an empty niche waiting to be filled. It took time for Fisher's early work to gain acceptance among the statistical and scientific communities. What allowed him to continue to innovate in statistical biology was the support of individuals, like Leonard Darwin, and institutions like Rothamsted, where his methodological insights more readily produced new results for field researchers.

FURTHER READING

Box, Joan Fisher. *R. A. Fisher: The Life of a Scientist*. New York: John Wiley, 1978.

Fisher, R. A. *The Genetical Theory of Natural Selection*. Oxford: Oxford University Press, 1930.

Fisher, R. A. "Statistical Methods in Genetics," *Heredity* 6 (1952), 1–12.

Mackenzie, Donald A. *Statistics in Britain, 1865–1930: The Social Construction of Scientific Knowledge*. Edinburgh: Edinburgh University Press, 1981.

NOTES

1. R. A. Fisher, "Statistical Methods in Genetics," *Heredity* 6 (1952), 1–12.

2. Fisher, "Statistical Methods in Genetics," 4.

3. R. A. Fisher, "The Renaissance of Darwinism," *Listener* 37 (1947), 1000, 1009.

4. Joan Fisher Box, *R. A. Fisher: The Life of a Scientist* (New York: John Wiley, 1978), p. 22.

5. Francis Galton, *Hereditary Genius: An Inquiry into Its Laws and Consequences* (London: Macmillan, 1869).

6. For more on British eugenics, see Donald Mackenzie, "Eugenics in Britain," *Social Studies of Science* 6 (1976), 499–532; Lyndsay A. Farrall, *The Origins and Growth of the English Eugenics Movement, 1865–1925* (New York: Garland Publishing, 1985); Pauline Mazumdar, *Eugenics, Human Genetics, and Human Failings: The Eugenics Society, Its Source and Its Critics in Britain* (London: Routledge, 1992).

7. Box, *R. A. Fisher*; Donald A. Mackenzie, *Statistics in Britain, 1865–1930: The Social Construction of Scientific Knowledge* (Edinburgh: Edinburgh University Press, 1981).

8. William B. Provine, *The Origins of Theoretical Population Genetics* (Chicago: University of Chicago Press, 1971); Kyung-Man Kim, *Explaining Scientific Consensus: The Case of Mendelian Genetics* (New York: Guilford Press, 1994).

9. R. A. Fisher, "Mendelism and Biometry (1911)," in J. H. Bennett, editor, *Natural Selection, Heredity, and Eugenics: Including Selected Correspondence of R. A. Fisher with Leonard Darwin and Others* (Oxford: Oxford University Press, 1983), p. 57.

10. R. A. Fisher, "On an Absolute Criterion for Fitting Frequency Curves," *Messenger of Mathematics* 41 (1912), 155–160.

11. John Aldrich, "R. A. Fisher and the Making of Maximum Likelihood, 1912–1922," *Statistical Science* 12 (1997), 162–176, 163.

12. Aldrich, "R. A. Fisher and the Making of Maximum Likelihood," 162.

13. Box, *R. A. Fisher*, 75–76.

14. Stephen Stigler, "How Ronald Fisher Became a Mathematical Statistician," *Mathematics and Social Sciences* 44 (2006), 23–30; R. A. Fisher, "On the Mathematical Foundations of Theoretical Statistics," *Philosophical Transactions of the Royal Society A* 222 (1922), 309–368.

15. Stigler, "How Ronald Fisher Became a Mathematical Statistician."

16. Box, *R. A. Fisher*, 49.

17. Box, *R. A. Fisher*, 51–52.

18. R. A. Fisher, *Statistical Methods for Research Workers* (Edinburgh: Oliver and Boyd, 1925).

19. Box, *R. A. Fisher*, 130–131; see Diane Paul and Barbara Kimmelman, "Mendel in America: Theory and Practice, 1900–1919," in *The American Development of Biology*, R. Rainger, K. Benson, and J. Maienschein, editors (New Brunswick, NJ: Rutgers University Press, 1988), pp. 281–310.

20. R. A. Fisher, "The Correlation of Relatives on the Supposition of Mendelian Inheritance," *Royal Society of Edinburgh* 52 (1918), 399–433, 399.

21. R. A. Fisher, "On the Dominance Ratio," *Proceedings of the Royal Society of Edinburgh* 42 (1922), 321–341, 334.

22. R. A Fisher, *The Genetical Theory of Natural Selection* (Oxford: Oxford University Press, 1930). Released in a revised second edition in 1958 from Dover Publications, and released in a variorum edition by Oxford University Press in 1999 edited by J. H. Bennett; J. B. S. Haldane, *The Causes of Evolution* (London: Longmans, 1932); Sewall Wright, "Evolution in Mendelian Populations," *Genetics* 16 (1931), 7–159; Sewall Wright, "The Roles of Mutation, Inbreeding, Crossbreeding and Selection in Evolution," *Proceedings of the Sixth Annual Congress of Genetics* (1932), 356–366.

23. Fisher, *Genetical Theory*, 37.

24. Warren Ewens, "An Interpretation and Proof of the Fundamental Theorem of Natural Selection," *Theoretical Population Biology* 36 (1989), 167–180; G. R. Price, "Fisher's 'Fundamental Theorem' Made Clear," *Annals of Human Genetics* 36 (1972), 129–140.

25. Fisher, "On the Dominance Ratio," 334.

26. Fisher, *Genetical Theory*, 11.

27. Provine, *The Origins of Theoretical Population Genetics*; Jean Gayon, *Darwinism's Struggle for Survival: Heredity and the Hypothesis of Natural Selection* (Cambridge:

Cambridge University Press, 1998); Jonathan Hodge, "Biology and Philosophy (Including Ideology): A Study of Fisher and Wright," in S. Sarkar, editor, *The Founders of Evolutionary Genetics* (Dordrecht: Kluwer, 1992), pp. 231–293.

28. Box, *R. A. Fisher.*

29. Stigler, "How Ronald Fisher Became a Mathematical Statistician"; Mackenzie, *Statistics in Britain.*

30. Paul and Kimmelman, "Mendel in America."

MAYA M. SHMAILOV

NICOLAS RASHEVSKY'S
PENCIL-AND-PAPER BIOLOGY

INTRODUCTION

When discussing disciplinary boundary-crossers, a question arises as to why and how one decides to transgress his or her comfort zone into an unknown land. Is this a discrete act of recognizing a need and merely introducing a methodology, a set of concepts or instruments, from one discipline to another in order to tackle a specific problem? Or is it a tendency inherent in the "outsider," which may be observed repeatedly throughout his or her scientific life? In the case of the mathematical physicist Nicolas Rashevsky (1899–1972), the latter seems to be the case. Rashevsky's outsiderness expressed itself in a wide range of disciplines, including biology, medicine, sociology, and psychology. "You name it; he had a theory on it," one of Rashevsky's first students, Alvin Weinberg, reminisced.[1] From the problem of cell division to the challenges of automobile driving, Rashevsky's forty-five years of scientific work provide more than their share of boundary crossing.[2]

This movement between disciplines coupled with Rashevsky's attempts to replace the biological problems presented by nature with mathematical investigations of simplified, hypothetical cases created antagonism and eventually neglect from biologists for the man who single-mindedly attempted to revolutionize their field. Rashevsky's career as mathematical biologist started while he was working as a mathematical physicist at the Westinghouse Research Laboratories in Pittsburgh (1924–1934), flourished at the Division of Biological Sciences at the University of Chicago (1934–1964), and dissipated with his resignation from Chicago and move to Ann Arbor, Michigan, where he worked at the Mental Research Institute (1964–1970).[3] In 1969, Rashevsky formed a nonprofit organization, Mathematical Biology Incorporated, the precursor of the currently active Society for Mathematical Biology, which provided an institutionalized venue for research in the field and an outlet for publishing.

While he established a new discipline in biology while coping with the social and political norms of science, Rashevsky's legacy, as his student Robert Rosen summarized in retrospect, stands "in stark contrast to the

161

fate of the man himself."[4] Antagonism towards Rashevsky was expressed even by those who had never met him or followed his research. In 1970, for instance, the applied mathematician Richard Bellman asserted that "if Rashevsky knew what he was doing, he would have been a charlatan."[5] Rashevsky was perceived as a Svengali propagating unorthodox views from within the

Figure 9.1: Nicolas Rashevsky, circa 1945.
Archival Biographical Files, Special Collections Research Center, University of Chicago.

realm of biology.[6] Perhaps his height, notorious long beard, confidence, and strong voice lent credence to such a judgment (see figure 9.1).

Rashevsky's crossing over into biology was not made in pursuit of a solution to a specific problem, nor was it an attempt to mathematically evaluate a domain in biology. He had a vision. Rashevsky had his mind set on the "building-up of a systematic mathematical biology similar in its structure and aims to mathematical physics."[7] He was in pursuit of fundamental laws governing the life processes and believed that only a persistent search for such laws using physico-mathematical reasoning would eventually unveil the mystery of life.

Francis Crick once commented that "cosmologists are . . . less inhibited than chemists in regard to scientific speculations."[8] When Rashevsky had a chance to discuss this statement with Crick in 1959, he jokingly pointed out to Crick that "mathematical biologists are much worse than cosmologists!"[9] Joke or not, it is precisely due to the speculative and abstract nature of their work that Rashevsky and his fellow mathematical biologists have often been accused of being entirely disconnected from biology.[10]

Rashevsky's decision to move from mathematical physics to biology occurred at a time when, as a commentator argued in 1925, "biology [was] fast approaching its scientific stage," if one considers "the amount of mathematical expression of a branch of science as a measure of how scientific that branch is."[11] Traditional methods of observation were being replaced by experimentation, as biologists revolted against the theoretical, speculative systems and were embracing the empirical methods of the laboratory.[12] Still, there were exceptions to the trend. While it was generally held that some biological problems were amenable to mathematical analysis, "fundamental physiological life processes did not . . . fall within the group."[13] With this as his background, equipped with pencil and paper as his instruments, Rashevsky embarked on his life's journey of turning his vision into reality.[14]

A BRIEF SKETCH OF RASHEVSKY'S LIFE

Nicolas (Nikolai) was born on September 20, 1899, to a wealthy *bourgeoisie* family, the eldest of two sons of Peter and Nadejda Rashevsky of the small Ukrainian town of Chernigov. Growing up, Rashevsky received an excellent education. By the age of nineteen, he obtained a doctorate in mathematical physics from the University of Kiev and was believed to be a man of "unusual ability in physics and mathematics, capable of . . . original work and . . . a skilled and resourceful experimenter."[15]

During the Russian Revolution, Rashevsky fought in the White Army, and in 1920 married the physicist Emilie (Emily) Zolotareff, a princess from the Caucasus.[16] Following the defeat of the Whites, Rashevsky was forced to flee to Constantinople, and by 1923 to Prague, where he taught physics and worked as a research assistant.

In the summer of 1924, Rashevsky immigrated to the United States, and the young family settled at Wilkinsburg, Pennsylvania. Rashevsky assumed a position as research physicist at the Westinghouse Research Laboratories (1924–1934), lectured on physics at the University of Pittsburgh, and translated scientific papers from German and Russian to English. His early works in mathematical physics dealt mainly with quantum mechanics and relativity theory, and towards 1925 he devoted himself to the problem of the thermionic effect and the thermodynamic properties of colloids and polydispersed systems. As historian Tara Abraham has documented, between 1927 and 1929 Rashevsky published seven papers on the dynamics of colloidal particles, with one of his first papers addressing the problem of size distribution of particles in a colloidal solution.[17]

CROSSING BOUNDARIES

Rashevsky's interest in biology began to crystallize in 1927 while he was still working as a mathematical physicist at Westinghouse. Alfred Lotka's *Elements of Physical Biology* (1925) triggered his interest in biology, and in particular his search for general mathematical principles that would apply to the entire realm of biology.[18] Rashevsky, however, advocated a different role for mathematics in ordering the "organic world." While, according to Rashevsky, "Lotka's interest lay . . . in mathematical theory of the interaction of different species," Rashevsky became interested in developing a "general method of approach" that would cover the whole field of biology while discovering the "physical mechanisms which underline the functioning of the individual organism."[19]

Rashevsky had his mind set on complementing experimental methods in biology with a methodology that would not necessarily lead to the solution of a specific problem but would rather provide variations of possible solutions, out of which at least one would be proven correct by subsequent experimentation. Thus, while physiologists would say to one another "experiment, experiment, experiment!", Rashevsky was introducing controversial speculative thinking. As Andrew Huxley indicated in 1950, Rashevsky was "attempting over a wide field, a synthesis for which an adequate experimental basis [did] not exist."[20]

In order to understand Rashevsky's relationship with the experimental community, we must understand his stand on what mathematical biology is and who are to be considered its practitioners. The cornerstone to Rashevsky's approach placed mathematical biology as a counterpart to mathematical physics. Just as mathematical physicists were not concerned with the statistical evaluation of experimental results, but rather with the *fundamentals* of physical phenomena, the same should apply to mathematical biologists. A mathematical biologist, by Rashevsky's own definition, "is *not* interested in finding an empirical equation which best fits a given set of experimental points. He is interested in *deriving from a set of assumptions* . . . an equation which will fit the set of experimental points"[21] (emphasis added).

Rashevsky's mathematical biologist was to work from *inside* the world of biologists rather than *outside*; the latter searches for correlation between measured parts, while the former embeds biological thought within mathematical schemes.[22] His approach was to transform biology from the descriptive, inductive stage to a stage where experimentalists are governed by deductively formulated theory. When he introduced his stand to the wider scientific community in 1935, Rashevsky characterized his methodology as "first studying . . . over simplified cases, which may even perhaps have no counterpart in reality" and only later examining "realistic" cases.[23] Simplification was used to predict *trends* rather than exact values, this sort of analysis being a standard practice in physics and applied mathematics.[24] Such a methodology, Rashevsky believed, would help to illuminate the *complexity* of all biological phenomena.

Realizing the importance of comprehending the domain he wished to transform, Rashevsky commenced by familiarizing himself with the ways of the "insider." Arguing that "even for a theoretician, familiarity with laboratory work was essential, in biology as well as in physics," Rashevsky studied biological literature, wetting his hands by doing informal laboratory work while still at Westinghouse with Davenport Hooker, a professor of anatomy; physiologist C. C. Guthrie from Pittsburgh University; and biologist Everett Kinsey, with whom he studied techniques of animal operations.[25] Nonetheless, Rashevsky never conducted experiments to support his research in mathematical biology. He solemnly believed that just as Lord Kelvin and James Clerk Maxwell never experimented with actual physical models in physics, but rather investigated the problem mathematically, so should the work of mathematical biologists be mathematical and *in abstracto*.

A first step towards unveiling the complexity of biological phenomena led Rashevsky to the "fundamental living unit"—the cell.[26] Inspired by his research on thermionic devices involving problems of spontaneous splitting of fluid drops, he realized that similarities might exist between the splitting of the drops and the division of cells.[27] Setting aside its complexity, Rashevsky abstracted the cell to the point where it was analogous to a small sphere suspended in a solution containing "food substances." This theoretical cell,

Rashevsky argued, would hold its spherical shape and would not divide until a disturbance is imposed thereupon, for example by interaction between the cell's interior and the surrounding substances. Based on this scheme, Rashevsky developed mathematical equations, what he called "pencil and paper models," stating that these have a value greater than actual "experimental" models.[28]

By 1933, it seemed that Rashevsky's approach was sprinting down the fast lane to success. Parallel to Rashevsky's interest in the physiology of the cell, he became interested in nerve excitation. By the end of the nineteenth century, attempts were made to mathematically express the effect stimuli had on excitation of nerves. In 1899 the theoretical chemist Walther Nernst, inspired by Jacques Loeb's work, introduced mathematical theories describing the effect the electric current had on electric excitation in living tissues. The relation between the intensity of the electric current, its duration, and the concentration of ions became a focal point of several theories developed on nerve excitation. Toward the end of the 1920s and throughout the 1930s, Rashevsky published several papers on the subject. But it was in 1933 that Rashevsky proposed a novel phenomenological theory (the "two-factor theory"), what he called "an essentially new point of view," in an attempt to embrace the "whole field of [nervous excitation and inhibition] phenomena" not accounted for by previous attempts. He published it in *Protoplasma*, with the physiological community as his primary audience.[29]

Rashevsky considered that an electrical stimulus applied to the axon would have two effects: it would cause a rise in an "excitatory" process, and a simultaneous rise in an "inhibitory" process. Each process increases at a rate proportional to the current flowing through the nerve and decays at a rate proportional to its own magnitude. Rashevsky's equations read as follows:

$$d_e/d_t = KI - k(e - e_0)$$
$$d_i/d_t = MI - m(i - i_0),$$

where K, k, M, and m are constants, e the excitatory process, and i the inhibitory one. Action occurs whenever e equals or exceeds i in value.

What made this theory interesting was the fact that less than three years later, the Nobel laureate Archibald Hill introduced a similar theory at which he arrived, apparently, without being aware of Rashevsky's work.[30] While Hill described the excitation itself by an equation similar to Rashevsky's, he thought of inhibition, which he termed "accommodation," due not to the stimulating current *per se,* but rather as a result of the rise in the excitatory process e. Again, excitation occurs when e exceeds i in value. Hill's equations (using Rashevsky's notations) read as follows:

$$d_e/d_t = KI - k(e - e_0)$$
$$d_i/d_t = M(e - e_0) - m(i - i_0).$$

The main difference between Rashevsky's and Hill's works lay in method-ology, and in particular the fact that Hill performed extensive experimental studies of the subject through which he designed his "model," while Rashevsky made assumptions, built a model, and verified it with available experimental data.[31] Thus the experimental "verification" that Hill had for his theory was not, in fact, any better verification for his model than for Rashevsky's.[32]

OUTSIDERS' SAD LOT

Rashevsky's phenomenological theory of nerve excitation, as Alvin Weinberg later noted, was the "most solid predecessor" of the Alan L. Hodgkin and Andrew F. Huxley model of action potential propagation, published in 1952.[33] Nonetheless, the theory was now being set aside by the insiders: in the fifth edition of *Recent Advances in Physiology*, published in 1937, the two-factor scheme was in fact described as Hill's theory "without any qualifications," as the University of Rochester neurophysiologist Henry Blair pointed out to Rashevsky later that year.[34] But an experimental analysis of the two theories had already been suggested earlier by the University of Chicago's neurophysi-ologist Ralph Gerard and was undertaken by his student Franklin Offner, who proposed to investigate which model corresponded best to experiment. Offner discovered that solutions of the mathematical equations suggested by Rashevsky and Hill led to exactly the same predictions and fit the experi-mental data equally.[35]

In lieu of this conclusion, Gerard and Offner submitted a short paper to the editor of *Nature* in 1936, who rejected it "on account of lack of space."[36] Hill never discussed the matter publicly, nor did he respond to "some embarrassing questions" raised by one of his graduate students, Donald Scott.[37] A copy of Rashevsky's paper was sent by Blair to Scott with the hope of receiving answers as to how Hill arrived at his theory without being aware of Rashevsky's work. Six months prior to submitting his own paper, after all, Hill was corresponding with Rashevsky regarding Rashevsky's 1933 article in *Physics* on nerve conduction.[38] When Blair wrote to Rashevsky about the "embarrassing questions," Rashevsky responded that Hill "is [not] a man to be embarrassed by such trifles" and decided to "take the matter philosophically."[39]

What factors could account for Rashevsky's failure to influence *insiders*? Why was his theory neglected while Hill's theory was put on the map of physio-logical research? Was it the fact that Hill based his theory on experimentation and Rashevsky on "speculations"? Was it because Hill was a Nobel laureate and had the kind of institutional standing Rashevsky was lacking? Or because Rashevsky proposed his theory as an intruder, while still working as a theoretical physicist at Westinghouse?

To quote the biologist and founder of the *Quarterly Review of Biology*, Raymond Pearl, in his review of Rashevsky's 1938 magnum opus *Mathematical Biophysics*:

> Somewhat unfortunately the pioneers in [mathematical biology] have a hard and discouraging row to hoe. The reaction of the biologists— including both those who are able to understand the mathematical procedures and the much larger number who are not—is apt to be that the initial postulates are always too much simplified to have any *significant* relation to biological reality as they know it (and the mathematicians do not).[40] [emphasis in original]

But our outsider's failure in making an impact cannot be attributed only to his inability to grasp the biological reality. Presenting his views while still at Westinghouse, Rashevsky's work was losing its power for the experimentalists in the shadow of his assertion of "the theorist's independence."[41] Since he approached biology from the perspective of a mathematical physicist, Rashevsky was not attempting to *trade* knowledge with physiologists; instead, he was using his perspective as a mathematical physicist to *dictate* how biology should work.

From the very beginning Rashevsky presented his work to insiders, admittedly "quite intentionally," as he wrote in a letter to a colleague, making presumptuous statements, which "might irritate *some* biologists."[42] He continuously advocated an abstract, theoretical approach to biology, comparing his work and vision to those of Kepler, Newton, and Einstein in physics.[43] A look at his published works reveals a rhetorical style filled with overstatements and exaggerations. In 1936 he wrote:

> Although we consider the development of mathematical [biology] . . . of greatest importance for interpretation of empirical biology, we do not consider this "utilitarian" aim as the principal driving motive for our study. . . . Mathematical [biology] has a right to existence of its own, and its interest lies not merely in the number of empirical facts which it can explain but in its . . . mathematical beauty. As a consolation for the "fact-seekers" we have many times pointed out that usually such pure theoretical studies bear . . . practical fruits. But this to us is really beside the point.[44]

Rashevsky was not advocating the use of the theoretical tools to explain empirical facts, but was rather asserting the independence of mathematical biology. Either due to his lack of command of English or his intentional attempts to irritate biologists, his tendency to pretension managed to put off and antagonize a fair share of the "insiders." [45]

When reviewing Rashevsky's early work, it becomes apparent that he was not looking for acceptance by insiders; he was trying to design a new kind of biologist, one who would work from *within* biology with a new approach. For him, mathematics was not to be made a mere handmaiden of the experimentalists; he was building up a new discipline that would require an expertise on the borderland between mathematics, physics, and biology. [46] Rashevsky's "outsiderness" soon unmasked not only the problem of acceptance of his science by insiders, but the problem of institutional acceptance.

MAKING AN "HONEST WOMAN" OF MATHEMATICAL BIOLOGY [47]

In March 1934, as a result of downsizing, Rashevsky was fired from his position as research physicist at Westinghouse. Around the same time, Rashevsky's application of methods from the physico-mathematical sciences in domains of the natural sciences attracted the interest of Warren Weaver, director of the Natural Sciences Division at the Rockefeller Foundation (1932–1955). Weaver's agenda was "to bring to reality a change in the . . . biological research that would open up if some of the most imaginative physical scientists turned their attention . . . to the examination of biological problems." [48] Weaver was in search of ideas that would produce "the intellectual ferment characteristic of much of the work in the physical sciences." [49] Rashevsky, it seemed, was just the physicist he was looking for. On April 5, 1934, Rashevsky was offered a one-year fellowship by the Rockefeller Foundation to develop an adventuresome project applying physico-mathematical methods to biological problems at the University of Chicago. [50]

The venue for this interdisciplinary project was not coincidental. Under the Robert Hutchins presidency, the University of Chicago was unique in having an administrative mechanism for promoting interdisciplinary studies. [51] With William H. Taliaferro as the dean of the Biological Sciences Division, interdisciplinarity was fostered with the notion that science should not be constrained by a demand for immediate application of its findings. [52]

Despite the administration's positive attitude towards interdisciplinary studies, Rashevsky initially had a hard time finding a place for his research. While a member of the Division of Biological Sciences, he spent his first year working under the auspices of the Department of Psychology. By the end of 1935, Rashevsky was moved to the Department of Physiology, despite the objection of its chairman, a devoted empiricist, the physiologist Anton J. Carlson. Carlson "actively disliked and mistrusted" Rashevsky, and eventually in 1936 forced the administration to move Rashesvky back to the psychology department.[53] In a way, Carlson's attitude was "self-defeating," as Taliaferro would later indicate to him, since rejection of Rashevsky's inclusion in his department forced the administration to "set R[ashevsky] up as a separate Department," providing Rashevsky with an institutionalized venue to pursue his work within the Division of Biological Sciences.[54] Thus, by late 1938 the Section of Mathematical Biophysics was formed.[55] By then Rashevsky, now an associate professor, was no longer a lone wolf and was working with a group of young men, including the physicists John Reiner, Gaylord Young, Alvin Weinberg, Herbert Landahl, and Alston Householder. This group formed what Rashevsky called "a permanent nucleus" around which the work in mathematical biology was crystallizing.[56]

While the group worked under the Division of Biological Sciences, they were isolated from its other members. Rashevsky and his team were provided with quarters at the outskirts of the university and away from the insiders. In isolation, his students teasingly pleaded with him to let them put on white coats to at least look like the "insiders," but Rashevsky refused, insisting on "a special niche for *mathematical* biology."[57]

Finding an appropriate venue for disseminating the group's research results proved to be a challenge, further isolating them from the insiders. While Rashevsky's earlier papers were published in the *Zeitschrift fur Physic, Physics, Protoplasma, Psychometrica,* and *Journal of General Physiology,* he was moving, as Weaver put it, "from one journal to another as difficulties have developed."[58] Rashevsky's research was falling between the cracks. As Weaver wrote in 1936 when he supported a grant for Rashevsky so that he could publish his work:

> Rashevsky has had to submit his papers either to physics journals or to biological journals. . . . In the physics journals he has had to suppress the biological interpretation and application. . . . Conversely, when publishing in biological journals he is required to eliminate a large share of the mathematical . . . argument. Thus . . . his researches have never been adequately presented in a form which gives the proper emphasis to the intimate relationship between the physico-chemical analysis and the biological problems.[59]

While the journal *Protoplasma* seemed to be a "relatively satisfactory" outlet, when the cellular physiologist Robert Chambers was appointed as its editor in 1938 it was becoming clear to Rashevsky and Weaver that "[the referees] did not really understand the analytical arguments in the papers of mathematical biologists" and that the editor was "not qualified to edit these papers," as summarized in one of Weaver's interviews.[60] These factors led to the formation of the *Bulletin of Mathematical Biophysics* in March 1939, with Rashevsky as its editor. The bulletin quickly became "a classical journal in mathematical biology and served as the principal publication outlet" for mathematical biologists.[61] During the first decade of its existence, almost all contributions were Rashevsky's and those of his students. However, from the early 1950s, work was contributed by over fifty scientists from the United States and about a score from abroad.[62] Most of the contributors were either graduates of Rashevsky's program or mathematical physicists interested in his approach.

With the university providing the means for supporting his research, by 1940 Rashevsky was establishing a name for himself. Rashevsky was primarily working on the problems of cell division, cellular growth, cancer, nerve conduction, and the central nervous system. He was invited three times to the Cold Spring Harbor Symposia on quantitative biology to discuss his mathematical theories on the division of cells (1934), excitation and conduction in nerves (1936), and the permeability of cells (1940). Rashevsky and his group were in close contact with on- and off-campus experimentalists, discussing the biological problems, gathering data, guiding their experiments, and verifying the mathematical theories—so much so that Carlson not only came to terms with Rashevsky, but in fact, as Rashevsky proudly reported to Weaver, "became quite an enthusiastic backer of Rashevsky's group."[63]

By now, Rashevsky's mathematical biology had grown out of its first, abstract stage of development, which, Rashevsky argued "must of necessity remain on . . . [a] purely theoretical level, without any apparent contact with actual data."[64] Having laid down what he believed to be the "theoretical foundations," his field was headed toward the "most important stage of development of a theoretical science." This stage, according to Rashevsky, occurs when the science "not only explains already known phenomena, but mathematically predicts new ones, and suggests ways for new experimentation."[65] With weekly Friday afternoon meetings devoted to reports of the current research, Rashevsky was establishing contact between his group and members of other departments, as well as off-campus and out of town experimentalists and theoreticians. With Rashevsky and his group making "considerably greater contact with the laboratory" than previously, in 1940 Rashevsky published a summary of their work in *Advances and Applications of Mathematical Biology* to show the progress their research was making.[66]

Into the 1940s, Rashevsky's *in abstracto* treatments of various phenomena were progressively graduating into more realistic cases. Thus in the case of cell division, his "spheres" were now incorporating some physiological characteristics of the living cell and accounting for the roles of cytoplasmic streaming and catalysts. Rashevsky was now working with experimentalists, designing and guiding their experiments, which at times lead to breakthroughs in their research.[67] For a while it even seemed that his "speculations" on cell division were bringing mathematical biology of the cell to its final stage of development, in which theory predicts reality. But the fortunes of the new field remained precarious as a result of occasional slips. In 1949, for example, the Swedish geneticist Gunnar Östergren pointed out to Rashevsky in a private letter that his theory of cell division overlooked mitosis, and Rashevsky was forced to admit that his new theory was "inadequate."[68] Such missteps presented setbacks that led to severe criticism of Rashevsky's approach and placed him yet again outside the world of experimentalists.

Rashevsky was now putting aside all his affiliation with the physico-mathematical world. By 1942 he had distanced himself from the mathematics and physics communities and even resigned from the American Physical Society, since he found it no longer appropriate to present his work to the physics community. Insisting that he be approached as a mathematical *biologist*, he considered his work "too biological" and out of place among papers in modern physics.[69] While the physico-mathematical world was mourning the loss of its member, members of the biological community were not welcoming him with open arms.[70]

By the early 1950s, Rashevsky was a tenured professor, and his department was, according to Weaver, attracting much attention as the only place "where the student is free to study mathematics *and* biology."[71] This attention eventually resulted in emancipation of the section from the Physiology Department, and in 1947 the Committee of Mathematical Biology was formed within the Division of Biological Sciences. The main function of the committee, Rashevsky wrote in a letter, was "to train research workers in this field so that when other institutions become interested in it, we could supply well trained mathematical [biologists]."[72]

The training program brought together scientists, mainly graduate and postdoctoral students, who were interested in doing mathematical biology. Rashevsky managed to develop a disciplined approach to honing the competencies of interdisciplinary researchers. The program was designed to provide the students with proficiency and competence in both biology and the physico-mathematical sciences. Students were required to complete a heavy curriculum that entailed at least one extra year of study. The list of available courses for students included over forty courses equally divided between the biological sciences and the physico-mathematical sciences, as well as nine research courses in mathematical biophysics. In response to the doubts raised about who would undertake such a heavy course schedule, Rashevsky responded that he "[was] not interested in quantity production of PhDs but rather in their quality."[73] Ultimately, between the years 1941 and 1963, the program granted over twenty-six PhDs under Rashevsky's mentoring.

By the 1950s, Rashevsky's own work was expanding and spreading to the problems of metabolism, brain functions, and cardiovascular and cardiopulmonary functions, as well as to a host of problems outside physiology and even biology, including sociology, history, and psychology. His work was receiving recognition by governmental agencies, foundations, and commercial bodies, and it seemed that his group was financially set to pursue their research.[74] But Rashevsky was growing uneasy.

LAST OF THE MOHICANS

By the 1950's Rashevsky came to realize that the reductionist treatment of physiology had led him to lose sight of the organisms themselves, and he proposed a new, grander scheme. In 1954, he wrote:

> A direct application of the physical principles, used in the mathematical models of biological phenomena, for the purpose of building a theory of life as an aggregate of individual cells is not likely to be fruitful. We must look for a principle which connects the different physical phenomena involved and expresses the biological unity of the organism and of the organic world as a whole.[75]

This new principle was coined as "relational biology" and would occupy most of Rashevsky's scientific work from that point on, for he had realized that separate models of biological phenomena or organisms cannot be patched to describe the entire organic world—instead, he was in search of a mathematical theory that would unveil the organization and function of the phenomena or organisms. He argued that the functions of complex organisms and their structural organization can be mapped onto lower organisms in a manner that certain relations are preserved between sets of different phenomena.

He was not entirely abandoning his hitherto optimistic vision of developing a mathematical theory of life starting at the cellular level. As he wrote in a letter dated 1969:

> When 35 years ago mathematical biology was still in an embryonic stage, my students and I were . . . universalists in that field. Now . . . a

specialization has set of necessity. . . . I consider that the field still must be developed as a whole and that its branches are closely interconnected at least methodologically. . . . I feel that the time has come to introduce into biology *new methods of thinking* [emphasis in original]. Because of their novelty they may appear to be . . . crazy to biologists. . . . But somebody has to make the first unorthodox step in a new direction in order to help a future Newton or Einstein of mathematical biology. . . . My introduction of the concept of "Relational Forces" . . . may appear crazy. But so did many new concepts in physics.[76]

While his students and followers were now each specializing in mathematizing a specific niche in biology, Rashevsky still considered himself a universalist. He was paving the road, laying foundation for those to follow him, hoping eventually to realize his vision of a mathematical biology that would correspond in its grandness to mathematical physics.

DEVIOUS ROADS OF SCIENCE

In "Legitimation is the Name of the Game," Richard Lewontin writes: "To understand the problem of establishing a new view . . . is to understand the problem of introducing that view into a collective consideration and final acceptance by the social and political organization that constitutes science."[77] Lewontin accounts for four interlocking structural elements that enforce the scientific orthodoxy, all of which scientists must cope with "if there is any hope of incorporating a heterodox view into the corpus of accepted scientific knowledge."[78] These four elements are controlled by peers and consist of public communication to a relevant scientific community; employment and promotion within the halls of science; professional dependents; and grants.[79] These elements in Rashevsky's career can be summarized as follows:

1. Rashevsky first published his work in *Protoplasma*, *Physics*, and the *Journal of General Physiology* and later established a new channel of information passage, the *Bulletin of Mathematical Biology*; he presented his views in *Nature* and *Science*; he was invited to various conferences and scientific meetings on the borderline of mathematics and biology, such as the Cold Spring Harbor Symposia and Gordon Research Conference (1965). He was invited by universities in the United States, Europe, and Russia as a guest lecturer. He was engaged as a consultant to the Federal Food and Drug Administration and was a member of the American Association for the Advancement of Science, Biometric Society, Biophysical Society, and International Brain Research Organization at the United Nations.

2. Rashevsky's employment and promotion path at the University of Chicago started with the Rockefeller Fellowship (1934–1935) and progressed to assistant professor (1935–1938), associate professor (1938–1946), and finally to professor (1947–1964).

3. Graduate students and postdoctoral fellows applied to work with Rashevsky knowing the path they chose would be a demanding one. Under Rashevsky's "sponsorship" over two dozen students received their PhD in Mathematical Biology. While some continued their scientific careers outside the discipline of Mathematical Biology, several of his dependents continued Rashevsky's path at other institutions. For instance, John Hearon headed a research group in mathematical biology at the National Institutes of Health, James Danielly directed a center for theoretical biology at the University of Buffalo, and George Karreman found a niche at the University of Pennsylvania and founded the Society of Mathematical Biology (1972).

4. Funding for Rashevsky's endeavors was available from various privately held foundations and governmental agencies such as the Rockefeller Foundation, Lucius N. Littauer foundation, the William T. Morris Foundation, General Motors, and the US Air Force, and received grants from the United States Public Health Service, National Institutes of Health, and National Science Foundation.

Examination of these interlocked elements in Rashevsky's case illustrates Rashevsky's importance in the history of biology and his contribution to its patchwork design. Equipped with heterodox views, Rashevsky was able to propagate his influence within the realm of biology on a tortuous path of successes and failures; with the academic world providing an institutionalized venue for his endeavors, he challenged the prevailing dogmas of biology and transformed a boundary-crossing event into a discipline in its own right. Rashevsky's story constitutes a piece in the puzzle of understanding the problem of introducing a new view into biology. It provides a view from within in understanding the "outsider's" struggle and his coping mechanisms in dealing with the elements enforcing the scientific orthodoxy. His struggle toward acceptance of his vision by the social and political organization that constitutes science provides new insights into the scholarship of outsiders and the history of biology.

Considering himself a biologist rather than a mathematical physicist, Rashevsky did not seek the acceptance of insiders but rather was engaged with the design of a new kind of biologist. His battles within the halls of science, which led to the establishment of the *Bulletin of Mathematical Biology* and the first degree-granting program in mathematical biology illustrate a unique and valuable facet of the dynamics of "outsider" transgression of boundaries.

And yet the story shows how outsiderness can often also remain to haunt a transgressor's scientific achievements. While driven by biological questions, his approach was and often still is lambasted, even by his own students, as having "nothing to do with the [biological] reality."[80] Rashevsky's constant search for universal mathematical principles in biology, akin to those of physics, sullied his reputation. He remains a notorious figure in the world of science in general and biology in particular.[81]

FURTHER READING

Abraham, Tara, "Nicolas Rashevsky's Mathematical Biophysics," *Journal of the History of Biology* 37, no. 2 (2004): 333–385.

Cull, Paul, "The Mathematical Biophysics of Nicolas Rashevsky," *Biosystems* 88, no. 3 (2007): 178–184.

Rashevsky, Nicolas, *Mathematical Biophysics Physico-Mathematical Foundations of Biology*, Dover Publications, New York, 1960.

Rosen, Robert, *Life Itself*, Columbia University Press, 1991.

NOTES

1. Lou Gross and Alvin Weinberg, video recorded interview, May 15, 2004 (hereafter GWI 2004), courtesy of Lou Gross.

2. For a comprehensive summary, see Nicolas Rashevsky, *Mathematical Biophysics Physico-Mathematical Foundations of Biology*, Vol. 1 and 2 (Dover Publications, New York, 1960).

3. Tara H. Abraham, "Nicolas Rashevsky's Mathematical Biophysics," *Journal of the History of Biology* 37 (2004): 333–385.

4. Robert Rosen, *Life Itself: A Comprehensive Inquiry into the Nature, Origin, and Fabrication of Life* (Columbia University Press, 1991), 112.

5. Ibid., 112–113.

6. Personal communications with Mrs. Gwen Rapoport, widow of Anatol Rapoport, Rashevsky's student and a close friend (2010–2011), in author's possession.

7. Rashevsky, *Mathematical Biophysics*, vii.

8. Correspondence with G. Gamov, June 26, 1959, Nicolas Rashevsky Papers (hereafter NRP), Box 10, Folder G, Special Collections Research Center, University of Chicago, IL, (hereafter SCRC).

9. Ibid.

10. Evelyn Fox-Keller, *Making Sense of Life: Explaining Biological Development with Models, Metaphors, and Machines* (Harvard University Press, 2003); Abraham, "Nicolas Rashevsky's Mathematical Biophysics."

11. O. W. Richards, "The Mathematics of Biology," *The American Mathematical Monthly* 32 (1925): 30–36.

12. E. B. Wilson, "Some Aspects of Progress in Modern Zoology," *Science* 41, no. 1044 (1915): 1–4; Garland Allen, *Life Sciences in the Twentieth Century* (John Wiley, New York, 1975); Jane Maienschein, "Shifting Assumptions in American Biology: Embryology, 1890–1910," *Journal of the History of Biology* 14, no. 1 (1981): 89–113.

13. Reginald G. Harris, "Mathematics in Biology," *The Scientific Monthly* 40 (1935): 504–510.

14. Paper and Pencil Biology, Research Reports, vol. 1, no. 5, 1950, University of Chicago, IL, Nicolas Rashevsky Biographical File, SCRC.

15. Letter of reference by Paul H. Dike, NRP, Box 12, Folder Misc., SCRC.

16. Personal Communication with Rashevsky's granddaughter, Dr. Vibeke Strand, December 20, 2010.

17. Abraham, "Nicolas Rashevsky's Mathematical Biophysics."

18. Correspondence with Mortimer Spiegelman, February 28, 1950, NRP, Box 8, Folder S, SCRC.

19. Nicolas Rashevsky, "Mathematical Approach to Fundamental Phenomena of Biology," n.d., NRP, Box 9, Folder Misc., SCRC.

20. D'Arcey Thompson, "Review: Nicolas Rashevsky, Mathematical Biophysics. Physicomathematical Foundations of Biology: Dr. Rashevsky Has a Way of His Own," *Nature* 142 (1938): 931–932; Andrew F. Huxley "Review: Nicolas Rashevsky, Mathematical Biophysics," *Nature* 165 (1950): 292.

21. Nicolas Rashevsky, Mathematical Approach to Fundamental Phenomena of Biology, n.d. NRP, Box 9, Folder Misc., SCRC.

22. Ibid.

23. Nicolas Rashevsky, "Mathematical Biophysics," *Nature* 135, (1935): 1938.

24. Paul Cull, "The Mathematical Biophysics of Nicolas Rashevsky," *Biosystems* 88, no. 3 (2007): 178–184.

25. Abraham, "Nicolas Rashevsky's Mathematical Biophysics"; Nicolas Rashevsky, "History of the Committee" (1963), Box 2, NRP, SCRC.

26. Rosen, *Life Itself*; Thompson, "Review: Nicolas Rashevsky, Mathematical Biophysics"; Nicolas Rashevsky, "Foundations of Mathematical Biophysics." *Philosophy of Science* (1934): 176–196, 181.

27. US patent number 1,840,130 and British patent number 271885 (1926); Robert Rosen and Dan P. Agin, eds., *Foundations of Mathematical Biology* (Academic Press, New York, 1972).

28. Nicolas Rashevsky, "Some Theoretical Aspects of the Biological Applications of Physics of Disperse Systems," *Physics* 1 (1931): 143–153.

29. Nicolas Rashevsky, "Outline of a Physico-Mathematical Theory of Excitation and Inhibition," *Protoplasma* 20, no. 1 (1933): 42–56.

30. Archibald V. Hill, "Excitation and Accommodation in Nerve," *Proceedings of the Royal Society of London, Series B, Biological Sciences* 119 (1936): 305–355.

31. C. Hodson and L. Y. Wei, "Comparative Evaluation of Quantum Theory of Nerve Excitation," *Bulletin of Mathematical Biology* 38, no. 3 (1976): 277–293.

32. Franklin Offner, "Excitation Theories of Rashevsky and Hill," *The Journal of General Physiology* 21 (1937): 89–91.

33. GWI, 2004.

34. Letter from H. A. Blair to Rashevsky, April 9, 1937, Folder "Correspondence," Box 8, NRP, SCRC.

35. Franklin Offner, "The Excitable Membrane-Biophysical Theory and Experiment," *Bulletin of Mathematical Biology* 35 (1973): 101–107.

36. Letter from Rashevsky to H. A. Blair May 7, 1937, Folder "Correspondence," Box 8, NRP, SCRC. The paper was eventually published in 1937 as an article by Franklin Offner: "Excitation Theories of Rashevsky and Hill," *The Journal of General Physiology* 21, no. 1 (1937): 89–91.

37. Letter from H. A. Blair to Rashevsky, November 14, 1934, Folder "Correspondence with H. A. Blair 1933–1939," NRP, SCRC.

38. Nicolas Rashevsky, "Some Physico Mathematical Aspects of Nerve Conduction," *Physics* 4, no. 9 (1933): 341–349.

39. Letter from Rashevsky to H. A. Blair, November 16, 1936, Folder "Correspondence," Box 9, NRP, SCRC.

40. Raymond Pearl, "Review: Nicolas Rashevsky, Mathematical Biophysics. Physicomathematical Foundations of Biology," *Bulletin of the American Mathematical Society (New Series)* 45, no. 3 (1939): 223–224.

41. Andrew F. Huxley, "Review: Nicolas Rashevsky, Mathematical Biophysics," *Nature* 165 (1950): 292.

42. Rashevsky to Weaver, September 24, 1936, RG 1.1, Series 216D, Box 11, Folder 147, RAC.

43. For example, Nicolas Rashevsky, "Foundations of Mathematical Biophysics," *Philosophy of Science* (1934).

44. Nicolas Rashesvky, "Mathematical Biophysics and Psychology," *Psychometrika* 1, no. 1 (1936): 1–26, 1

45. Weaver to Rashevsky, September 19, 1936, RG 1.1, Series 216D, Box 11, Folder 147, RAC.

46. Nicolas Rashevsky, "Mathematical Biophysics: Physico-Mathematical Foundations of Biology," *Bulletin of the AMS* 45 (1939): 223–224.

47. Memorandum from president in charge of administrative affairs at the University of Chicago, paleographer Ernst Colwell, to his vice president, chemist R.G. Gustavson, December 10, 1945, Box 214, Folder 6, Office of the President, Hutchins Administration Records, Special Collections Research Center, University of Chicago, IL.

48. Mina Rees, "Warren Weaver: July 17, 1894–November 24, 1978," *Biographical Memoirs, National Academy of Sciences* 57 (1987): 493–529.

49. Ibid.

50. Abraham, "Nicolas Rashevsky's Mathematical Biophysics."

51. R. B. Emmett, "Specializing in Interdisciplinarity: The Committee on Social Thought as the University of Chicago's Antidote to Compartmentalization in the Social Sciences," *History of Political Economy* 42 (2010): 261–287.

52. William H. Taliaferro, "Science in the Universities," *Science* 108 (1948): 145–148.

53. Weaver Interviews, January 19, 1939, RG1.1, Series 216D, Box 11, Folder 148, RAC.

54. Ibid.

55. By the beginning of the 1938–1939 academic year, the Department of Psychology was shifted from the Division of Biological Sciences to the Division of Social Sciences, and a new place had to be found for some of its members, including Rashevsky.

56. Rashevsky's letter to Weaver, March 26, 1938, RG 1.1, Series 216D, Box 11, Folder 148, RAC.

57. Anatol Rapoport, *Certainties and Doubts: A Philosophy of Life* (Black Rose Books Ltd., Montreal, 2000), 90.

58. Weaver Interviews, July 3, 1938, RG 1.1, Series 216D, Box 11, Folder 148, RAC.

59. Grant in Aid, October 14, 1936, RG 1.1, Series 216D, Box 11, Folder 147, RAC.

60. Weaver Interviews, July 3, 1938, RG 1.1, Series 216D, Box 11, Folder 148, RAC.

61. Philip K. Maini, et al., "Bulletin of Mathematical Biology—Facts, Figures and Comparisons," *Bulletin of Mathematical Biology* 66, no. 4 (2004): 595–603.

62. Alvin Weinberg, personal communication, May 27, 2004, in author's possession.

63. Weaver Interviews, June 18, 1940, RG1.1, Series 216D, Box 11, Folder 148 RAC.

64. Rashevsky to R. G. Gustavson, September 24, 1945, Box 214, Folder 6, HOP-SCRC. This approach was reflected in the first edition of Rashevsky's magnum opus, *Mathematical Biophysics: Physico-Mathematical Foundations of Biology* (University of Chicago Press, 1938).

65. Rashevsky to R. G. Gustavson, September 24, 1945, Box 214, Folder 6, Office of the President Hutchins, SCRC.

66. Ibid.; Alston S. Householder, "Review: Advances and Applications of Mathematical Biology," *National Mathematics Magazine* 15 (1941): 384–386.

67. One example is Rashevsky's work with biologist F. C. Besic on the effects of corrosion of the teeth enamel by various acids.

68. Gunnar Östergren to Rashevsky, December 5, 1949, Folder "O," Box 8, NRP, SCRC; Rashevsky, "Mathematical Approach to Fundamental Phenomena of Biology."

69. Correspondence with Arthur Compton, 1942, Folder "Correspondence," Box 8, NRP, SCRC.

70. Ibid.

71. Weaver Interviews, February 22, 1951, RG1.1, Series 216D, Box 11, Folder 150 RAC.

72. Letter from Rashevsky to Gustavson, September 24, 1945, Office of the President Hutchins Administration Records, Box 137, Folder 6, SCRC.

73. Memorandum on "History of the Committee," 1963, Box 2, NRP, SCRC.

74. Rashevsky and his group received grants from the National Institute of Health, National Science Foundation, and the U.S. Air Force, including a training grant of over half a million U.S. dollars for training mathematical biologists.

75. Nicolas Rashevsky, "Topology and Life: In Search of General Mathematical Principles in Biology and Sociology," *Bulletin of Mathematical Biology* 16 (1954): 317–348.

76. Rashevsky to Irving Gerring, February 26, 1969, Folder "NIGMS-RSS," NRP, SCRC.

77. Richard C Lewontin, "Epilogue: Legitimation Is the Name of the Game," in Oren Harman and Michael Dietrich, eds., *Rebels, Mavericks, and Heretics in Biology* (Yale University Press, 2008), 372–380, 372.

78. Ibid.

79. Ibid.

80. GWI, 2004.

81. Keller, *Making Sense of Life*, and personal communications with Lee Segel and Alvin Weinberg in 2004, in the author's possession.

JAY ODENBAUGH

SEARCHING FOR PATTERNS,
HUNTING FOR CAUSES
ROBERT MACARTHUR, THE
MATHEMATICAL NATURALIST

INTRODUCTION

Robert Helmer Macarthur was born in Toronto, Canada, on April 7, 1930, and died of renal cancer on November 1, 1972 (see figure 10.1). However, his legacy as an ecologist is not adequately represented by his mere forty-two years of life. MacArthur received an undergraduate degree in mathematics from Marlboro College in Marlboro, Vermont, where his father, John Wood MacArthur, was a geneticist. From there, MacArthur received a master's degree in mathematics from Brown University, and in 1957 he began a PhD program at Yale University starting in mathematics but quickly moving to ecology. At Yale, he studied with George Evelyn Hutchinson, the most important ecologist of the twentieth century, who influenced him both in style and substance.[1] During 1957 and 1958, MacArthur worked with ornithologist David Lack at Oxford University. From 1958 to 1965, he went from assistant to full professor at the University of Pennsylvania and finally became the Henry Fairfield Osborn Professor of Biology at Princeton University. In 1969, he was elected to the National Academy of Sciences.

In their memorial volume *Ecology and Evolution of Communities*, Martin Cody and Jared Diamond write:

> In November 1972 a brief but remarkable era in the development of ecology came to a tragic, premature close with the death of Robert MacArthur at the age of 42. When this era began in the 1950s, ecology was still mainly a descriptive science. It consisted of qualitative, situation-bound statements that had low predictive value, plus empirical facts and numbers that often seemed to defy generalization. Within two decades new paradigms had transformed large areas of ecology into a structured, predictive science that combined powerful quantitative theories with the recognition of widespread patterns in nature. This revolution in ecology had been largely due to the work of Robert MacArthur.[2]

When ecologists consider MacArthur's work, they emphasize the use of simple analytic models with interspecific competition as the primary

Figure 10.1: Robert MacArthur.
Photograph by Orren Jack Turner.

mechanism structuring ecological communities, accompanied by an approach to hypothesis testing consisting of qualitatively comparing models and patterns.[3] MacArthur worked as a mathematical "outsider," and not the customary biologist, identifying ecological patterns and providing simple models representing the causes of such patterns. As Eric Pianka notes, when MacArthur was with mathematicians he claimed to be a biologist, and when he was with biologists he claimed to be a mathematician.[4]

Some "insiders" denied these patterns exist, and some denied simple models could explain them.[5] Nevertheless, MacArthur's mathematical approach was historically important. As the writer David Quammen recognized, his influence has been profoundly methodological; he changed the way ecologists asked and answered questions about populations and communities.[6] MacArthur, as an applied mathematician with a love of fieldwork, changed the face of ecology.

In this essay, we first look at an influential view of the nature of mathematics espoused by G. H. Hardy; namely, that it is a science of patterns. Second, we consider the views of MacArthur's teacher G. E. Hutchinson and his emphasis on theory and pattern. Third, we explore the methodologies of co-members in the "Marlboro Circle," Richard Lewontin and Richard Levins. Fourth, we directly engage MacArthur's own views as a "mathematical naturalist." Finally, I offer some speculative reflections on MacArthur's role as an outsider.

HETEROGENEOUS UNSTABLE POPULATIONS

During the 1950s, vociferous but unproductive stultifying debates were occurring in ecology. In population ecology, ecologists argued intensely over whether populations are "regulated." Many populations persist through time and vary moderately in abundance. The "biotic school," of which David Lack was a member, argued that this was so because of density-dependent processes such as intraspecific competition.[7] Contrarily, the "climatic school" argued that populations minimally vary because of the abiotic environment (e.g., weather).[8] Confusions and complications appeared at every juncture in the debate. The facts were unclear, mechanistic explanations were often merely assumed, experiments were unrealistic, different groups used different model organisms, and terminology was ill defined. At the 1957 Cold Harbor Symposium, where the two sides epically clashed, G. E. Hutchinson suggested that the symposium itself was a "heterogeneous unstable population."[9] To MacArthur, the debate appeared to be mired in terminological difficulties over terms like "carrying capacity," "competition," and "density."[10]

He noted in his review of the volume in which Hutchinson's "Concluding Remarks" appears:

> Science usually progresses faster when theory is able to keep abreast of facts; when the array of facts is as complicated as human demography already is, one despairs of finding a theorist who can set up a complete, adequate theory. It may be very true that demographers know too much.[11]

Likewise, he opined that an "almost religious fervor replaces objectivity in the symposium."[12]

Similarly, in community ecology, Frederic Clements and Henry Gleason, along with their respective followers, disagreed vigorously over succession and the nature of ecological communities. Clements argued that disturbed communities followed a very specific sequence of stages to a single "climax" community. In fact, he considered communities to be "superorganisms." Gleason argued Clements's views were empirically unfounded and that community properties were the result of individual species's "physiological" requirements; "every species of plant is a law unto itself" with no climax community.[13]

Like population ecology, MacArthur saw the state of community ecology as deeply problematic. With regard to Clements, he was very skeptical of the notion of superorganisms and their "emergent properties," since most scientists believe that properties of wholes are the result of the behavior and interactions of their parts.[14] With regard to Gleason, he argued that ecologists primarily interested in separate species "have never made any progress in unraveling community patterns."[15] More generally, he writes:

> The question is not whether such communities exist but whether they exhibit interesting patterns about which we can make generalizations. This need not imply that communities are superorganisms or have properties not contained in the component parts and their interactions. Rather it implies simply that we see patterns of communities and that, at this stage of ecology, the patterns may be more easily related than the complex dynamics of the component species.[16]

Why couldn't ecologists leave behind fraught debates about population regulation and the nature of communities? In both disciplines, something new was needed. As we shall see, something new in ecology is exactly what MacArthur proposed.

MATHEMATICS AS A SCIENCE OF PATTERNS

To appreciate MacArthur's mathematical approach, we should consider the nature of mathematics. E. O. Wilson and G. E. Hutchinson note that MacArthur "resembled very much in temperament and philosophy" the pure mathematician G. H. Hardy and shared his "conviction" that mathematics was a science of patterns.[17]

Hardy begins with the idea that a mathematician is, like a painter or a poet, "a maker of patterns."[18] As an example of pure mathematics, Hardy considers pure geometries. He suggests that pure geometries are models; they are maps or pictures that are partial and imperfect copies of mathematical reality.[19] Pure mathematics is an attempt to describe mathematical objects, but not the physical world. Applied mathematics is an attempt to describe the patterns exemplified by spatiotemporal objects. He continues:

> Applied mathematicians, mathematical physicists, naturally take a different view, since they are preoccupied with the physical world itself, which also has its structure or pattern. We cannot describe this pattern exactly, as we can that of a pure geometry, but we can say something significant about it. We can describe, sometimes fairly accurately, sometimes very roughly, the relations which hold between some of its constituents, and compare them with the exact relations holding between constituents of some system of pure geometry. We may be able to trace a certain resemblance between the two sets of relations, and then the pure geometry will become interesting to physicists; it will give us, to that extent, a map which "fits the facts" of the physical world. The geometer offers to the physicist a whole set of maps from which to choose. One map, perhaps, will fit the facts better than others, and then the geometry which provides that particular map will be the geometry most important for applied mathematics.[20]

Thus, for Hardy, mathematical and physical objects exemplify patterns respectively that can be "roughly" compared.

So, pure mathematicians study and prove theorems regarding patterns independent of any physical exemplification. Applied mathematicians study patterns too, but particularly ones that "fairly accurately, sometimes very roughly" are exemplified by spatiotemporal systems. For example, a pure mathematician studying differential equations is concerned with how a point "changes position" in a geometric space. However, an applied mathematician studies differential equations in order to model, say, how birds feed in a forest. Mathematics, then, is the science of patterns.

G. E. HUTCHINSON, PATTERNS, AND NICHES

Remarkably, G. E. Hutchinson also wrote about the importance of patterns in his "The Concept of Pattern in Ecology."[21] According to Hutchinson, the concept of "pattern" is fundamental to science because the "completely disordered is unimaginable," and "if we are going to say anything at all, some structure is certain to be involved."[22] Put simply, intelligibility in science requires that objects be patterned. He defines "pattern" in ecology as follows: "The structure which results from the distributions of organisms in, or from, their interactions with, their environments, will be called pattern."[23]

Hutchinson also offered theoretical frameworks for investigating patterns in ecology; specifically, the concept of a "niche." In his 1957 "Concluding Remarks," he provided a commentary on the population regulation debates, with the hope that some clear theory could profitably redirect ecologists. In this essay, Hutchinson formalized the competitive exclusion principle: roughly, species with identical niches cannot coexist. Suppose that we have an n-dimensional hypervolume composed of independent variables, each affecting the abundance of species. This hypervolume has a nonempty area where the species persists; this is its "fundamental niche." Similarly, the "realized niche" is the volume where a species persists given interspecific interactions. If the realized niches of species in a community overlap, but not completely, then they will coexist. Hutchinson controversially claimed that there were groups of species, such as the European insect species *Corixa affinis*, *C. macrocephala*, and *C. punctate,* that differed in size by a factor of 1.3 to avoid competitive exclusion.

Hutchinson also claimed the principle of competitive exclusion might be falsified by territorial birds whose population size was too low for competition to occur.[24] In his dissertation, MacArthur did extensive fieldwork on five warbler species in Maine.[25] In areas of feeding where competition would be most likely to occur, he found an amazing degree of niche specificity. Each bird species fed in different parts of the trees, thus avoiding competitive exclusion. Hutchinson impressed upon MacArthur both the importance of well-chosen patterns and mathematical theory. Ironically, Macarthur noted regarding Hutchinson that his most significant achievements occurred in part by using procedures from other sciences on ecological questions.[26]

THE MARLBORO CIRCLE

The 1960s were an exciting time in theoretical biology. MacArthur, along with Ebgert Leigh (another Hutchinson student), Richard Levins, Richard Lewontin, and E. O. Wilson, were conceiving of a new integrative mathematical biology.[27] The group informally met at MacArthur's home in Marlboro,

Vermont. E. O. Wilson called this group the "Marlboro Circle" (Wilson 1994, 253). In his autobiography, Wilson notes that this group of five biologists met in July of 1964 to discuss their individual research agendas and to how they might jointly contribute to the future of population biology.[28] The question was how would they mathematically integrate population genetics, ecology, biogeography, and ethology into population biology:

> For two days between walks in the quiet northern woodland, we expanded upon our common ambition to pull evolutionary biology onto a more solid base of theoretical population biology. Each in turn described his particular ongoing research. Then we talked together about the ways in which that subject might be extended toward the central theory and aligned with it.[29]

Interestingly, Wilson reports they considered publishing under the pseudonym "George Maximin" (254). However, though the group did not, Levins and Lewontin did write under a pseudonym, "Isadore Nabi," criticizing systems ecology.[30] In the end, the group did not meet thereafter, instead pursuing their various interests in smaller groups.[31]

In the seventies, E. O. Wilson reminisced that they were interested in "simple theory." They deliberately attempted to simplify natural systems in order to articulate mathematical principles.[32] Wilson compared their work with that of other systems ecologists like Kenneth Watt, C. S. Holling, and Paul Ehrlich, who devised "complex theory."

> They say that because ecosystems are so vastly complex, you must be able to take all the various components into account. You really must feed in a lot of the stuff that we simple theorists leave out, like sunsets and tides and temperature variations in winter, and the only way you can do this is with a computer. To them, in other words, the ideal modern ecologist is a computer technologist, who scans the whole environment, feeds all the relevant information into a computer, and uses the computer to simulate problems and make projections into the future.[33]

This juxtaposition between "simple" and "complex theory" was important for the Marlboro Circle's members. Richard Lewontin similarly recollects,

> Dick Levins and I had hooked up with Robert MacArthur, who was then at Penn but then went to Princeton, and the three of us had this idea that we ought to be able to build a science of population biology that would fuse the intrapopulation genetic variation aspect of biology with demography and population ecology, and so on. . . . Dick Levins and I and Robert MacArthur used to meet, and we had a sense of really building some new

science of population biology. We had contact with Ed Wilson, who was also interested in that, and with Lee Van Valen. We met a couple of times in Vermont at Robert's "in-laws" place and, in general, had a kind of zeal for founding a new field.[34]

To understand this group's methodological opinions, we can start with geneticist Richard Levins in his 1966 classic essay, "The Strategy of Model Building in Population Biology" and his 1968 classic *Evolution in Changing Environments*. As MacArthur himself notes, they shared ideas so continuously that it was difficult to trace their individual history.[35]

In Levins's 1966 essay, there are several issues in play. First, Levins, along with others in the "Marlboro circle," was convinced that ecological and evolutionary processes must be modeled together. Clearly, ecological processes like plant succession can occur over centuries, and evolutionary processes like pesticide resistance can occur over a few years. Thus, given their entwinement, they should be jointly modeled, contrary to traditional theory. This is the "fusing" of the "intrapopulation genetic aspect" with population ecology, Lewontin notes. Second, mathematical models that represent the dynamics of the ecological-evolutionary multispecies ensembles could not be "photographically exact."[36] Maximally general, realistic, and precise models would be analytically insoluble, their parameters would be meaningless, and they could not be measured. This sort of "FORTRAN ecology," which "Isadore Nabi" poked fun at, was being advocated by "complex theorist" Kenneth Watt.[37] Third, Levins claimed one could build models that maximized any two factors among generality, realism, and precision, but not all three. MacArthur and Levins preferred general and realistic models at the expense of precision.[38] He concluded famously that theories in biology were collections of models with "robust theorems"; he writes, "Our truth is the intersection of independent lies."[39] Curiously, according to Wilson and Hutchinson, MacArthur often claimed to quote Picasso when he would say, "Art is the lie that helps us to see the truth."[40] What Pablo Picasso actually said was,

> We all know that art is not truth. Art is a lie that makes us realize truth, at least the truth that is given us to understand. The artist must know the manner whereby to convince others of the truthfulness of his lies.[41]

It seems likely that Levins's famous quote derives from Picasso via MacArthur.

In *The Genetics of Evolutionary Change* (1974), evolutionary geneticist Richard Lewontin articulates his own perspective on modeling systems, with a strong resemblance to that of Levins'.[42] At some time t, the system is in state E, and we want to predict the system's state E' at $t + n$. To do so, we must construct laws that contain the relevant variables and parameters. For example, we might predict gene frequencies using parameters describing fitness, mutation rates, and so forth. Lewontin crucially notes that there may be states E and E' such that no law of transformation can be constructed to obtain $E'(t+n)$ from $E(t)$.[43] For example, given merely the present position of a space capsule at some time, we cannot successfully predict its future position. However, given information regarding its velocity and acceleration in three orthogonal directions, a *dynamically sufficient* description can be given; that is, a set of laws such that given $E(t)$, we can successfully predict $E'(t+n)$.

Crucially, Lewontin notes dynamic sufficiency is relative to chosen variables and "tolerance limits." That is, for each state E a *tolerance set e* is provided such that states in e are regarded as "indistinguishable"; we do not care about differences among states within e.[44] With broad tolerance limits, the required model dimensionality (i.e., the number of variables needed for dynamic sufficiency) will be low, and if the limits are narrow, the required dimensionality will be high. For example, if population ecologists explain the changes in population abundance "to one order of magnitude," then net fecundity and mortality rates are sufficient. On the other hand, if fisheries biologists want to predict abundances to an accuracy of 10%–20%, this requires complete age-specific life tables. Human demographers' attempt to predict population size within 1% requires knowledge of age, sex, socioeconomic class, education, geography, and so on.[45] Echoing Levins's view of theories, Lewontin writes,

> The building of a dynamically sufficient theory of evolutionary processes will really entail the simultaneous development of theories of different dimensionalities, each appropriate to the tolerance limits acceptable in its domain of explanation.[46]

Of course, models must be more than dynamically sufficient; they must be *empirically sufficient* too.[47] The variables and parameters must be measured. Otherwise, the theory becomes a "vacuous exercise in formal logic."[48]

Lewontin extends his methodological discussion to ecology as well:

> It is not always appreciated that the problem of theory building is a constant interaction between constructing laws and finding an appropriate set of descriptive state variables such that laws can be constructed. We cannot go out and describe the world in any old way we please and then sit back and demand that an explanatory and predictive theory be built on that description. The description may be dynamically insufficient. Such is the agony of community ecology. We do not really know what a sufficient

description of a community is because we do not know what the laws of transformation are like, nor can we construct those laws until we have chosen a set of state variables. That is not to say that there is an insoluble contradiction. Rather, there is a process of trial and synthesis going on in community ecology, in which both state descriptions and laws are being fitted together.[49]

Community ecologists must find the right state variables for their laws to be dynamically sufficient. Only then, can they be empirically sufficient.

Thus, the methodological views of the evolutionary geneticists Levins and Lewontin have several features in common. First, both recognize the importance of theorizing in biology. Second, theories are judged relative to the task at hand. When we are not primarily interested in precision (i.e., "narrow tolerance limits"), we add to the generality and realism (i.e., dimensionality) of our models. Third, we need multiple models for adequate population biology. As we shall see, MacArthur accepts these points too.

MACARTHUR'S MATHEMATICAL MIND

After considering the most prominent influences, let's finally examine MacArthur's mathematical approach to ecology. His approach has four components: the search for patterns, the construction of simple theory, the testing of such theory with natural experiments, and a disregard for conceptual disagreements. Let's consider each in turn. MacArthur infamously writes:

> To do science is to search for general patterns, not simply to accumulate facts, and to do the science of geographical ecology is to search for patterns of plant and animal life that can be put on a map. The person best equipped to do this is the naturalist who loves to note changes in bird life up a mountainside, or changes in plant life from mainland to island, or changes in butterflies from temperate to tropics. But not all naturalists want to do science; many take refuge in nature's complexity in a justification to oppose any search for patterns. This book is addressed to those who do wish to do science.[50]

According to MacArthur, like Hardy and Hutchinson, science is the study of patterns. Ecologists study patterns that species and communities exhibit in space and time. Like Hardy, he claims that ecological patterns are importantly different from pure mathematical ones. First, ecological patterns admit of variation; they are seen with "blurred vision":

> Ecological patterns, about which we construct theories, are only interesting if they are repeated. They may be repeated in space or time, and they

may be repeated from species to species. A pattern which has all of these kinds of repetition is of special interest because of its generality, and yet these very general events are only seen by ecologists with rather blurred vision. The very sharp-sighted always find discrepancies and are able to say that there is no generality, only a spectrum of special cases.[51]

The "sharp-sighted" naturalist finds exceptions. Second, not only are the patterns "blurry," they are sensitive to the morphological, economical, and dynamical properties of the species.

Science should be general in its principles. A well-known ecologist remarked that any pattern visible in my birds but not in his *Paramecium* would not be interesting, because, I presume, he felt it would not be general. . . . [A] bird pattern would only be expected to look like that of *Paramecium* if birds and *Paramecium* had the same morphology, economics, and dynamics, and found themselves in environments of the same structure.[52]

Ecology involves natural history, but it inescapably requires the construction of theory too.

Unraveling the history of a phenomenon has always appealed to some people and describing the machinery of the phenomenon to others. In both processes generalizations can be made and tested against new information so both are scientific, but the same person seldom excels at both. The ecologist and the physical scientist tend to be machinery oriented, whereas the paleontologist and most biogeographers tend to be history oriented. They tend to notice different things about nature. The historian often pays special attention to *differences* between phenomena, because they may shed light on the history. . . . [The machinery person] tends to see *similarities* among phenomena, because they reveal regularities.[53]

Reviewing Lawrence Slobodkin's 1961 publication, *Growth and Regulation of Animal Populations*, MacArthur suggests that ecologists can be placed into two groups. One group pays the utmost attention to nature's complexity, documenting it through observations at endless lengths. The other group proposes theories that are continuously patched up to account for as much data as possible. This latter group of theoreticians will sometimes have to ignore important observations in order to articulate generalizations that will have to be revised considerably. However, it is only through this constant revision of principles that ecology can have any hope of becoming a "respectable branch of science."[54] The existence of "blurry patterns" undergirds MacArthur (and

Levins's) preference for general, realistic, but imprecise theories (and is similar to Lewontin's notion of broad "tolerance limits"). "Simple" as opposed to "complex" theory represents simple causal mechanisms that make qualitative differences, such as "fine vs. coarse-grained, pursuers vs. searchers, jacks of all trades vs. masters of one, r selection vs. K selection."[55]

Simple theory coupled with "blurry" patterns leads MacArthur to his "qualitative" view of theory evaluation.

> The concept of pattern or regularity is central to science. Pattern implies some sort of repetition, and in nature it is usually an imperfect repetition. . . . The imperfection of the repetition gives us the means of making comparisons. We witness an event A, occurring under conditions C, then, under slightly altered conditions, C', we witness a slightly altered event, A'. Now we have the seed of a scientific hypothesis: "the difference between C and C' causes (i.e., is always associated with) the difference between A and A'," which we test by further observations. In geographic ecology, we study patterns repeated in space, not time, and natural comparisons are those of events occurring in different places. Over and over again in what follows we compare the species on the mainland to those on an island, the species on one mountain to those on another, the species high on a mountain to those lower on the mountain, the communities of the tropics to those of the temperate, and so on.[56]

Thus, this approach lends itself to "natural experiments" where differences between mainland and island, temperate regions and the tropics, etc. allow one to look for these simple difference makers.

Interestingly, MacArthur was very skeptical of laboratory or "bottle experiments," considering their "dramatic failures" due to the difficulty of adding environmental heterogeneity.[57] But he also claimed we do not need them, since astronomy was a respected science even though Copernicus and Galileo "never moved a star."[58] As with Levins, he was skeptical of computer modeling, since he thought computers could never replace the good judgment provided by the training of a field naturalist.[59] As Levins suggested, using computers prematurely could "confuse numbers with knowledge."[60]

Finally, we have MacArthur's disregard for conceptual debates. Consider again his views on the nature of ecological communities (which highlight the concerns regarding empirical sufficiency and the choice of variables raised by Lewontin).

> Humpty Dumpty told Alice, "When *I* use a word, it means just what I choose it to mean—neither more nor less." Irrespective of how other people use

the term "community"—and there are almost as many uses as there are ecologists—I use it here to mean any set of organisms currently living near each other and about which it is interesting to talk. . . . The question is not whether such communities exist but whether they exhibit interesting patterns about which we can make generalizations.[61]

MacArthur chose community-level properties like species diversity because there are interesting patterns to be found concerning interacting species, and, on his view, populations exhibit too much variation for simple theory (having learned this lesson in part from the population regulation debates).[62] MacArthur's mathematical approach is now clear. First, he was an ardent defender of the search for patterns. Second, he forcefully proposes simple theories representing simple causes of those patterns. Third, qualitative theory capturing said mechanisms should be evaluated by natural experiments where the relevant causal factors naturally vary and produce the blurry patterns of interest. Finally, we should put to the side obscuring debates over conceptual issues.

CONCLUSION

MacArthur's approach, though borrowing elements from an ecologist like Hutchinson and from geneticists Levins and Lewontin, was applied with his preeminent mathematical powers to population and community ecology, where they never had made an appearance. He chose a variety of ecological topics and offered remarkably novel mathematical theories (e.g., species abundance distributions, island biogeography, optimal foraging theory, limiting similarity). "Simple theory" appeared to have the resources to turn ecology into a quantitatively successful science like physics, and this prospect was extremely alluring. This was especially tantalizing for him and others against the background of stagnating debates over population regulation and the nature of communities, as we have seen. Given the dreary state of ecology, MacArthur envisioned a radical ecology of patterns, simple qualitative theory, natural experiments, and pragmatism about conceptual disputes.

Unfortunately, MacArthur's "simple theories" ultimately had to be tested against ecological patterns, and here is where the critics demurred. First, in some instances, ecological patterns discovered were challenged as being mere statistical artifacts or explainable without interspecific competition.[63] Second, what started as analytically tractable theory eventually had the problems of the "complex theory" MacArthur originally opposed (e.g., the theory of limiting similarity).[64] Third, even in those cases where there are patterns and simple theory, the qualitative methodology sometimes fails. Schematically, suppose C causes A and C' causes A'; then we would assume observing A confirms C is the relevant mechanism. However, if sometimes C' causes A, then our simple theory and natural experiments can fail us. Different theories can lead to the same observations and thus precise predictions are required to discriminate between the theories (e.g., "broken stick" distributions and the equilibrium theory of island biogeography).[65] Of course, there were theoretical and empirical successes, but there have been challenges and failures too.

Presciently, MacArthur believed his theories, even if false, could be of great importance, writing:

> A theory attempts to identify the factors that determine a class of phenomena and to state the permissible relationships among the factors as a set of verifiable propositions. A purpose is to simplify our education by substituting one theory for many facts. A good theory points to possible factors and relationships in the real world that would otherwise remain hidden and thus stimulates new forms of empirical research. Even a first, crude theory can have these virtues. (MacArthur and Wilson 1967, 5)

MacArthur thought that ecology "can never have too much theory."[66] Likewise, the worst sin of a scientist is not to be wrong but to be trivial.[67] Robert MacArthur's work, even if wrong, was never trivial. In Hutchinson's "Concluding Remarks," he wrote:

> It is not necessary in any empirical science to keep an elaborate logicomathematical system always apparent, any more than it is necessary to keep a vacuum cleaner conspicuously in the middle of a room at all times. When a lot of irrelevant litter has accumulated the machine must be brought out, used, and then put away.[68]

In my estimation, MacArthur's outsider approach attempted to remove "a lot of irrelevant litter" from ecology. Some of the litter was surely removed but some was merely swept under the rug.

FURTHER READING

Cody, M. and J. Diamond, *Ecology and Evolution of Communities.* Cambridge: Harvard University Press, 1974.

Kingsland, S. *Modeling Nature.* Chicago: University of Chicago Press, 1995.

MacArthur, R. H. *Geographical Ecology.* Princeton: Princeton University Press, 1972.

MacArthur, R. H. and E. O. Wilson. *The Theory of Island Biogeography.* Princeton: Princeton University Press, 1967.

Odenbaugh, J. "Philosophical Themes in the Work of Robert MacArthur," in *Handbook of the Philosophy of Science, vol. 11: Philosophy of Ecology*, eds. Bryson Brown, Kevin de Laplante, and Kent Peacock. North Holland: Elsevier, 2011.

Wilson, E. O. and G. E. Hutchinson. "Robert Helmer MacArthur 1930–1972: A Biographical Memoir," *National Academy of the Sciences Biographical Memoirs* 58 (1982): 319–327.

NOTES

1. Nancy Slack, *G. Evelyn Hutchinson and the Invention of Modern Ecology* (New Haven: Yale University Press, 2011).

2. Martin Cody and Jared Diamond, eds., *Ecology and Evolution of Communities* (Cambridge: Harvard University Press, 1997), vii.

3. Sharon Kingsland, *Modeling Nature* (Chicago: University of Chicago Press, 1995); Thomas Schoener and Scott Boorman, "Mathematical Ecology and Its Place Among the Sciences," *Science* 178 (1972): 389–394; John Wiens, *The Ecology of Bird Communities* (Cambridge: Cambridge University Press, 1992).

4. Henry Horn and Eric Pianka, "Ecology's Legacy from Robert MacArthur," in *Ecological Paradigms Lost: Routes of Theory Change*, eds. Beatric Beisner and Kim Cuddington (Amsterdam: Academic Press, 2005), 212–232.

5. Roger Lewin, "Santa Rosalia Was a Goat," *Science* 221 (1983): 636–639; George Salt, ed., *Ecology and Evolution Biology: A Round Table on Research* (Chicago: University of Chicago Press, 1983); Donald Strong et al., eds., *Ecological Communities: Conceptual Issues and the Evidence* (Princeton: Princeton University Press, 1984).

6. David Quammen, *The Song of the Dodo: Island Biogeography in the Age of Extinction* (New York: Scribner, 1997), 177.

7. A. J. Nicholson and V. A. Bailey, "The Balance of Animal Populations, Part I," *Proceedings of the Zoological Society* 3 (1935): 551–598; H. S. Smith, "The Role of Biotic Factors in the Determination of Population Densities," *Journal of Economic Entomology* 28 (1935): 873–898.

8. H. Andrewartha and C. Birch, *The Distribution of and Abundance of Animals* (Chicago: University of Chicago Press, 1954).

9. G. E. Hutchinson, "Concluding Remarks," in *Cold Spring Harbor Symposia on Quantitative Biology* 22 (1957): 415–427, 415.

10. V. Diether and Robert H. MacArthur, "A Field's Capacity to Support a Butterfly Population," *Nature* 21 (1964): 728–729, 728. Apparently, Theodosius Dobzhansky felt similarly; he wrote regarding the symposium, "To a non-ecologist, the controversy which has made our session so lively is, I confess, somewhat bewildering. I have had a feeling for several years now that this is a controversy chiefly about words, about 'semantics,' to use a fashionable word. Having tried to the best of my ability to understand the issue involved, I still continue to feel that way" (Gordon Orians, "Natural Selection and Ecological Theory," *The American Naturalist* 96 [1962]: 257–263, 257).

11. Robert MacArthur, "Population Studies: Animal Ecology and Demography," review of *Cold Spring Harbor Symposium Vol. XXII*, *Quarterly Review of Biology* 35 (1960): 82–83, 82.

12. MacArthur, "Population Studies," 83.

13. As Henry Horn humorously notes, "The many fragments of the traditional analysis of succession have been christened and described in detail, but the resulting jargon has added more to Freudian imagery than it has to a genuine understanding of successional patterns" (Henry Horn, "Markovian Properties of Forest Succession," in *Ecology and Evolution of Communities*, eds. Martin Cody and Jared Diamond [Cambridge: Harvard University Press, 1997], 196–211).

14. Robert H. MacArthur, *Geographical Ecology* (Princeton: Princeton University Press, 1972), 154.

15. MacArthur, *Geographical Ecology*, 55.

16. Robert H. MacArthur, "Patterns of Terrestrial Bird Communities," *Avian Biology* 1 (1971): 189–221, 190.

17. E. O. Wilson and G. E. Hutchinson, "Robert Helmer MacArthur 1930–1972: A Biographical Memoir," *National Academy of the Sciences Biographical Memoirs* 58 (1982): 319–327.

18. G. H. Hardy, *A Mathematician's Apology* (Cambridge: Cambridge University Press, 1992), 13.

19. Hardy, *A Mathematician's Apology*, 35.

20. Hardy, *A Mathematician's Apology,* 36–37.

21. G. E. Hutchinson, "The Concept of Pattern in Ecology," *Proceedings of the Academy of Natural Sciences of Philadelphia* 105 (1953): 1–12.

22. Hutchinson, "The Concept of Pattern in Ecology," 2.

23. Hutchinson, "The Concept of Pattern in Ecology," 3.

24. Kingsland, *Modeling Nature*, ch. 8.

25. Robert H. MacArthur, "Population Ecology of Some Warblers of Northeastern Coniferous Forests," *Ecology* 39 (1958): 599–619.

26. Slack, *G. Evelyn Hutchinson and the Invention of Modern Ecology*, 290.

27. Anne Chishom, *Philosophers of the Earth: Conversations with Ecologists* (New York: Dutton, 1972); R. C. Singh et al., eds., *Thinking About Evolution* (Cambridge: Cambridge University Press, 2001); E. O. Wilson, *Naturalist* (New York: Harper Collins, 1994).

28. Wilson, *Naturalist*, 253.

29. Wilson, *Naturalist*, 253.

30. Richard Levins and Richard Lewontin, *The Dialectical Biologist* (Cambridge: Harvard University Press, 1987).

31. The reasons the group did not meet again are complicated. At the time, Wilson was beginning his work on sociobiology and Levins and Lewontin would later oppose it. Levins writes, "Robert MacArthur, E.O. Wilson and I had planned a division of labor in which they would ask 'How many species are there on islands?' while I would ask, 'How many islands does a species occupy?' It was intended that both

approaches would converge in a continental biogeography. Although this program was thwarted by Robert's early death and disagreements with Ed over socio-biology, it remains a valid strategy" (Gry Oftedal, Jan Kyrre Berg O. Frillis, Peter Rossel, eds., *Evolutionary Theory: 5 Questions* [New York: Automatic Press, 2009], 120).

32. Chisholm, *Philosophers of the Earth*, 177.

33. Chisholm, *Philosophers of the Earth*, 181–182.

34. Singh et al., *Thinking About Evolution*, 37.

35. MacArthur, *Geographical Ecology*, vii.

36. Richard Levins, *Evolution in Changing Environments* (Princeton: Princeton University Press, 1968), 304.

37. Levins, *Evolution in Changing Environments*, 504.

38. Model properties generality, realism, and precision are controversial and complicated. Put simply, a model is general if widely applicable, is realistic to the extent that it represents causal information, and is more precise than another model if the set of predicted values of the former is smaller than that of the latter.

39. Richard Levins, "The Strategy of Model Building in Population Biology," *American Scientist* 54 (1966): 421–431, 423.

40. Wilson and Hutchinson, "Robert Helmer MacArthur 1930–1972," 321.

41. P. Selz, and H. Chipp, *Theories of Modern Art* (Berkeley: University of California Press, 1982), 264.

42. MacArthur had a great respect for Lewontin. E. O. Wilson writes, "Robert Macarthur told me, when we three were young men, that Lewontin was the only person who could make him sweat" (Wilson, *Naturalist*, 342).

43. Richard Lewontin, *The Genetic Basis of Evolutionary Change* (New York: Columbia University Press, 1974), 7.

44. Lewontin, *The Genetic Basis of Evolutionary Change*, 9.

45. Lewontin, *The Genetic Basis of Evolutionary Change*, 9.

46. Lewontin's and Levins's views are similar though importantly different given that Lewontin compresses generality and realism into the dimensionality of models. Lewontin merely commits himself to the claim that given different tolerance limits, models with different dimensionalities will be needed, and this is distinct from Levin's claim about tradeoffs.

47. Lewontin, *The Genetic Basis of Evolutionary Change*, 12.

48. Lewontin, *The Genetic Basis of Evolutionary Change*, 12.

49. Lewontin, *The Genetic Basis of Evolutionary Change*, 8.

50. MacArthur, *Geographical Ecology*, 1. Strangely, MacArthur writes, "It is a pity that several promising young ecologists have been wasting their lives in philosophical nonsense about there being only one way—their own way, of course—to do science. Anyone familiar with the history of science knows that it is done in the most astonishing ways by the most improbable people and that its only real rules are honesty and validity of logic, and that even these are open to public scrutiny and correction" (Robert H. MacArthur, "Coexistence of Species," in *Challenging Biological Problems*, ed. J. Benke [New York: Oxford University Press, 1972], 259).

51. Robert H. MacArthur, "The Theory of the Niche," in *Population Biology and Evolution*, ed., R. C. Lewontin (Syracuse: Syracuse University Press, 1968), 159–176, 159.

52. MacArthur, *Geographical Ecology*, 1.

53. MacArthur, *Geographical Ecology*, 238.

54. Robert H. MacArthur, "Reviewed Work: *Growth and Regulation of Animal Populations*," *Ecology* 43 (1962): 579.

55. Schoener and Boorman, "Mathematical Ecology and Its Place Among the Sciences," 390.

56. MacArthur, *Geographical Ecology*, 77.

57. MacArthur, *Geographical Ecology*, 193.

58. James Brown, "The Legacy of Robert MacArthur: From Geographical Ecology to Macroecology," *Journal of Mammalogy* 80 (1999): 333–344, 335.

59. MacArthur, "Patterns of Terrestrial Bird Communities," 220.

60. Levins, *Evolution in Changing Environments*, 504.

61. MacArthur, "Patterns of Terrestrial Bird Communities," 190.

62. One might be perplexed, given MacArthur's skepticism regarding population-level "repetitive" patterns, by his pursuit of theoretical population biology and that he was in fact a founder of the journal *Theoretical Population Biology*. Probably, "population biology" included much more than merely population ecology and genetics but community ecology and biogeography as well.

63. Strong et al., *Ecological Communities*, 316–331.

64. Robert H. MacArthur and Richard Levins, "The Limiting Similarity, Convergence and Divergence of Coexisting Species," *American Naturalist* 101 (1967): 377–385; Robert H. MacArthur and Robert May, "Niche Overlap as a Function of Environmental Variability," *Proceedings of the National Academy of the Sciences* 69 (1972): 1109–1113.

65. Robert H. MacArthur, "On the Relative Abundance of Bird Species," *Proceedings of the National Academy of Sciences* 43 (1957): 293–295; Robert H. MacArthur and E. O. Wilson, *The Theory of Island Biogeography* (Princeton: Princeton University Press, 1967).

66. Brown, "The Legacy of Robert MacArthur," 335.

67. Brown, "The Legacy of Robert MacArthur," 335.

68. Hutchinson, "Concluding Remarks," 1957.

IV

OUTSIDERS FROM
THE HUMAN SCIENCES

W. TECUMSEH FITCH

NOAM CHOMSKY AND THE
BIOLOGY OF LANGUAGE

INTRODUCTION

It is dangerous to discuss the contributions of a scholar who is still alive and active, and this is particularly true for a scholar as prolific as Noam Chomsky. Chomsky's revolutionary contributions to linguistics are well known, as is his vigorous and consistent opposition to U.S. foreign policy.[1] But his contributions to the cognitive sciences more broadly are equally fundamental, and Chomsky's ideas continue to play a formative role in all of those disciplines today. In all of these fields, Chomsky's contribution is ongoing, and it would thus be foolish to attempt to encapsulate this contribution within the neat confines (and comfortable distance) of scientific history. The intellectual terrain is still being reworked under our feet, and both noisy ongoing battles and quiet marches of normal science are prominent. While it is safe to say that Chomsky's ideas continue to loom large, it is equally safe to observe that many of his ideas remain highly controversial.

In this essay, I will attempt to characterize Chomsky's contributions to biology by focusing on three central debates, three ongoing areas of investigation in which Chomsky's ideas have been extremely influential. First, Chomsky was the first to clearly highlight the richness and complexity of the unconscious knowledge underlying our use of language, and particularly the generative aspects of syntax. By explicitly delineating the rule-governed nature of syntax, Chomsky demonstrated the apparent complexity of these unconscious syntactic rules, along with the fundamentally creative role syntactic rules play in everyday language use. These insights played a crucial role in the "cognitive revolution," helping to topple American behaviorism and replace it with the new paradigms of cognitive science, and had further effects in developmental biology and immunology.[2] Here, Chomsky's role as an "outsider" influencing biology and cognitive science is clear.

Second, under the heading of "biolinguistics," Chomsky was a leading figure in the reawakening of nativist perspectives in linguistics and the cognitive sciences. Inspired by Continental ethologists like Konrad Lorenz,

Chomsky suggested that the human capacity to acquire language might be "special" to our species, in just the same way that a bat's capacity to sense objects via sound is special to its. Here, Chomsky functioned more as an "insider" within cognitive science, championing "outside" ideas already established within biology. While it is safe to say that nativist perspectives have gained credibility and respectability, nativism remains a locus of ongoing battles in the contemporary cognitive sciences.

Third and finally, Chomsky's interest in the biological foundations of human language has fed a long-simmering debate about their evolutionary origins. Chomsky has consistently defended a nonadaptationist viewpoint, suggesting that many aspects of language, and specifically syntax, may result from mechanistic, developmental, or phylogenetic constraints rather than being adaptations driven by natural selection. Chomsky's ideas on this count converged closely with those of molecular geneticists Francois Jacob and Salvador Luria, and evolutionary biologists Steven Jay Gould and Richard Lewontin, representing a direct link between Chomsky's ideas and modern evolutionary biology, particularly for the new disciplinary merging of developmental and evolutionary biology ("evo-devo").[3] Disciplinary boundaries are harder to draw in this case, and again the controversy is ongoing, but Chomsky's contributions are both clear and consistent.

In the rest of this essay, after a concise biographical sketch, I will explore each of these three themes in more detail. It should already be clear that, throughout his career, Chomsky has been a maverick, consistently championing neglected facts (e.g., the complexity and creativity of our unconscious rules of syntax) and unpopular hypotheses (e.g., innate constraints on human learning, or evolutionary constraints on natural selection). In each of these choices he has tended to favor Continental European perspectives over then-current Anglophone orthodoxy, and in each case he has helped resurrect issues of fundamental and transdisciplinary importance. As the leader of major revolutions in linguistics and psychology, he began his career as an unconventional outsider, developing a more rigorous formal approach, but today this approach is well-ensconced within the fields. Thus I will focus mainly on the early days of these revolutions, when his "outsider" nature was most clear.

BIOGRAPHY

Avram Noam Chomsky was born in Philadelphia in 1928 (see figure 11.1). His mother was Elsie Simonofsky, who was born in Belarus, but grew up in the United States, and his father was William Chomsky, an eminent Hebrew scholar and professor at Gratz College in Philadelphia, who had immigrated

Figure 11.1: Noam Chomsky.
Courtesy of the Austrian Press Agency.

from the Ukraine. Although both his parents spoke Yiddish, Chomsky grew up speaking English and studying classical Hebrew, and was discouraged from speaking Yiddish.

He began his university studies at the University of Pennsylvania in 1945. Chomsky was interested in language, mathematics, and politics, but was unsure of what he would study. He soon came into contact with Penn professor Zellig Harris as a result of their shared political interests. Harris was the prominent founder of the Department of Linguistics at Penn, and under his influence Chomsky became interested in linguistics in 1947. After completing an honor's thesis on Hebrew morphophonemics in 1949, he began graduate studies in that field. In the same year he married Carol Schatz (1930–2008), a fellow student studying French and phonetics. During work on his master's thesis on Hebrew (completed in 1951), Chomsky recognized that the attempt to formulate mathematically rigorous rules for linguistic phenomena raised some fascinating fundamental issues, and he began to explore the application of mathematical tools to these problems. This work was rapidly recognized as groundbreaking, and in 1951, Chomsky began a fellowship at the Harvard Society of Fellows, where he worked independently on these issues for four years. Chomsky completed his Penn doctoral dissertation in 1955. The thesis, entitled "Transformational Analysis," laid out the fundamental principles of transformational generative grammar; although copies were widely circulated and read, it remained unpublished until a revised version came out in 1975.[4]

Immediately upon completing his PhD in 1955, he was offered a job at the Massachusetts Institute of Technology (MIT) and an office at the famous Building 20. His funding, in retrospect surprisingly, came from military sources. When Chomsky arrived for his interview, the Research Lab of Electronics was directed by Jerome Wiesner (1915–1994), later a presidential science advisor, then provost and finally president of MIT. Under Wiesner's directorship, Building 20 became the heart of many of the most risky and innovative projects at MIT—including the new department of linguistics (formed in 1960). Previous linguistic work at the RLE led by Victor Yngve had focused on automatic computer translation. Wiesner was looking to hire additional linguists, and asked Chomsky what he thought of this work, whereupon Chomsky replied that he thought it was fundamentally misguided and "a complete waste of time." Presumably aware of the value of such honest skepticism, Wiesner immediately offered Chomsky the job.

In 1957 Chomsky published *Syntactic Structures*, inaugurating the generative revolution in linguistics. This book was innovative in combining linguistic insights with a formal apparatus borrowed from mathematics, and is a founding document in formal language theory (see below). In 1959 he published his famous critique of behaviorism and related associationist approaches to human language, which was framed as a review of the book "Verbal Behavior" by eminent Harvard psychologist B. F. Skinner (1904–1990).[5] This critique has been widely recognized as a key event in the "cognitive revolution" and the resurgence of scientific discussion of the mind.[6] In 1961 Chomsky was appointed full professor in the Department of Modern Languages and Linguistics (now the Department of Linguistics and Philosophy). In 1965 he published *Aspects of the Theory of Syntax*, in which most of the principles of modern generative syntax were established.[7] From 1966 to 1976 he held the Ferrari P. Ward Professorship of Modern Languages and Linguistics, and in 1976 he was appointed Institute Professor. He has remained at MIT ever since.

COMPLEXITY AND CREATIVITY IN THE UNCONSCIOUS RULES OF LANGUAGE

The "cognitive sciences" are today construed broadly as those disciplines that contribute to the study of human and animal minds, including psychology, neuroscience, linguistics, computer science, ethology, philosophy, and anthropology. Chomsky's first and most broadly accepted contribution to the cognitive sciences was his recognition that the (unconscious) rules underlying human language in general, and syntax in particular, appear to be dauntingly complex. This insight came from linguistics, and is an important reason that linguistics has been part of cognitive science from its beginnings. The complexity of unconscious cognition is a central axiom of modern cognitive science, and it is thus easy to forget why Chomsky's observations were so incendiary in the intellectual climate of the United States in the late 1950s, when Skinner's radical behaviorism still ruled in psychology.

Behaviorism rode a wave of euphoric post-war American triumphalism, where the general tenor in the United States was that modern science and technology illustrated an almost unlimited flexibility and power of the human mind. The negative side of this optimism was a firm rejection of many older traditions in science and philosophy, particularly those of Continental Europe, as hidebound and speculative. Within psychology, "modern" was taken to imply a sole focus on observable behavior, and a rejection of mental constructs such as "beliefs," "desires," or "goals" that typified earlier approaches (e.g., German introspective psychology). Although a few behaviorists such as Edward Tolman had stressed the need for mental constructs to explain even rat behavior, radical behaviorists like Skinner rejected the need for "intervening variables" and believed that a few simple and highly general principles of stimulus-response learning and operant conditioning could fully explain the behavior of rats running mazes and pressing levers in highly controlled laboratory conditions.[8] Furthermore, Skinner argued that these principles could be naturally extended to all aspects of human behavior, including the acquisition of language.

The general applicability of a behaviorist framework for language was supported by leading linguists like Leonard Bloomfield and analytic philosophers like W. V. O. Quine and Bertrand Russell. It was this flavor of psychology that Chomsky encountered in 1951 when he came to Harvard as a graduate student, just after Skinner had given the William James lectures on language learning.[9] Chomsky also came into close contact with Quine, who was a prominent supporter of radical behaviorism and its applicability to philosophy (Chomsky soon found that he disagreed with Quine about virtually everything). Finally, information theory and cybernetics were sweeping the world at this time, closely followed by computers, and reinforcing the sense that scientists (and American scientists in particular) were on the brink of solving the major problems of language and mind that previous generations of scholars in other countries had spilled so much, mostly useless, ink upon. It was this broadly US-centered, behaviorist, empiricist perspective that Chomsky encountered as the dominant strain of thinking at the time.

To judge from his own writings, Chomsky found this triumphalism distasteful and the claims of the radical behaviorists vastly overblown.[10] He suggested instead that "the behavioral sciences have commonly insisted upon certain arbitrary methodological restrictions that make it virtually impossible for scientific knowledge of a nontrivial character to be attained" (p. xiv). He further argued that "the contributions of earlier thought and speculation cannot be safely neglected, and in large measure they provide an indispensable basis for serious work today." He also cited the "brilliant" 1948 critique of radical behaviorist approaches by early neuroscientist Karl Lashley, but noted that this critique went unacknowledged and virtually ignored.[11] While these were general arguments about the nature of contemporary psychological theory, it was Chomsky's own research into language and syntax that gave teeth to these critiques, for Chomsky's assertion of the centrality of creativity in language use, and the complexity of linguistic rules, posed insuperable difficulties for strict behaviorist models of language learning.

Chomsky was not, of course, the first scholar to note the complexity and creativity of language use (he credits Descartes and subsequent rationalist philosophers with this insight).[12] But many previous structural linguists had tended to follow Saussure in seeing the syntactic construction of sentences as distant from the core foci of linguistics: as part of "parole" rather than "langue."[13] Part of the relative lack of research on syntax at this time also resulted from the lack of a clear and accepted methodology to deal with sentence structure: while phonemes or words can be listed, the set of possible sentences is clearly unlimited in size. Chomsky's interest in mathematics led him to recognize that the formal mathematical apparatus of infinite sets, developed mostly by formal logicians (such as Russell and Whitehead) and mathematicians (such as Gödel, Church, and Turing), could be applied productively to natural languages like English. This apparatus, today termed "formal language theory," was a critical interdisciplinary bridge between linguistics, mathematics, and computer science, and Chomsky played a central role in creating it.[14]

Formal language theory was important in the early days of the cognitive revolution for at least three reasons. First, it provided an adequately explicit formalism within which to discuss the creative aspects that Chomsky saw as central to human language (echoing Descartes long before him).[15] The infinite set of English sentences no longer seemed to be hopelessly beyond reach of careful and rigorous investigation. This is one element of what is now called the "generative" aspect of Chomsky's approach to syntax. "Generative" connotes creativity both in a technical sense, in which functions generate sets or axioms generate theorems, and in the psychological sense that speakers frequently generate novel sentences, and listeners understand such novel sentences effortlessly. Second, this generative formal turn furthered an ongoing trend in linguistics of embracing explicit algorithms and attempting to design nonsubjective scientific methodologies applicable to natural languages.[16] Chomsky's early mathematical work, especially *Syntactic Structures*, helped solidify the foundations and broaden the scope of such algorithmic research in a way that was extremely influential in the development of computer science (where the eponymous "Chomsky hierarchy" is taught today to every undergraduate).[17] For example, the influential computer scientist Donald Knuth notes the crucial influence of this book, read during his honeymoon, on his subsequent research, stating, "Here was a marvelous thing: a mathematical theory of language in which I could use a computer programmer's intuition!"[18]

Finally, and most importantly, Chomsky argued in this early work that some relatively simple generative algorithms, such as the so-called finite state automata being intensively investigated at that time, were unable to cope with various important syntactic phenomena in natural languages.[19] For example, an unlimited number of words might intervene between the semantically connected "if" and a "then" in a sentence, including other "if/then" statements, and this fact cannot be handled naturally in finite-state grammars. Such constructions require a more powerful class of grammars called context-free grammars. Furthermore, Chomsky argued that even these were not enough. Although the "deep structure" of a sentence—the representation closest to its logical representation—might be generated by a context-free grammar, constructions such as the passive required an additional layer of complexity, termed "transformations." In a sense, Chomsky used the rigor of the nascent theory of automata to clearly demonstrate its own inadequacies vis à vis natural languages. Interestingly, however, Chomsky's earlier formal work on the Chomsky hierarchy had little influence on later generative work on syntax, despite its foundational influence in computer science, and by 1965 had mostly disappeared from Chomsky's own research agenda.[20]

Nonetheless, this early formal work had an immediate, central implication: that the general associationist rules typical of behaviorist approaches to language learning were grossly inadequate to deal with the actual observed complexity of natural language. This work was already quite visible at the time—for example in a symposium presentation Chomsky made in 1956, a symposium that has been argued to mark the birth of modern cognitive science—and the rigorous formal form of Chomsky's arguments made them hard to dismiss.[21] Thus, when Chomsky published his critique of Skinner a few years later, it served as a death knell for strict associationist models of language learning.[22] At the same time, Chomsky's tentative suggestions about the correct model, involving generative algorithms and transformations of deep structure, seemed concordant with a spreading conviction that explanations of human behavior required models of mental activity, and not just superficial behavioral activities and stimulus reinforcement and operant conditioning.

By the time Chomsky published *Aspects* in 1965, generative approaches to linguistics and cognitive approaches to human behavior were already widely seen as the flavor of the day. The notion of generative models was quickly and enthusiastically taken up not only by linguists, but also by many psychologists, musicologists, and scholars in a variety of other disciplines, and served as a nexus for much of the excitement that characterized the first decade of the cognitive revolution.[23] Within biology, the generative perspective was explicitly adopted in research on the immune system by Nils Jerne, who ultimately won a Nobel Prize for this work, and on development by Brian Goodwin.[24] Thus, the first and most widely accepted of Chomsky's contributions was the demonstration that the complexity of linguistic rules both necessitated new models, and illustrated the vacuity of attempting to explain human language using currently reigning behaviorist dogma. This not only played a crucial role in the success of the cognitive revolution, but also spurred considerable research in many other disciplines, from computer science to philosophy to biology.

NATIVISM AND HUMAN LANGUAGE:
THE BIOLINGUISTIC PERSPECTIVE

The second core focus of Chomsky's work is neatly captured by the term "biolinguistics," the approach to language that he, Morris Halle, and Eric Lenneberg began formulating in the early 1950s.[25] Although the biolinguistic approach was already relatively mature in Lenneberg's *Biological Foundations of Language*, published in 1967 (a book to which Chomsky contributed a chapter), it was not named until Massimo Piatelli-Palmarini used the term for a workshop in 1974.[26] The central tenet of biolinguistics is that the capacity for language rests upon an inherited biological basis, and that the principles underlying language are to at least some degree specific to it (just like those underlying vision, spatial navigation, or motor control). This biological perspective on language rejected the notion that simple associationist rules, fed by environmental contingencies, were enough for a child to acquire language with speed and precision. Furthermore, the failure of cats or dogs raised in a human home to acquire human language was not simply a result of their lack of general intelligence, but rather resulted from a lack of human-specific mechanisms tailored to language acquisition.

The biolinguistic perspective emerged in clear opposition to the dominant empiricist perspective of American psychology and linguistics at the time, and was strongly influenced both by the ethological approach to animal behavior, pioneered by Konrad Lorenz and Niko Tinbergen, and the new research on regulatory genetics led by Francois Jacob and Salvador Luria (with whom Chomsky co-taught a seminar on the biology of language in the 1970s).[27] It also developed in close connection with the then relatively young field of neuroscience. The biolinguist's focus on language as a biological entity contrasted with anthropology, where most believed that human languages and cultures, beyond a few very basic constraints, were almost infinitely flexible. The clear recognition of the complexity of syntactic systems immediately raised the question: how could complex rules about covert structures be acquired so quickly by the developing child, when the evidence for them at the perceptual surface was both rare and equivocal? The answer, suggested Lenneberg and Chomsky, was that they are not learned at all. Rather, they are acquired via unconscious inference by the child's language acquisition system, in a process that is tightly constrained by the child's biology. In other words, humans are biologically equipped to acquire language, using a system that fails to even consider most of the possible interpretations of the utterances they perceive. The biolinguistic perspective found support from Continental ethology, where it remained commonplace to recognize instinctual behavior (as it had been in Darwin's time).[28] In contrast, the word "instinct" had become a dirty word among Anglophone psychologists, despite increasing evidence that the recognition and study of instinct was a necessity as soon as one ventured outside the laboratory and beyond the white rat. Despite this increasingly discrepant research from within behaviorism, talk of instinct was still vigorously opposed by many in America and England, and after a long, hard battle Tinbergen himself caved to this pressure and stopped using the word "instinct."[29] Thus, by adopting a biolinguistic perspective, Chomsky and Lenneberg were sharply bucking the trend of their times.

In retrospect, much of the debate over innateness was a result of miscommunication and different scientific goals, rather than disagreements over substantive empirical issues.[30] For those primarily interested in the evolution of behavior, like the ethologists, inherited patterns of behavior, including instincts to learn, are a conceptual necessity. As Lorenz pointed out, Darwinian evolution cannot occur without inheritance of behavior patterns, and specifically without heritable differences in behavior within a population.[31] If all organisms possessed identical instinctual behavior and identical instincts to learn, all of the observed behavioral differences within and between species must result from the environment, a patent absurdity if we contrast an ant, a lizard, and a monkey raised in the same environment. Even the most radical behaviorist had to acknowledge that the few learning mechanisms they postulated must be inborn, and that difference in learning abilities among taxonomic groups like insects and vertebrates result partially from inborn differences in such abilities. Inborn inherited behaviors are a necessary part of any evolutionary theory of behavior, regardless of what one chooses to call them.

There were two substantive disagreements at the heart of the debate between behaviorists and ethologists. The first was the object of study. The behaviorists were primarily interested in development in the individual, while the ethologists focused on evolution.[32] For developmentalists, arguing that a behavior results from instinct simply puts off the question of interest: namely, development. For evolutionists, a focus on individual development begs the question of how the capabilities of that individual of that species got there in the first place. Which one of these questions a scientist chooses to focus on is a matter of taste, of course, but in a classic essay, Tinbergen correctly pointed out that *both* of these questions need to be answered for complete understanding, and emphasized that there is no conflict or con-tradiction between them (a message that bears repeating even today).[33] The second source of conflict was definitional.[34] For the ethologists "instinc-tual" simply meant "having an inherited, reliably developing component" and did *not* imply that the behavior in question was invariable or present at birth, or that it did not incorporate learning.[35] In contrast, for many psy-chologists "innate" or "instinctual" connotes a fixed, reflexive, invariant behavior, with no learning possible.[36] Thus psychologists saw learned *versus* innate as a dichotomy, while the ethologists argued that most behaviors have *both* a learned and an innate component.[37]

The early biolinguists' hypothesis of an innate, human-specific language acquisition system must be viewed against the backdrop of this ongoing battle. But this debate between biologists and psychologists cannot alone explain the ire elicited by Chomsky's arguments about an innate language acquisition system. When applying "innate" to the human mind and lan-guage, a second and much older strand of philosophical argument was equally prominent. The behaviorists adopted a broadly empiricist worldview, while the biolinguists sided with the rationalist perspective of Descartes and Kant, even citing an obscure paper by Lorenz on the biological basis for a Kantian a priori.[38] The basic argument is simple, and ancient: that we humans know things that we have not learned from our environments in any direct sense, and that such knowledge is therefore inborn.

Chomsky, characteristically, put a much sharper point on this old argu-ment, again using the complexity of syntactic rules as his primary evidence, and resuscitating an old argument by the American philosopher Charles Sanders Peirce concerning "abduction." Deduction allows a thinker, given premises and rules, to derive inferences, while induction represents an attempt, given observations, to generalize from and extend those observa-tions. Abduction is the term Peirce gave to the process of generating hy-potheses given observations, and according to Peirce is the core deliberative process by which scientific theories are built.[39] Chomsky observed a direct analogy between the (unconscious) task of a child acquiring language and that of the scientist attempting to build a theory: in both cases the goal is to generate an underlying theory, based on observations, that adequately copes with an indefinite number of future observations. The child's internal set of grammatical principles (which Chomsky termed "I-language") must be inferred from a smattering of data that, in principle, are compatible with a huge variety of different models and rules. The differences between the two situations are obvious: the child does this unconsciously, in a few short years, on the basis of the data of her senses, while the scientist has a lifetime of conscious deliberation that includes experimental data augmented by technology, and builds on the findings of previous scientists. While Peirce emphasized the role of abduction in scientific discovery—a conscious process—in the spirit of the cognitive revolution, Chomsky extended this notion to the unconscious inferences drawn reflexively by the child's language acquisition system.[40] But these differences simply emphasize the almost miraculous speed and reliability with which the child acquires the language of their community.

Crucially, Peirce observed that an astronomical number of hypotheses could account for any set of data, and therefore that abduction can only be effective at producing true theory if the hypotheses considered by the scientist are sharply limited in number. Chomsky, applying this insight to child language acquisition, argued that similar constraints on the hypothesis space are necessary for the child to develop an accurate model of the grammar of her language in a few years.[41] He argued that the space of possible grammars that are even conceivable to the child must be quite small compared to those that would theoretically account for the observed data. Chomsky revived an old rationalist term, "universal grammar," to refer to this set of constraints on possible hypotheses. Despite persistent misconstrual, Chomsky's notion of universal grammar was never intended as a set of universal grammatical rules, true of all languages.[42] Instead, the biolinguists proposed a set of constraints on the language acquisition system that allowed the child, in a short time given impoverished data, to almost unerringly hone in on the "correct" model (that of the others in its community).

Chomsky's arguments about the innate basis for human language acquisition galvanized the cognitive sciences and helped reorient linguistics from an exclusive focus on individual languages (like English or Mohawk) to the problems of language acquisition. Chomsky's wife, Carol, performed early detailed studies on child language acquisition, and the field has grown into a vibrant one today.[43] In parallel, the broader study of infant cognition was revitalized by Chomsky's nativist arguments.[44] Such arguments are now widespread and seen as respectable in this literature, though vociferous debate continues.[45]

Although the ongoing nature of this debate makes it difficult to draw firm conclusions about the future of nativist ideas in psychology and linguistics, I will conclude this section with a brief retrospective and outlook. Viewed against the backdrop of the behaviorist 1950s, Chomsky's nativist perspective can be seen as bringing relatively uncontroversial biological assertions of genetics and ethology into linguistics, anthropology, psychology, and philosophy. The idea is simply that humans have a specific drive to acquire language, and that they are innately equipped with the cognitive apparatus to do so. This idea was anathema to the dominant American discourse at the time Chomsky introduced it, but now is very broadly accepted. For example, in their avowedly "anti-nativist" collection of readings on child language, Elizabeth Bates and Michael Tomasello state that "each human child must learn the specific linguistic conventions used by the people around her. Children are biologically prepared for this prodigious task, of course."[46] The debate, today, has shifted to questions of how specific to our species or to language this innate equipment is.[47] Even fierce foes of the terms "innate" or "universal grammar" readily accept that the human capacity to acquire language rests on a biologically given capacity that is, at least in part, unique to our species.[48] Thus it seems that Chomsky's role, in the 1960s and 1970s, as an emissary for a biological perspective in the cognitive sciences was successful. Certainly, the biolinguistic perspective is growing in prominence: several books and a journal incorporating this term have appeared in recent years.[49]

In 1965, when Chomsky resurrected the notion of universal grammar, the data suggested that the innate basis for language acquisition must be quite complex and detailed, immediately raising problems regarding its evolution.[50] But through a gradual process of simplification and streamlining, his view has evolved into a much simpler and minimalistic perspective on the innate basis for language. This "minimalist" perspective has interesting implications for language evolution, to which we now turn.[51]

CONSTRAINTS ON ADAPTATION IN LANGUAGE

Despite occasional off-hand remarks, Chomsky had little to say about language evolution until 2002, when he co-authored a paper published in the journal *Science* with psychologist Marc Hauser and biologist Tecumseh Fitch.[52] Prior to that, most of his comments had simply expressed skepticism about the idea that the details of syntax were to be explained by the slow hand of natural selection gradually "tweaking" a primate communication system into the form of spoken language.[53] Instead, Chomsky suggested, many important aspects of language might result from physical or developmental limitations, suggesting that the constraints of cramming 10^{10} neurons "into something the size of a basketball" might have played a role in generating the constraints underlying universal grammar.[54] Furthermore he suggested, along with Luria and Jacob, that the precursors of these constraints may have had little or nothing to do with communication, but rather stem from internal cognitive systems for thought.[55] In the last decade he has written several papers that clarify and extend this perspective, which again exemplifies a set of ideas that are well known to biologists, but have been sharply opposed by some linguists and psychologists. Oversimplifying somewhat, these ideas can be broadly characterized as exaptationist.

In a seminal 1979 paper, geneticist Richard Lewontin and paleontologist Stephen Jay Gould suggested that many evolutionary biologists, and sociobiologists in particular, had become overly enamored with adaptive explanations.[56] They caricatured this enthusiasm as "Panglossian" after Dr. Pangloss in Voltaire's Candide, who believed that everything had a god-given purpose and must be precisely as it is. In contrast, Gould and Lewontin argued that many aspects of living organisms are not adaptations, in the technical sense of having been shaped by natural selection for their current function. Instead, they suggested, a host of nonadaptive explanations are also necessary to explain aspects of living things, including architectural constraints derived from physics or geometry (dubbed "spandrels"), developmental constraints imposed by the developmental process, and historical constraints derived from phylogenetic history. Although this article was highly rhetorically charged, these ideas were not particularly new, and echoed the foundational message of George C. Williams that adaptation is "an onerous concept," not to be invoked or accepted lightly. Darwin himself had emphasized his recognition of such nonadaptive aspects of the evolutionary process in the introduction of *On the Origin of Species*, with increasing vehemence in each edition, and made frequent mention of "laws of growth" and change in function during evolution.[57] Gould continued to develop these ideas, in what he termed a "pluralist" perspective (opposing this to a "panadaptationist" alternative), throughout his career.[58] In a particularly relevant article with Elisabeth Vrba, Gould introduced the term "exaptation" to denote those features of organisms that originated as nonadaptive byproducts, but were "co-opted" and put to use later in evolution.[59] An exaptation is thus a trait, or aspect of a trait, that is useful in some current function but was not built for that function. Following Williams, Gould and Vrba suggested that the term "adaptation" in evolutionary biology be restricted to those features that have been shaped by a long period of natural selection for their current function. Features that simply have fortuitous effects at present should be termed exaptations. Gould and Vrba furthermore argued that exaptations are common, and that many features of living organisms are better characterized as spandrels, exaptations, or both than as adaptations.

A closely related body of research dates back to D'Arcy Wentworth Thompson's work on the role of physical constraints on body form and Alan Turing's work on the mathematics of development.[60] Both authors stressed the important role played by physics and chemistry in determining the form of organisms, arguing that natural selection by its nature can only select among those forms that are both physically possible and developmentally reachable. These were among the most prominent examples of a long tradition in developmental biology of emphasizing so-called internal forces in evolution that resulted from the nature of the developmental process rather than the shaping of natural selection.[61] The co-discoverer of natural selection, Alfred Russell Wallace, was also skeptical about the value of adaptive explanations for the human mind.[62] Although such ideas have had a checkered past, by the 1980s they were increasing in prominence, and among biologists it had become widely accepted that adaptation by natural selection was only one of the forces required for evolutionary explanation, and that physical, phylogenetic, and developmental constraints also had important roles to play.[63] Much more recently, in the impressive development of the new field of evolutionary developmental biology, or "evo-devo," the invocation of such constraints has become mainstream.[64] Nonetheless, in psychology, and in particularly evolutionary psychology, the importance of exaptation, spandrels, and constraints remains controversial.[65]

The adaptationist-constraint debate is, at its core, a biological controversy with implications for psychology and linguistics. Chomsky's arguments about the evolutionary origins of the human language faculty are drawn from biology and fall firmly in the exaptationist camp (although he does not himself use the terms "exaptation" or "spandrel"). It is important to note that Chomsky fully accepts the Darwinian principle of evolution by natural selection, and has never doubted that "there is an obvious selective advantage in the ability to discover the language of one's speech community."[66] His argument is about whether or not the details of this faculty can be explained as evolutionary adaptations for communication, as is widely accepted by many other scholars.[67] Such scholars assume that communication and interaction drove the evolution of language, from its beginnings to the present.

An additional aspect of Chomsky's evolutionary argument is that language evolved initially and primarily as a system for augmenting thought (rather than as communication), and that communication developed secondarily.[68] Whatever constraints operated during the evolution of this internal cognitive system are either preexisting aspects of the other cognitive systems with which this system must interact (e.g., for vision, audition, memory, or motor control) and are therefore exaptations, or are inevitable results of the computational requirements such a system must have to be usable at all (and thus "spandrels" in Gould and Lewontin's sense). Specifically, Chomsky has advanced the argument that the key innovation leading to modern human language was the appearance of a simple but powerful computational operator, termed "Merge."[69] This operator simply combines two preexisting elements to make a new, hierarchically structured element, and it thus has the minimal characteristics necessary to generate the kind of hierarchical structures at the heart of human language. For this operation to be able to generate a discrete infinity of hierarchical structures, it must be recursive (that is, able to operate on its own previous outputs). Chomsky argues that it is therefore unnecessary to propose a gradualistic scenario in which Merge slowly extended its scope from two-word combinations to three-elements and so on, since it is no more difficult to make the computational transition from an operator limited to two-word elements to discrete infinity, as from seven or ten words to infinity. In either case the transition to discrete infinity is required, and the simplest model is the sudden one, from no Merge to full Merge in a single evolutionary step. Chomsky argues that, if this thesis is correct, many of the complexities of human language (e.g., the distinction between argument structure and discourse structure) may be inevitable results of the computational nature of the Merge operator (in this case the distinction between external and internal Merge). Adopting Gould and Lewontin's terminology, such complexities would be spandrels, rather than adaptations per se.

Chomsky's argument that important aspects of the human language faculty may result from physical, developmental, or computational constraints was strongly seconded by Gould, who suggested that many of the most interesting aspects of the human mind may be exaptations.[70] His argument that much of the complexity of human language is the result of cognitive constraints that predated language evolution (e.g., on sensory, motor, and conceptual systems) is congenial to biologists, in that it increases the degree to which the biological basis of language is shared with other species, and thus reduces the evolutionary burden of explanation.[71] It also finds a natural place in the broader and ongoing biological debate about the relative role of constraints and adaptation in evolution as well as in the burgeoning evo-devo literature.[72] These ideas have nonetheless again met with strong resistance from linguists and psychologists, ironically including many who strongly support Chomsky's generative and biolinguistic perspectives.[73] Again, these are ongoing debates, and no firm prognosis is justified. However, a number of recent critiques make clear that traditional evolutionary psychology has erred on the side of panadaptationism, making room for an increasingly pluralistic awareness of the possible role of exaptation and spandrels in the evolution of the human mind.[74] Again, Chomsky's willingness to consider unorthodox possibilities, and his unparalleled ability to frame precise hypotheses about the nature of language, has led the way.

CONCLUSION

In conclusion, Noam Chomsky has been an important innovator in a number of fields, often functioning as a true outsider (e.g., in psychology or philosophy) in the sense that he championed ideas seen as unorthodox or even heretical. His early interest in and familiarity with mathematics led to his consistent use of precise, and often formal, characterizations of human language that was both innovative in linguistics, and influential in neighboring fields. This formal precision helps explain why a relatively abstract set of linguistic observations played a central role in the cognitive revolution and the downfall of radical behaviorism. Beyond cognitive science, Chomsky's early work on formal language and automata theory proved seminal in the creation of computer science, and had ripple effects in developmental biology and immunology. His advocacy, along with Lenneberg, for a fundamentally biological perspective on human language profoundly influenced linguistics and had important and ongoing knock-on effects in psychology, anthropology, and philosophy. Finally, his interest in nonadaptive explanatory factors in understanding the evolution of language, and the human mind more generally, has been increasingly influential in discussions of language evolution. Although he himself has been at pains to find and cite historical precursors, sometimes at book length, there can be little question that in the context of the intellectual climate of his own time Chomsky was a maverick. And despite the repeated storms of scholarly debate unleashed by his ideas, there can be equally little doubt that Chomsky's profound though controversial influence on a wide diversity of disciplines has no parallel among other living scholars.

FURTHER READING

Barsky, R. F. *Noam Chomsky: A Life of Dissent*. Cambridge: MIT Press, 1998. Chomsky, N. *Language and Mind*, third edition. Cambridge: Cambridge University Press, 2006.

NOTES

I thank Robert Berwick, Oren Harman, Gesche Westphal Fitch, Robert Freiden, and Noam Chomsky for comments on previous versions of this manuscript, and Jerry Fodor, Lila Gleitman, Morris Halle, Ray Jackendoff, and particularly Noam Chomsky for discussions of the early days of generative linguistics.

1. R. Freidin, "A brief history of generative grammar," *A Companion to the Philosophy of Language*, Gillian Russell and Delia Graff Fara, editors (London: Routledge, 2012), 895–916; R. F. Barsky, *Noam Chomsky: A Life of Dissent* (MIT Press, 1998), 255.

2. H. Gardner, *The Mind's New Science: A History of the Cognitive Revolution* (Basic Books, 1985); G. A. Miller, "The cognitive revolution: A historical perspective," *Trends in Cognitive Sciences* 7 (2003), 141–144.

3. S. J. Gould, "The exaptive excellence of spandrels as a term and prototype," *Proceedings of the National Academy of Sciences* 94 (1997), 10750–10755; S. J. Gould, and R. C. Lewontin, "The spandrels of San Marco and the panglossian paradigm: A critique of the adaptationist programme," *Proceedings of the Royal Society B* 205 (1979), 581–598; R. C. Lewontin, "The evolution of cognition: Questions we will never answer," *An Invitation to Cognitive Science: Methods, Models, and Conceptual Issues*, D. Scarborough and Saul Sternberg, editors (MIT Press, 1998), 107–131.

4. N. Chomsky, *The Logical Structure of Linguistic Theory* (Plenum Press, 1975).

5. B. F. Skinner, *Verbal Behavior* (Appleton-Century-Crofts, 1957); N. Chomsky, "Review of 'Verbal Behavior' by B. F. Skinner," *Language* 35 (1959), 26–58.

6. H. Gardner, *The Mind's New Science*.

7. N. Chomsky, *Aspects of the Theory of Syntax* (MIT Press, 1965).

8. E. C. Tolman, "The determinants of behavior at a choice point," *Psychological Review* 45 (1938), 1–41; E. C. Tolman, "Cognitive maps in rats and men," *Psychological Review* 55 (1948), 189–208.

9. N. Chomsky, *Language and Mind*, first edition (Harcourt, Brace & World, 1968).

10. N. Chomsky, *Language and Mind*, third edition (Cambridge: Cambridge University Press, 2006).

11. K. Lashley, "The problem of serial order in behavior," *Cerebral Mechanisms in Behavior: The Hixon Symposium*, Lloyd A Jeffress, editor (Wiley, 1951), 112–146.

12. N. Chomsky, *Cartesian Linguistics: A Chapter in the History of Rationalist Thought* (Harper & Row, 1966).

13. F. d. Saussure, *Course in General Linguistics* (McGraw-Hill, 1916).

14. N. Chomsky, "Three models for the description of language," *I.R.E. Transactions on Information Theory* IT-2 (1956), 113–124; N. Chomsky, "On certain formal properties of grammars," *Information and Control* 2 (1959), 137–167. N. Chomsky and G. A. Miller, "Finite state languages," *Information and Control* 1 (1958), 91–112; N. Chomsky and G. A. Miller, "Introduction to the formal analysis of natural languages," *Handbook of Mathematical Psychology*, vol. II, R Duncan Luce, Robert R. Bush, and Eugene Galanter, editors (John Wiley & Sons, 1963), 269–322.

15. N. Chomsky, *Cartesian Linguistics: A Chapter in the History of Rationalist Thought* (Harper & Row, 1966).

16. Z. S. Harris, *Methods in Structural Linguistics* (University of Chicago Press, 1951).

17. N. Chomsky, *Syntactic Structures* (The Hague: Mouton, 1957).

18. D. E. Knuth, "Preface," in *Selected Papers on Computer Languages*, D. E. Knuth, editor (Center for the Study of Language and Information, Stanford University, 2003), 1–5.

19. Chomsky, "Three models for the description of language."

20. J. L. Gersting, *Mathematical Structures for Computer Science*, fourth edition (W. H. Freeman, 1999); G. K. Pullum, "Formal linguistics meets the Boojum," in *The Great Eskimo Vocabulary Hoax and Other Irreverent Essays on the Study of Language*, Geoffrey K. Pullum, editor (University of Chicago Press, 1991), 47–55; N. Chomsky, "On

formalization and formal linguistics," *Natural Language and Linguistic Theory* 8 (1990), 143–147.

21. Gardner, *The Mind's New Science*; Miller, "The cognitive revolution."

22. Chomsky, "Review of 'Verbal Behavior' by B. F. Skinner."

23. P. N. Johnson-Laird and R. Stevenson, "Memory for syntax," *Nature* 227 (1970), 412; L. Bernstein, *The Unanswered Question: Six Talks at Harvard (Charles Eliot Norton Lectures)* (Harvard University Press, 1981); Miller, "The cognitive revolution."

24. N. K. Jerne, "Antibodies and learning: Selection versus instruction," *The Neurosciences: A Study Program*, G. C. Quarton, T. Melnechuk, and F. O. Schmitt, editors (Rockefeller University Press, 1967), 200–205; N. K. Jerne, "The generative grammar of the immune system," *Science* 229 (1985), 1057–1059; B. Goodwin, *Form and Transformation: Generative and Relational Principles in Biology* (Cambridge: Cambridge University Press, 1996); B. Goodwin, "Development and evolution," *Journal of Theoretical Biology* 97 (1982), 43–55.

25. E. H. Lenneberg, *New Directions in the Study of Language* (MIT Press, 1964).

26. E. H. Lenneberg, *Biological Foundations of Language* (Wiley, 1967); L. Jenkins, *Biolinguistics: Exploring the Biology of Language* (Cambridge: Cambridge University Press, 1999).

27. Chomsky, *Language and Mind*, third edition.

28. C. Darwin, *The Descent of Man and Selection in Relation to Sex*, first edition (London: John Murray, 1871); C. Darwin, *The Expression of the Emotions in Man and Animals* (London: John Murray, 1872); N. Tinbergen, *The Study of Instinct*, first edition (Oxford: Oxford University Press, 1969), 228; K. Lorenz, *Evolution and Modification of Behavior* (University of Chicago Press, 1965).

29. Tinbergen, *The Study of Instinct*; Lorenz, *Evolution and Modification of Behavior*; D. S. Lehrman, "A critique of Konrad Lorenz's theory of instinctive behavior," *Quarterly Review of Biology* 28 (1953), 337–363; N. Tinbergen, "On aims and methods of ethology," *Zeitschrift für Tierpsychologie* 20 (1963), 410–433.

30. W. T. Fitch, "Innateness and human language: A biological perspective," *The Oxford Handbook of Language Evolution*, M. Tallerman and K. R. Gibson, editors (Oxford: Oxford University Press, 2011), 143–156.

31. Lorenz, *Evolution and Modification of Behavior*.

32. Cf. Lehrman, "A critique of Konrad Lorenz's theory of instinctive behavior."

33. N. Tinbergen, "On aims and methods of ethology," *Zeitschrift für Tierpsychologie* 20 (1963), 410–433.

34. Cf. W. T. Fitch, "Innateness and human language"; P. Bateson and M. Mameli, "The innate and the acquired: Useful clusters or a residual distinction from folk biology?" *Developmental Psychobiology* 49 (2007), 818–831.

35. Lorenz, *Evolution and Modification of Behavior*.

36. E. Sober, "Innate knowledge," *Routledge Encyclopedia of Philosophy*, E. Craig, editor (London: Routledge, 1998), 794–797; Samuels, "Innateness in cognitive science."

37. L. Jenkins, *Biolinguistics: Exploring the Biology of Language* (Cambridge: Cambridge University Press, 1999).

38. N. Chomsky, *Cartesian Linguistics: A Chapter in the History of Rationalist Thought* (Harper & Row, 1966); K. Lorenz, "Kants lehre vom apriorischen im lichte gegenwärtiger biologie," *Blätter für Deutsche Philosophie* 15 (1941), 94–125.

39. C. S. Peirce, "The logic of abduction," *Pierce's Essays in the Philosophy of Science* (Liberal Arts Press, 1957).

40. A. W. Burks, "Peirce's theory of abduction," *Philosophy of Science* 13 (1946), 301–306.

41. Chomsky, *Aspects of the Theory of Syntax*; Chomsky, *Language and Mind*, first edition.

42. Chomsky, *Language and Mind*, third edition; R. Jackendoff, *Foundations of Language* (Oxford: Oxford University Press, 2002); W. T. Fitch, "Unity and diversity in human language," *Philosophical Transactions of the Royal Society B* 366 (2011), 376–388.

43. C. Chomsky, *The Acquisition of Syntax in Children from 5 to 10* (MIT Press, 1969).

44. J. Mehler and T. G. Bever, "Cognitive capacities of young children," *Science* 158 (1967), 141–142; E. S. Spelke, "Infants' intermodal perception of events," *Cognitive Psychology* 8 (1976), 533–560; S. Carey and R. Diamond, "From piecemeal to configurational representation of faces," *Science* 195 (1977), 312–314.

45. S. Pinker, *Language Learnability and Language Development* (Harvard University Press, 1984); M. Tomasello and E. Bates, "General introduction," *Language Development: The Essential Readings* (Oxford: Wiley-Blackwell, Oxford, 2001), 1–12; S. Pinker, "Why nature and nurture won't go away," *Daedalus* 133 (2004), 5–17; G. Sampson, *Educating Eve: The "Language Instinct" Debate* (London: Cassell, 1997), 184; G. K. Pullum and B. C. Scholz, "Empirical assessment of stimulus poverty arguments," *Linguistic Review* 19 (2002), 9–50.

46. Tomasello and Bates, "General introduction," 1.

47. M. D. Hauser, N. Chomsky, and W. T. Fitch, "The language faculty: What is it, who has it, and how did it evolve?" *Science* 298 (2002), 1569–1579; W. T. Fitch, M. D. Hauser, and N. Chomsky, "The evolution of the language faculty: Clarifications and implications," *Cognition* 97 (2005), 179–210; S. Pinker and R. Jackendoff, "The faculty of language: What's special about it?" *Cognition* 95 (2005), 201–236.

48. Sampson, *Educating Eve*; M. Tomasello, *The Cultural Origins of Human Cognition* (Harvard University Press, 1999).

49. L. Jenkins, *Biolinguistics: Exploring the Biology of Language* (Cambridge: Cambridge University Press, 1999); L. Jenkins, editor, *Variation and Universals in Biolinguistics* (Elsevier, 2004), 446; T. Givón, *Bio-Linguistics: The Santa Barbara Lectures* (Amsterdam: John Benjamins, 2002).

50. N. Chomsky, " On the Nature of Language," *Annals of the New York Academy of Sciences* 280 (1976), 46–57.

51. Chomsky, *Language and Mind*, third edition; N. Chomsky, "Some simple evo-devo theses: How true might they be for language?" in *The Evolution of Human Language: Biolinguistic Perspectives*, Richard Larson, Viviane Deprez, and Hiroko Yamakido, editors (Cambridge: Cambridge University Press, 2010), 45–62.

52. Hauser, Chomsky, and Fitch, "The Language Faculty."

53. Chomsky, *Language and Mind*, first edition.

54. N. Chomsky, *Reflections on Language* (Pantheon, 1975).

55. Chomsky, *Language and Mind*, third edition; Chomsky, "Some simple evo-devo theses."

56. Gould and Lewontin, "The spandrels of San Marco and the panglossian paradigm."

57. Darwin, *On the Origin of Species*.

58. S. J. Gould, *Ontogeny and Phylogeny* (The Belknap Press, 1977); S. J. Gould, *The Structure of Evolutionary Theory* (Harvard University Press, 2002).

59. S. J. Gould and E. S. Vrba, "Exaptation—A missing term in the science of form," *Paleobiology* 8 (1982), 4–15.

60. D. A. W. Thompson, *On Growth and Form* (Cambridge: Cambridge University Press, 1948); A. M. Turing, "The chemical basis of morphogenesis," *Philosophical Transactions of the Royal Society of London B* 237 (1952), 37–72.

61. Gould, *Ontogeny and Phylogeny*; Gould, *The Structure of Evolutionary Theory*; L. L. Whyte, *Internal Factors in Evolution* (London: Tavistock, 1965), 81.

62. A. R. Wallace, "Limits of natural selection as applied to man," *Contributions to the Theory of Natural Selection* (Macmillan, 1871).

63. J. Maynard Smith, et al. "Developmental constraints and evolution," *The Quarterly Review of Biology* 60 (1985), 265–287; J. C. Ahouse, "The tragedy of a priori selectionism: Dennett and Gould on adaptationism," *Biology and Philosophy* 13 (1998), 359–391.

64. M. Kirschner and J. Gerhart, "Evolvability," *Proceedings of the National Academy of Sciences* 95 (1998), 8240–8427; M. Kirschner and J. Gerhart, *The Plausibility of Life: Resolving Darwin's Dilemma* (Yale University Press, 2005); E. M. De Robertis, "Evo-devo: Variations on ancestral themes," *Cell* 132 (2008), 185–195; B. C. Goodwin, *How the Leopard Changed Its Spots: The Evolution of Complexity* (Princeton University Press, 2001).

65. D. M. Buss, M. G. Haselton, T. K. Shackelford, A. L. Bleske and J. C. Wakefield, "Adaptations, exaptations, and spandrels," *American Psychologist* 53 (1998), 533–548 (1998); S. Pinker, *How the Mind Works* (Norton, 1997); J. A. Fodor, *The Mind Doesn't Work That Way* (MIT Press, 2000).

66. Chomsky, *Reflections on Language*, 252.

67. C. F. Hockett, "The origin of speech," *Scientific American* 203 (1960), 88–96; S. Pinker and P. Bloom, "Natural language and natural selection," *Behavioral and Brain Sciences* 13 (1990), 707–784; Jackendoff, "Possible stages in the evolution of the language capacity"; R. Jackendoff and S. Pinker, "The nature of the language faculty and its implications for evolution of language (reply to Fitch, Hauser, & Chomsky)," *Cognition* 97 (2005), 211–225; F. J. Newmeyer, "On the supposed 'counterfunctionality' of universal grammar: Some evolutionary implications," *Approaches to the Evolution of Language*, James R. Hurford, Michael Studdert-Kennedy, and Chris Knight, editors (Cambridge: Cambridge University Press, 1998), 305–319.

68. Chomsky, "Some simple evo-devo theses."

69. W. T. Fitch, "Evolutionary developmental biology and human language evolution: Constraints on adaptation," *Evolutionary Biology* 39 (2012), 613–637.

70. S. J. Gould, "Exaptation: A crucial tool for evolutionary psychology," *Journal of Social Issues* 47 (1991), 43–65.

71. Fitch, Hauser, and Chomsky, "The evolution of the language faculty"; W. T. Fitch, *The Evolution of Language* (Cambridge: Cambridge University Press, 2010), 203; T. Deacon, "Universal grammar and semiotic constraints," *Language Evolution*, Morten Christiansen and Simon Kirby, editors (Oxford: Oxford University Press, 2003), 111–139.

72. Maynard Smith et al., "Developmental constraints and evolution"; Kirschner and Gerhart, *The Plausibility of Life*; De Robertis, "Evo-devo"; F. Jacob, *The Possible and the Actual* (Pantheon, 1982); Y. Narita and S. Kuratani, "Evolution of the vertebral formulae in mammals: A perspective on developmental constraints," *Journal of Experimental Zoology* 304B (2005), 91–106; S. B. Carroll, "Genetics and the making of *Homo sapiens*," *Nature* 422 (2003), 849–857; S. B. Carroll, "Evo-devo and an expanding evolutionary synthesis: A genetic theory of morphological evolution," *Cell* 134 (2008), 25–36; S. F. Gilbert, "The morphogenesis of evolutionary developmental biology," *International Journal of Developmental Biology* 47 (2003), 467–477; W. T. Fitch, "Evolutionary developmental biology and human language evolution: Constraints on adaptation," *Evolutionary Biology* 39 (2012), 613–637.

73. Pinker and Jackendoff, "The faculty of language;" Newmeyer, "On the supposed 'counterfunctionality' of universal grammar."

74. J. C. Ahouse, "The tragedy of a priori selectionism: Dennett and Gould on adaptationism," *Biology and Philosophy* 13 (1998), 359–391; Fodor, *The Mind Doesn't Work That Way*; D. J. Buller, *Adapting Minds: Evolutionary Psychology and the Persistent Quest for Human Nature* (MIT Press, 2005); R. C. Richardson, *Evolutionary Psychology as Maladapted Psychology* (MIT Press, 2007).

DUNKING THE TARZANISTS
ELAINE MORGAN AND THE
AQUATIC APE THEORY

12

INTRODUCTION

Elaine Morgan had sass. In chapter four of *Descent of Woman*, published in 1972, she asked her readers to take science into their own hands. "Try a bit of fieldwork," she suggested. "Go out of your front door and try to spot some live specimens of Homo sapiens in his natural habitat. It shouldn't be difficult because the species is protected by law and in no immediate danger of extinction." After completing observations of twenty random people, she suggested, substitute them when you are reading statements about universal human nature. The result?

That window cleaner is one of the most sophisticated predators the world has ever seen.

The weapon is my grocer's principal [*sic*] means of expression, and his only means of resolving differences.

The postman's aggressive drive has acquired a paranoid potential because his young remain dependent for a prolonged period.

Morgan then added that you might imagine you were observing the wrong species. "But," she continued, "if you're going to be any good as an ethologist, you must learn to trust the evidence of your own senses above that of the printed word and the television image. Remember, you have been living among thousands of these large carnivores all your life, on more intimate terms than those on which Jane Goodall lived among the chimpanzees."[1]

In positing such a scenario, Morgan was engaged in serious work. All popular theories of human evolution to date, she insisted, were based on a male-centered notion of human evolution. Where were the evolutionary scenarios that began, "When the first ancestor of the human race descended from the trees, she had not yet developed the mighty brain that was to distinguish her so sharply from all the other species . . ."?[2] This formed one of two major points she wished to make in *Descent of Woman*. The second was to advance a theory of aquatic adaptation that preceded life on the savannah.

In our semi-adaptation to a watery world, we lost our body hair, gained a layer of subcutaneous fat to keep us warm, learned to walk upright (keeping our head above water while foraging for tasty snacks in the shallows), came to use stones and manufacture tools for breaking open shells, and developed the ability to control our breathing when diving beneath the surface—a precondition for true spoken language and an obvious boon to any individual trying to communicate with most of her body submerged. These activities, Morgan noted, were associated with gathering, not hunting, an activity in which women's contributions were becoming acknowledged. In short, we acquired precisely those traits that distinguish us from the rest of the animal world while living around water, not in the arid grasslands.

Descent of Woman fit well into the contemporary genre of public science that aimed to present scientific arguments in ways accessible to people with common sense but little scientific training. Such books provided scientific authors with a way of rendering their ideas in a more conversational tone than they might allow themselves in standard academic venues and provided nonscientifically trained authors, like Morgan, with opportunities to contribute to highly visible debates over scientific ideas.[3] By relying on arguments and evidence intelligible to nonscientific audiences, these books sold extremely well but were also left open to marginalization by the scientific community. Morgan was Oxford-educated (in English), the mother of three children, lived in Wales, and had been writing screenplays and dramas for the BBC for decades (see figure 12.1). When she turned her attention to nonfiction, it was largely out of irritation with the popular books of the time that emphasized the evolution of *man*.[4] The aggressive masculinity they advanced left little room for women, and in stark opposition to these volumes, Morgan was read as a radical feminist with a sense of humor.

The legacy of her book today exists in both the gender and science literature, where she is cited as one of the first authors to call attention to the widespread male bias in anthropological theories, and in the small, but steadily growing literature on the idea that humans may have undergone an aquatic phase in our evolutionary past. On the one hand, her work highlights a historical moment when anthropologists and evolutionary biologists self-consciously mobilized to change the dominant theories of human evolution by including women as self-determining actors. Even as an outsider with no scientific training, Morgan's book contributed to this groundswell. On the other hand, advocates of the aquatic ape increasingly sought to distance the theory from its historical origins. The sarcastic wit of Morgan's *Descent of Woman*, combined with her lack of scientific training, eventually hurt their

Figure 12.1: Elaine Morgan.
Courtesy of Elaine Morgan.

cause more than it helped. As a result, in the decades following the book's publication, these legacies have become increasingly divergent.

THE TARZANISTS

To understand Morgan's arguments, we first need to explore the era in which she was writing. In the late 1960s, many evolutionary theorists and popular writers believed that early humanity was defined by the ability of men to manufacture weapons and engage in cooperative hunting.[5] The movie *2001: A Space Odyssey*, for example, envisioned the dawn of humanity as the moment at which a human ancestor picked up a bone and realized it could be weaponized to hunt tapir.[6] Writer Arthur C. Clarke, together with director Stanley Kubrick took this idea from Australian paleoanthropologist Raymond Dart, whose ideas had been popularized by best-selling author (and Hollywood writer) Robert Ardrey.[7] Kubrick's depiction of the dawn of humanity reflected the dominant conception of the forces defining human evolution at the time. It was the hunt that required intragroup cooperation and communication, and only men could hunt.

When Ardrey published *African Genesis* in 1961, it was merely the first in a series of books depicting man as an aggressive hunter.[8] Ardrey, an early advocate of Dart's "killer ape" theory, suggested that "even in the first long days of our beginnings we held in our hand the weapon, an instrument somewhat older than ourselves."[9] Weapons preceded humanity, weapons fathered humanity: "A rock, a stick, a heavy bone—to our ancestral killer ape it meant the margin of survival. But the use of the weapon meant new and multiplying demands on the nervous system for the co-ordination of muscle and touch and sight. And so at last came the enlarged brain; so at last came man."[10] In his next foray into anthropology—*The Territorial Imperative*—it became clear that by mankind Ardrey really meant men, and the violence with which he was preoccupied was associated with male members of a group defending their common territory against outsiders.[11] Evolutionary pressures on groups of men, intent on hunting to support the women and children safe in camp and cooperating to defend their home from hostile incursions, drove the increased intelligence that came to characterize all humanity.

Sociologist Lionel Tiger similarly built on the work of primatologists and concentrated on the importance of male bonding and competition. In *Men in Groups*, Tiger argued that as males became more specialized as hunters, their behavior became increasingly dissociated from the ways females act. Three main behavioral "links" that held the community together, he posited, the male-female reproductive link, the female-offspring generative link, and the male-male cooperative hunting link. Of these, it was the latter all-male interactions that drove evolutionary changes in "perception, brain-size, posture, hand formation, locomotion, etc."[12]

Zoologist Desmond Morris, on the other hand, was initially more preoccupied with sex than violence or male-male bonding. In the first paragraph of the introduction to *The Naked Ape*, Morris suggested that *Homo sapiens* "is proud that he has the biggest brain of all the primates, but attempts to conceal the fact that he also has the biggest penis."[13] In contrast to Ardrey and Tiger, he emphasized the development of a pair bond between adult males and females.[14] The pair bond ensured that females remained faithful to their individual males when hunting duties called males away from camp; it reduced serious sexual rivalries between males which aided in the development of male cooperation; and the young benefited from a cohesive family unit. Such a process, he was careful to add, "was never really perfected," but was a crucial component of his new ecological role as a "lethal carnivore."[15]

It was while reading a scant two pages that Morris devoted to a possible aquatic phase in human history that Elaine Morgan first came across the idea.[16] She eventually contacted the theory's originator—Sir Alister Hardy—and asked whether he would mind if she worked on a book responding to the contemporary theories of human evolution and developing his suggestion. Like Ardrey, she may have believed that artists and writers were well-developed students of human nature, despite their lack of anthropological training.[17] Hardy, who began wrestling with the aquatic theory thirty years earlier, had initially refused to publish anything on it lest he ruin his nascent academic career (at the time, he had plenty of experience at sea and in marine biology, but lacked a permanent post).[18] In 1960, as a recently knighted and well-established professor of zoology at Oxford, he agreed to write a brief article in *The New Scientist* proposing an aquatic past for humanity.[19] Although he intended to publish his own book on the topic, he was more than happy to let Morgan run with the idea.

WOMEN'S LIB PREHISTORY

Elaine Morgan's frustration with "the Tarzanists" is palpable when reading *Descent of Woman*. She began by noting, "The legend of the jungle heritage and the evolution of man as a hunting carnivore has taken root in man's mind. . . . He may even genuinely believe that equal pay will do something terrible to his gonads." The scientific facts used to buttress such arguments were unassailable, she insisted, but not the interpretations of those facts. She further noted that the term "man" was ambiguous, denoting both the entire species and also the males of the species. The trick was not to confuse the two. Adding carefully that although this might sound like "a piece of feminist petulance" to some, Morgan hoped to convince her readers that her semantic point vitiated much of the "speculation" concerning the evolutionary origins of humanity.[20]

Further, Morgan contended that based on the available evidence it was impossible to distinguish weapons from tools and impossible to know which sex invented them. "A knife is a weapon or a tool according to whether you use it for disemboweling your enemy or for chopping parsley." In fact, she reasoned, it was likely that early humans used tools for both purposes—the strict dichotomy between herbivores and carnivores in our past was never clean-cut.[21] Morgan also noted that although male hunters were widely acknowledged as being the inventors of tools and pottery, there was no solid evidence supporting this assumption. She caricatured the dominant theory of pottery invention as follows:

> One day he noticed with secret chuckle that the little woman was wearing herself out trotting to and fro carrying seed home by the handful. He quietly laid aside his beautiful symmetrical weapons and forsook his male-bonded companions for a few weeks while he devoted himself to the problem, and finally invented the pot. He gave her a few prototypes and a crash course of instruction, patted her on the head, and sped away across the savannah to rejoin the hunting party.[22]

That was equally likely, she contended, as assuming that a woman invented her partner's weapons, saying, "Play quietly among yourselves today, children: I'm busy inventing the bow and arrow for your father." Both tales were equally plausible and equally unsubstantiated.[23]

Why, then, were such assumptions about the importance of men to the evolutionary progress of humanity so widespread? Morgan chalked this up to a combination of male pride and egotism. She extensively quoted Ardrey's *Territorial Imperative* in order to rebut his characterization of men as baboons. A male baboon, he had written, "is born a bully, a born criminal, a born candidate for the hangman's noose. He is as submissive as a truck, as inoffensive as a bulldozer, as gentle as a power-driven lawnmower. He has predator inclinations and enjoys nothing better than killing and devouring the newborn fawns of the delicate gazelle." Morgan imagined that a male reader would naturally find this characterization appealing. She pictured him polishing his glasses, thinking: "Yeah, that's me all right. Tell me more about the bulldozer and how I ravaged that delicate gazelle." The reason such theories continued to be popular was because Ardrey's typical reader, and the author, got "no end of a kick out of thinking that all that power and passion and brutal virility is seething within him, just below the skin, only barely held in leash by the conscious control of his intellect."[24]

It was high time, she thought, to expose these arguments for what they really were—myths, yes, but politically useful myths. Morgan argued that man-the-hunter theories were used to "bolster up with pseudo-history and pseudo-anthropology the belief that it is 'against nature' for women to play a part in economic life; that 'from time immemorial' men have said 'she shall have no other food and that will make her my slave'; and that we are descended from females whose sole function was to placate the hunters and keep them happy and mind the babies."[25] In particular, Morgan cited the wide acclaim and recognition that Tiger received for *Men in Groups*. His idea of "male bonding" dominated discussions of human social interactions, with almost no attention paid to equally important (for Morgan) female-female relations.[26] Additionally, based on evidence from modern hunter-gatherer societies, she noted that the bulk of the total diet of early humans was probably the result of gathering vegetable matter, not hunting meat.[27] Women *were* important to the evolutionary history of humanity, but had been ignored by the "blood-and-thunder boys."[28]

Descent of Woman sold extremely well, and reviews of the book quickly picked up on her none-too-subtle revisionist agenda. *Playboy*, for example, hailed it as a "stunning tour de force." They hypothesized that, "even the most militant male chauvinist will find it difficult to cling to all his prior convictions in the face of the evidence marshaled here."[29] Similarly, *Life* magazine described *Descent of Woman* as a lively "women's-lib prehistory."[30] Several reviewers missed that Morgan didn't have any anthropological training, describing her as "a female anthropologist," and "a scholarly woman, educated at Oxford in the fields of paleontology, ethology, and anthropology."[31] Yet even when people recognized her outsider status, they still acknowledged the force of her arguments against the standard evolutionary history of humanity. Her reviewer in the *New York Times*, for example, characterized *Descent of Woman* as "a potent commentary on the state of the social sciences in general, and anthropology in particular."[32]

By the late 1970s, scientists also cited Morgan's *Descent of Woman* as an early critique of "man the hunter" theories of human evolution.[33] Yet these citations often wielded a double-edged sword. Although Morgan's critique was useful, it came packaged along with her interest in Hardy's aquatic ape. Physical anthropologist Adrienne Zihlman, for example, argued that Morgan's sensible and "substantial critique of existing evolutionary dogma did not get the attention and credibility it deserved," because it had been "contaminated by Morgan's own elaboration and support of a very dubious theory of human origins, the 'Aquatic Ape' hypothesis."[34] Zihlman regretted Morgan's advocacy for two reasons. First, the myths against which Morgan had been fighting were still present over a decade later. Second, readers might get the mistaken impression that Morgan's book was the best feminist anthropology could offer. "A feminist revision of human evolution," Zihlman insisted, "does not require life in the water."[35]

THE AQUATIC APE

Why then, you might wonder, was Elaine Morgan so invested in the idea that humans had an aquatic past? Morgan wanted a viable theory to fill the hole left by her banishment of "man the hunter." She insisted that, in concentrating solely on the males of the species, evolutionary narratives like those advanced by Ardrey, Morris, and Tiger overstated the capacity of females (and therefore the species as a whole) to survive. From a female perspective, without protective weapons of any sort or easy hiding places, weighted down with a nursing infant, "the only thing she had going for her was the fact that she was one of a community, so that if they all ran away together a predator would be satisfied with catching the slowest and the rest would survive a little longer." In other words, when this sweet "generalized vegetarian prehominid hairy ape" experienced the "first torrid heat waves of the Pleistocene," she would have been unable to avoid being eaten by predators.[36] Both our putative ancestor and Morgan needed an escape route, and that's what the water provided.

A watery environment, Morgan posited, could explain a wide variety of our uniquely human characteristics. As our putative ancestor ran into the water up to her waist or even her neck to escape predators, she was forced into upright, bipedal walking. Lakeshores and seashores provided much easily accessible food, but hard shells needed to be broken open to access the tasty tidbits they contained—the use and eventual manufacture of tools thus started with gathering food, and were only later adapted as weapons for hunting. Living in caves along the seashore would have provided shelter and also an explanation for the origins of family structure. The loss of our body hair would then be the result of the first stages of adaptation to an aquatic environment.[37] Even our capacity for speech could have emerged as a result of learning breath control for swimming and diving. Communicating while wading can be difficult, Morgan suggested, because our limbs would be covered with water (preventing active gesturing or the use of body language) and the water would additionally mask the chemical particles that form part of our olfactory communication.[38]

Morgan presented evidence for her claims that nonscientifically trained readers could easily understand and relate to. For example, she explained the incredible sensitivity of our fingertips as resulting from the need to grope for food under the water, and the development of a subcutaneous layer of fat as a way to keep us warm. She noted that our heartbeat slows down when we dive to great depths and further pointed out that newborn babies can swim. She also explained long hair on the heads of women as an adaptation allowing babies to grab ahold of their mother in the water and her "pendulous, dollop breasts" as easy hand-holds for breastfeeding babies—not as traits that were controlled by the sexual preference of men.[39] In a broad survey of the animal kingdom, she argued, only aquatic mammals exhibit traits like breath control, a slowed heartbeat while diving, a layer of subcutaneous fat, and functionally hairless skin.

After this aquatic phase of our prehistory, she posited, the rains once again returned to Africa, the Pliocene merged into the Pleistocene, and man the hunter learned to roam the savannah in search of prey.[40]

Morgan was aware that her interest in the aquatic ape might be perceived as outside the normal bounds of science. In an interview she remarked that perhaps it was easier for her than for an established scientist (one can only assume she meant Hardy), because she had "nothing to lose, no high academic position to think of." She added, "if you talk about flying saucers you're branded a kook. I don't believe in flying saucers but I suppose this kind of thing looks flying-saucerish to the Establishment."[41] In fact, this was exactly what "Establishment" science initially thought. Her reviewer in the *New York Times*, for example (although he had lauded her critique of contemporary anthropology), suggested that her characterization of human ancestors as including "a breed of sea beasts" should be consumed along "with a grain or two of salt."[42] Several years later, physician Jerold Lowenstein and anthropologist Adrienne Zihlman wrote a brief article for *Oceans* magazine in which they compared the "Aquatic Ape Theory" to "the existence of bigfoot" and "visitors from outer space."[43] After analyzing key aspects of the evidence, they concluded, "the Aquatic Ape Theory does not hold water, anatomically, biochemically, behaviorally or archaeologically. With a similar combination of imagination, a grab bag of unrelated 'facts' and a popular literary style, one could make an equally convincing case that our ancestors evolved in the air—as von Daniken has more or less done in his cult book *Chariots of the Gods?*"[44]

Despite this dismissal, and with some urging from an American fan, Morgan wrote a second book, *The Aquatic Ape*, in 1982, in which she re-presented her evidence in more scientific terms.[45] This time, she included figures illustrating her points and, notably, her prose no longer contained the sarcastic critique which had been so well received in her first book. Scientists remained skeptical. One reviewer noted, "Until some hard evidence is found though, I fear we are left with several equally convincing theories floating in a sea of speculation."[46] Lowenstein, too, remained unconvinced, remarking, "It is fun to make up evolutionary fables."[47] Unfazed by such criticisms, she published *The Scars of Evolution* in 1990 and *The Descent of the Child* five years later, each updating her evidence to reflect more recent findings.[48] And although most scientists continued to dismiss her ideas as speculative or pseudoscientific,[49] a small number became intrigued.

The first sign of interest came in the form of a 1987 conference organized to debate the theory, which included both advocates and detractors.[50] Most of the supporters, however, lacked anthropological training, including, for example, Derek Ellis (a marine ecologist from the University of Victoria in Canada) and Marc Verhaegen (a Belgian physician). Ellis argued that the aquatic ape was ecologically viable and called for scientists to test the theory, rather than dismissing it out of hand (anthropologists, of course, felt it *had* been evaluated).[51] Verhaegen, along with a handful of other scientists, began to amass new kinds of evidentiary support for an aquatic phase in human history.[52] They argued that chimpanzees and bonobos were known to walk upright while wading in shallow water and when carrying objects in their arms,[53] that recent paleontological evidence suggested that early hominins spent more time in swampy or coastal forests than on the savannah,[54] and that foods high in fatty acids (like fish) could have been important to the increased brain development of our ancestors.[55]

Anthropologists, with a few notable exceptions, have been far more reticent. The first convert was Philip Tobias, former student of Raymond Dart. Like Ellis, he came to insist that the savannah hypothesis didn't hold water, and that the aquatic ape needed a fair hearing.[56] More recently, well-respected professors Richard Wrangham (biological anthropologist), Dorothy Cheney (primatologist), and Robert Seyfarth (psychologist) published a paper in which they suggested that early human ancestors used aquatic environments to forage for fallback foods, like water lilies and floodplain herbs, when their preferred foods were unavailable, and the aquatic environment in which these foods are found could have helped human ancestors become bipedal.[57]

There are several important points to note about this history of the aquatic ape. The theory itself remains extremely controversial; all of these claims have been challenged in the scientific literature.[58] Additionally, none of the recent papers supporting an aquatic (or semi-aquatic) environment for human ancestors mention or cite Morgan or Hardy by name—to find the connection, you must look in the footnotes of the papers they do cite. Scientific theories can thus become separated from their most ardent supporters, even (perhaps especially) in cases where the equation is almost iconic.[59]

Historians of science have spilled much ink on the demarcation question, asking how scientists and philosophers have tried to cleanly differentiate legitimate scientific inquiry from "non-science," "pseudoscience," "pathological" science, or even simply "bad" science.[60] In considering Morgan's case, however, we can see that in addition to these hard demarcations, scientists in the 1970s also used feminist science as a tool of soft demarcation. By labeling Morgan's *Descent of Woman* "feminist anthropology" they both sidestepped her critique and used the aquatic ape to delegitimize feminist science as a whole (precisely the reason Zihlman attacked the book as pseudoscientific). Emulating the efforts of many fringe scientists, Morgan responded by re-articulating her argument and her evidence with each new publication, in an attempt to rehabilitate her claims by presenting them in a style and language scientists would take seriously.[61] Notably, as the first step in this process, she removed all traces of her feminist critique of other theories of human evolution.

In order for Morgan's ideas to be accepted as scientific, they had to be stripped of her sharp feminist wit, dissociated from Morgan herself, and repackaged as legitimate science. In this form, her ideas are now getting the professional hearing she always wanted. In 2005, Morgan was asked whether or not she still felt like an "outsider" in biology. "Not nearly as much as I used to," she replied.[62]

FURTHER READING

Haraway, Donna. *Primate Visions: Gender, Race, and Nature in the World of Modern Science*. New York: Routledge, 1989.

Milam, Erika. "Making Males Aggressive and Females Coy: Gender Across the Animal-Human Boundary," *Signs: Journal of Women in Culture and Society* 37 (2012): 935–959.

Montagu, Ashley, ed. *Man and Aggression*. Oxford: Oxford University Press, 1968.

Morgan, Elaine. *Descent of Woman*. New York: Stein and Day, 1972.

Sperling, Susan. "The Troop Trope: Baboon Behavior as a Model System in the Postwar Period," in *Science Without Laws: Model Systems, Cases, Exemplary Narratives*, ed. Angela Creager, Elizabeth Lunbeck, and M. Norton Wise. Durham, NC: Duke University Press, 2007, 73–89.

Weidman, Nadine. "Popularizing the Ancestry of Man: Robert Ardrey and the Killer Instinct," *Isis* 102 (2011): 269–299.

NOTES

1. Elaine Morgan, *Descent of Woman* (New York: Stein and Day, 1972): 57.

2. Morgan, *Descent of Woman*, 3.

3. James A. Secord, "Knowledge in Transit," *Isis* 95/4 (2004): 654–672; Katherine Pandora, "Popular Science in National and Transnational Perspective: Suggestions from the American Context," *Isis* 100/2 (2009): 346–358.

4. Morgan, *Descent of Woman*, 56.

5. Robert Ardrey, *African Genesis* (New York: Atheneum, 1961); Robert Ardrey, *Territorial Imperative: A Personal Inquiry into the Animal Origins of Property and Nations* (New York: Atheneum, 1966); Konrad Lorenz, *On Aggression*, trans. Marjorie Kerr Wilson (New York: Harcourt Brace Jovanovich, 1966); Desmond Morris, *The Naked Ape: A Zoologist's Study of the Human Animal* (New York: McGraw Hill, 1967); Lionel Tiger, *Men in Groups* (New York: Random House, 1969).

6. Stanley Kubrick, director, *2001: A Space Odyssey* (MGM, 1968), 141 min.

7. Eric Johnson, "Ariel Casts Out Caliban," *Times Higher Education* (April 21, 2011); E. G. Marshall and Nicolas Noxon, *The Man Hunters*, 16mm (MGM, 1970), 52 min.

8. The most scholarly of these volumes was Richard B. Lee and Irven DeVore, eds., *Man the Hunter* (Chicago: Aldine, 1968). On Ardrey and pop anthropology, see Nadine Weidman, "Popularizing the Ancestry of Man: Robert Ardrey and the Killer Instinct," *Isis* 102 (2011): 269–299.

9. Ardrey, *African Genesis*, 1.

10. Ardrey, *African Genesis*, 29.

11. Ardrey, *Territorial Imperative*, 6–7.

12. Tiger, *Men in Groups*, 44–49, 44. Tiger aimed at a more academic readership than the other authors discussed so far, as evidenced, for example, by his extensive use of footnotes.

13. Morris, *Naked Ape*, 9.

14. In his next book, Morris changed his tune, writing that businessmen in boardrooms, like both baboons and men of the Stone Age, fought to maintain their status in the eyes of their peers; Morris, *Human Zoo*, "Status and Super Status," 15–38. For a more extreme example of the same trend, see Antony Jay, *Corporation Man: Who He Is, What He Does, Why His Ancient Tribal Impulses Dominate the Life of the Modern Corporation* (New York: Random House, 1971).

15. Morris, *Naked Ape*, 37–39, 38.

16. Morris, *Naked Ape*, 43–45; Morgan, *Descent of Woman*, 24–25.

17. Weidman, "Popularizing the Ancestry of Man," 288–289.

18. N. B. Marshall, "Alister Clavering Hardy, 10 February 1896–22 May 1985," *Biographical Memoirs of Fellows of the Royal Society* 32 (1986): 222–273.

19. Sir Alister Hardy, "Was Man More Aquatic in the Past?" *The New Scientist* (March 17, 1960): 642–645.

20. Morgan's sematic point was later adopted by anthropologists and linguists. See, for example, Virginia L. Warren, "Guidelines for the Nonsexist Use of Language," *Proceedings and Addresses of the American Philosophical Association* 59/3 (1986): 471–484.

21. Morgan, *Descent of Woman*, 156.

22. Morgan, *Descent of Woman*, 165.

23. Morgan, *Descent of Woman*, 165.

24. Morgan, *Descent of Woman*, 180–181.

25. Morgan, *Descent of Woman*, 159.

26. Morgan, *Descent of Woman*, 190–193.

27. Morgan, *Descent of Woman*, 160–163.

28. Morgan, *Descent of Woman*, 179.

29. "Books," *Playboy* (June 1972): 20.

30. Hugh Kenner, "Reviewer's Choice," *Life* 73/2 (July 14, 1972): 20; see also Mary Morain, "The Endocentric [sic] Bias," *The Humanist* 33/2 (1973): 44.

31. Jurate Kazickas, "Is the Missing Link a Mermaid?" *The Stars and Stripes* (July 6, 1972): 14–15; Morain, "The Endocentric Bias," 44.

32. John Pfeiffer, "The Descent of Woman," *New York Times* (June 25, 1972): BR6.

33. Richard B. Lee, "Politics, Sexual and Non-Sexual in an Egalitarian Society," *Social Science Information* 17/6 (1978): 871–895, 871; Marie Withers Osmond, "Cross-Societal Family Research: A Macrosociological Overview of the Seventies," *Journal of Marriage and Family* 42/4 (1980): 995–1016, 1004; Martin J. Waterhouse, [review Nancy Tanner, *On Becoming Human* (Cambridge, 1981)], *Man* 17/2 (1982): 352; Margaret W. Conkey and Janet D. Spector, "Archeology and the Study of Gender," *Advances in Archeological Method and Theory* 7 (1984): 1–38, 7 and 14; Hilary Callan, "The Imagery of Choice in Sociobiology," *Man* 19/3 (1984): 404–420, 418; Donna J. Haraway, "Primatology Is Politics by Other Means," *PSA: Proceedings of the Biennial Meeting of the Philosophy of Science Association*, vol. 2 (1984): 489–524, 503; Robert Attenborough, "Between Men and Women" *Nature* 319 (January 23, 1986): 271–272; Nancy Tuana, "Re-Presenting the World: Feminism and the Natural Sciences," *Frontiers: A Journal of Women Studies* 8/3 (1986): 73–75; Matt Cartmill, "Paleoanthropology: Science or Mythological Charter?" *Journal of Anthropological Research* 58/2 (2002): 183–201.

34. Adrianne Zihlman, "Gathering Stories for Hunting Human Nature," *Feminist Studies* 11/2 (Summer 1985): 365–377, 367; see also Robert Attenborough, "Between Men and Women," *Nature* 319 (January 23, 1986): 271–272, and Helen E. Longino, "Science, Objectivity, and Feminist Values," *Feminist Studies* 14/3 (1988): 561–574.

35. Jerold M. Lowenstein and Adrienne L. Zihlman, "The Wading Ape: A Watered-Down Version of Human Evolution," *Oceans* 13 (May/June 1980): 3–6, 6.

36. Morgan, *Descent of Woman*, 17, 14, 15, 17.

37. Morgan, *Descent of Woman*, 21–22.

38. Morgan, *Descent of Woman*, 113–130.

39. Morgan, *Descent of Woman*, 28, 30.

40. Morgan, *Descent of Woman*, 131.

41. Jordan Bonafonte, "The Naked Ape Is All Wet, Says a Liberated Lady," *Life* (July 21, 1973): 77.

42. Pfeiffer, "The Descent of Woman," BR6.

43. Jerold Lowenstein and Adrienne Zihlman, "The Wading Ape: A Watered-Down Version of Human Evolution," *Oceans* 13 (May/June 1980): 3–6, 3.

44. Lowenstein and Zihlman, "The Wading Ape," 6; Erich von Däniken, *Chariots of the Gods? Unsolved Mysteries of the Past*, trans. Michael Heron (New York: Putnam, 1968). A couple of years later, Zihlman compared the aquatic ape to Immanuel Velikovsky's *Worlds in Collision* (New York: Macmillan, 1950); Adrienne Zihlman [review of John Gribbin and Jeremy Cherfas, *The Monkey Puzzle: Reshaping the Evolutionary Tree* (New York: Pantheon Books, 1982)], *American Anthropologist* 85/2 (June 1983): 458–459. See also Ian Tattersall and Niles Eldredge, "Fact, Theory, and Fantasy in Human Paleontology," *American Scientist* 65/2 (1977): 204–211, 207.

45. Elaine Morgan, *The Aquatic Ape* (New York: Stein and Day, 1982); Barbara Miner, "Author Waters Down Anthropological Views," *Hutchinson News* (April 2, 1983): 10; Kate Douglas, "Natural Optimist" [interview with Elaine Morgan], *New Scientist* 186/2496 (2005): 50–53.

46. Jennifer Rees, "Trigger of Change," *New Scientist* 94/1307 (May 27, 1982): 592.

47. Jerold M. Lowenstein, "Swimmers or Swingers," *Oceans* 17 (July/August 1984): 72.

48. Elaine Morgan, *Scars of Evolution* (London: Souvenir Press, 1990); Elaine Morgan, *Descent of the Child: Human Evolution from a New Perspective* (Oxford University Press, 1995).

49. Andrew Hill, "Scientists' Bookshelf," *American Scientist* 72/2 (1984): 188–189; Adrienne Zihlman, "Review: Evolution, a Suitable Case for Treatment," *New Scientist* 1752 (January 19, 1991); Nina Jablonski, "Naked Truth," *Scientific American* (February 2010): 42–49, 45; Paul Griffiths, "The Historical Turn in the Study of Adaptation," *The British Society for the Philosophy of Science* 47/4 (1996): 511–532, 523.

50. Machteld Roede, Jan Wind, John Patrick, and Vernon Reynolds, eds., *Aquatic Ape: Fact of Fiction: Proceedings from the Valkenburg Conference* (London: Souvenir Press, 1991).

51. Derek V. Ellis, "Is an Aquatic Ape Viable in Terms of Marine Ecology and Primate Behaviour?" in *Aquatic Ape*, 36–74; Derek V. Ellis, "Wetlands or Aquatic Ape? Availability of Food Resources," *Nutrition and Health* (Journal of the McCarrison Society) 9 (1993): 205–217. Such calls for additional testing are a standard tactic among "fringe" science groups. Ellis should not be confused with the marine conservationist and painter, Richard Ellis, who has also written on the topic: "Everybody Back into the Water," in *Aquagenesis: The Origin and Evolution of Life in the Sea* (Viking, 2001): 243–267.

52. Marc Verhaegen, "Aquatic versus Savannah: Comparative and Paleoenvironmental Evidence," *Nutrition and Health* 9 (1993): 165–191; Marc Verhaegen, "Aquatic Ape Theory, Speech Origins, and Brain Differences with Apes and Monkeys," *Medical Hypotheses* 44 (1995): 409–413; Marc Verhaegen and Pierre-François Puech, "Hominid Lifestyle and Diet Reconsidered: Paleo-Environmental and Comparative Data," *Human Evolution* 15 (2000): 151–162; Marc Verhaegen, Pierre-François Puech, and Stephen Munro, "Aquarboreal Ancestors?" *Trends in Ecology and Evolution* 17/5 (2002): 212–217.

53. K. D. Hunt, "The Evolution of Human Bipedality: Ecology and Functional Morphology," *Journal of Human Evolution* 26 (1994): 183–202; E. Esteban Sarmiento, "Generalized Quadrupeds, Committed Bipeds and the Shift to Open Habitats: An Evolutionary Model of Hominid Divergence," *American Museum of Natural History Novitates* 3250 (1998): 1–78, 36.

54. Verhaegen, Puech, and Monroe, "Aquarboreal Ancestors?"

55. Stephen Cunnane, Laurence Harbige, and Michael Crawford, "The Importance of Energy and Nutrient Supply in Human Brain Evolution," *Nutrition and Health* 9/3 (1993): 219–235; C. Leigh Broadhurst, Stephen Cunnane, and Michael Crawford, "Rift Valley Lake Fish and Shellfish Provided Brain-Specific Nutrition for Early *Homo*," *British Journal of Nutrition* 79/1 (1998): 3–21; Stephen Cunnane, "Hunter-Gatherer Diets—A Shore-Based Perspective," *American Journal of Clinical Nutrition* 72/6 (2000): 1584–1585; C. Leigh Broadhurst, Yiqun Wang, Michael Crawford, Stephen Cunnane, John Parkington, and Walter Schmidt, "Brain-Specific Lipids from Marine, Lacustrine, or Terrestrial Food Resources: Potential Impact on Early African *Homo Sapiens*," *Comparative Biochemistry and Physiology B, Biochemistry & Molecular Biology* 131/4 (2002): 653–673.

56. Kate Douglas, "Taking the Plunge," *New Scientist* (November 25, 2000).

57. Richard Wrangham, "The Delta Hypothesis: Hominoid Ecology and Hominin Origins," in *Interpreting the Past: Essays on Human, Primate and Mammal Evolution in Honor of David Pilbeam*, eds. D. E. Lieberman, R. J. Smith, and J. Kelley (Boston: Brill Academic, 2005): 231–242; Richard Wrangham, Dorothy Cheney, Robert Seyfarth, and Esteban Sarmiento, "Shallow-Water Habitats as Sources of Fallback Foods for Hominins," *American Journal of Physical Anthropology* 149 (2009): 630–642.

58. Debates over the aquatic ape are also alive and well on the Internet. See, for example, www.riverapes.com (pro) and www.aquaticape.org (con). Accessed on March 13, 2013.

59. On the association of "fringe" scientific theories with charismatic gurus, especially in the 1960s and '70s, see David Kaiser, *How the Hippies Saved Physics: Science, Counterculture, and the Quantum Revival* (New York: Norton, 2011).

60. For recent publications on the question, see Thomas Geiryn, *Cultural Boundaries of Science: Credibility on the Line* (Chicago: University of Chicago Press, 1999); James Rodger Fleming, *Fixing the Sky: The Checkered History of Weather and Climate Control* (New York: Columbia University Press, 2010); Naomi Oreskes and Erik Conway, *Merchants of Doubt: How a Handful of Scientists Obscured the Truth on Issues from Tobacco Smoke to Global Warming* (New York: Bloomsbury, 2010); Michael Gordin, *The Pseudoscience Wars: Immanuel Velikovsky and the Birth of the Modern Fringe* (Chicago: University of Chicago Press, 2012). In terms of evolutionary theory specifically, see Philip Kitcher's *Abusing Science: The Case Against Creationism* (Cambridge, MA: MIT Press, 1982).

61. For similar efforts, see Michael Gordin, "Experiments in Rehabilitation," in *The Pseudoscience Wars*, 106–134.

62. Douglas, "Natural Optimist," 52.

DAVID HULL'S PHILOSOPHICAL CONTRIBUTION TO BIOLOGY

INTRODUCTION

On the face of it, philosophy might well be expected to support and, if only indirectly, contribute to science. That certainly was the assumption in the era in which philosophers of science—the "logical positivists" who dominated Anglo-American philosophy in the first half of the twentieth century—believed they could map out the structure of all scientific disciplines on the model of physics, mathematics, and logic, and also define "scientific method." The classic statement of the orthodox position was Nagel's *The Structure of Science* (1961). And yet, typically, scientists—then as now—have successfully gone about their business entirely without benefit of being formally taught the "scientific method."

The aim of this chapter is to examine the interactions of science and philosophy through the uniquely revealing contribution of David Hull (1935–2010). According to Michael Ruse, "David Hull is the father of modern studies of biology from a philosophical viewpoint"[1] and "today's leading philosopher of biology"[2] (see figure 13.1). What impact did this eminent philosopher have on biologists?

THE EMERGENCE OF PHILOSOPHY OF BIOLOGY

Hull studied philosophy at Illinois Wesleyan University (1956–1960). He was in the first cohort of students in the History and Philosophy of Science graduate program, initiated at Indiana—alongside Princeton and Yale—in 1960.[3] The founding staff were the historians of the scientific revolution, Rupert and Marie Hall, and philosophers with biological interests, Roger Buck and Michael Scriven, both trained in Britain.[4] The driving force was the ex-physicist Russell Hanson, already well known for his pioneering study of the process of discovery in physics.

Philosophy of science in the 1960s was "ripe for a revolution."[5] For Hull and many of his contemporaries, "Thomas Kuhn's *The Structure of Scientific Revolutions* (1962) provided the spark that ignited the imaginations of historians, philosophers, and scientists alike."[6]

Figure 13.1: David Hull.
Photograph by Lizzie Ruse.

In 1974, the thirty-eight-year-old Hull published one of the first texts under the new rubric of "philosophy of biological science."[7] In his words, "My underlying concern in writing this book is to investigate whether there is a single philosophy of science adequate for all science or whether there are many, each appropriate in its own domain"[8] Nagel had argued that biology might be fundamentally different from physics (e.g., in its invoking of teleology and vitalism).[9] By explicating the molecular basis of genetics Hull countered any such position, but went on to argue that the (ontological) reduction of one to the other is logically inappropriate and actually impracticable: this undermines Hempel's logical positivist concept of "covering laws" and (nomological) "theory reduction."

"Anyone who has ever taught a course in Philosophy of Biology," Hull wrote, "knows how difficult it is to teach philosophy to biology students while at the same time teaching biology to philosophy students."[10] One notable feature of the book was that more than half of it was devoted to explaining biology itself, reflecting the growing realization among philosophers of the need to catch up on now-established scientific understanding based on molecular biology and the evolutionary synthesis. The second striking feature is the extent to which the agenda for Hull still reflected the topics of concern of the past; namely, scientific change as Kuhnian revolutions and Hempelian reduction.

An influential early paper by Hull reveals his critical attitude to what was available in the literature at the time (e.g., by authors Bunge, Flew, Goudge, Grene, Woodger, and Beckner).[11] In this paper he tries to define the aims of the discipline: "A philosopher might uncover, explicate, and possibly solve problems in biological theory and methodology. He might even go on to communicate these results to other philosophers, to scientists, and especially to biologists. . . . These are some of the things which philosophers of biology might do. With rare exceptions, they have not. . . . It must be admitted that thus far it is not very relevant to biology, nor biology to it."[12]

Hull's rejoinder came with his classic book of 1988, *Science as a Process*, the result of some fifteen years' work. In its preface, he explained that it was triggered by his increasing dissatisfaction with logical positivism, with the latter's remoteness from the actual processes of science, and also by "critics . . . bent on debunking science at all costs." Here Hull was referring to the "externalist" perspective favored by sociologists and historians (most influentially the "Edinburgh strong program"),[13] which explains science as the product of socioeconomic, class, and political factors—the ultimate extension of the charge of "theory-ladenness" of evidence and observations arising out of the tradition of Feyerabend, Lakatos, and Popper—an "epistemological relativism" that undermines any "objectivity," and therefore veridical standing, in ("internalist") scientific claims.[14]

Hull's book is a study of the development of three new schools of taxonomic method that emerged in the 1970–1980s (233–76, 504–5). Since 1970 Hull had become involved in the editing of *Systematic Zoology*, the only journal covering taxonomy at the time.[15] Military service had given him information technology skills.[16] His doctoral work had started his long-standing interest in "species" definitions.[17] His initial intention[18] was to study the Kansas school of taxonomic method, "numerical taxonomy" or "numerical phenetics"—so named because it sought to use numerical (indeed, computer-based) tabulations of large numbers of identifying features to characterize any given biological species, aiming to maximize "objectivity": earlier methods relied on subjectively selected ideal characters. But Hull was lucky; at that very time a rival method called "cladism," originally formulated by Willi Hennig in the 1950s, became known in the United States.[19] Also known as "phylogenetic classification," this method formalized the identification of relationships between species (each defined by a few, selected, and distinctive shared traits) while also linking species classes to their positions in a phylogenetic tree according to strict rules. Of even more interest was the subsequent splitting-off of a school of "pattern (or transformed) cladism," which reasserted the need to avoid the theory-ladeness of considering trees at all in the process of characterizing a species (262–63, 375, 490–91).

Hull was thus directly involved in the battles that ensued in the period between 1970 and 1985, not only in his editorial role but also because he had direct access to the participants. During 1973 through 1983 he conducted fifty-two structured interviews with all the leading personalities. A good half of Hull's complex book is taken up with detailed empirical and historical documentation. The result is a remarkably frank record of the social (and underlying personal) forces in the rise and fall of groupings of scientists (i.e., rivalry, competition, cooperation, utilization of publications and conferences, professional bodies, etc). His ("ambitious," 284) aim is to define a general explanatory principle (evolutionary selection): it is spelled out in the key chapter, "The Need for a Mechanism." "I present a general analysis of selection processes which I take to be applicable equally to biological, social and conceptual change," Hull writes (284).

Hull's book is the product of a time at which the full impact was felt of the Kuhn-inspired methodological liberalization in philosophy of science. Hull avoids discussing such traditional philosophical themes as the foundations of knowledge, the nature of truth, and matters of justification or belief. Studies of science in the 1980s became polarized into "internal" and "external" explanations for scientific change. "I strongly suspect," he predicted, ". . . that once the current 'spirited' dispute over internalism and externalism has run its course, neither program will remain unchanged" (16).

Regarding Hull's use of taxonomy as his case example, there is hardly any more suitable subject for a philosopher: it relates directly to fundamental philosophical concerns, such as "essentialism," "natural kinds" (and related scientific realism),[20] mereology, "universals," language and the role of terminology. Hull's position is that, as with species definitions (213–23) and social systems, concepts could not be open to change if they were defined by "essences": scientific change must involve theories and concepts that have flexible defining properties.

It becomes clear that Hull sees himself as mounting a critique of many previous approaches to the philosophy of science.[21] He builds up his case in the following terms; he insists that the work of philosophers must be judged in the light of empirical evidence and in the same way as the work of scientists: "There is nothing wrong with philosophers providing general definitions of metascientific terms . . . [Hull is referring here to terms such as explanation, rationality, and justification] if they go on to *do* something with them. In science the chief role of such general terms is to function in scientific laws. What analogous function are philosophical analyses of metascientific terms [referring here to concepts such as covering laws and theory reduction] . . . to perform? If none, then I see no point to all this effort" (79–80). For Hull, "the only way to evaluate these analyses is in the context of a general theory about the nature of science. The next step is to confront these general theories with empirical data" (298). He adds, "anyone proposing a theory about the scientific process itself must treat it as seriously as Darwin and other scientists treat their theories. Vague gestures . . . are . . . not good enough" (281).

Hull's philosophical stance is "naturalistic": avoiding traditional epistemological concerns, he claims to provide explicit, empirical evidence as a way to address the philosophical theme of the nature of science. Hull frequently argued for the use of real-life examples in philosophy rather than thought experiments or "other world" examples.[22] Like a scientist, he bases his arguments on his own observational, documentary, and sometimes quantitative evidence. "I . . . do something that is usually not considered appropriate for a philosopher. I present a mechanism which, I argue, is adequate to explain a great deal about the way in which science works" (285).

THE NATURE OF HULL'S BRIDGING OF DISCIPLINES

In a wide-ranging review of Hull's book, historian of geology David Oldroyd points out Hull's remarkable qualifications for his project: "Hull's work regularly appears in scientific as well as philosophical and historical journals, and his clarifications of conceptual problems in biology are evidently appreciated by scientists themselves. Indeed, he has the unique distinction of having been President of both the Philosophy of Science Association (in 1986) and the Society of Systematic Zoology (in 1984–1985). With such credentials, then, Hull has found it possible to gain entry to the 'control centre' of the community of systematic zoologists in North America."[23]

Hull was insistent on the potential for real interaction between science and philosophy, and he continued to do all he could to put this into practice; as in, for example, his 2001 paper co-authored with two scientists.[24] As he wrote, "An interplay between science and philosophy has . . . characterized science since its inception. . . . As science developed, [the disciplines] became increasingly detached. Even so, their connections were never totally severed. . . . For example, both Woodger and Gilmour were attracted to the tenets of logical positivism. . . . Cladists in turn were attracted to the philosophy of Karl Popper" (516). And twenty years later: "Increasingly, philosophers of biology and biologists began to coauthor papers, as they pooled their conceptual resources."[25] Hull listed a number of such combined publications.[26]

Still, there is surprisingly little evidence in the book of any effects he may have had on the taxonomy battles.[27] Despite his expertise in the matters at stake—Hull was an occasional author in *Systematic Zoology* from 1964 and, as mentioned, ex-president of the Society of Systematic Zoology—he does not claim, or imply, that he personally influenced the scientific debates. How come? The style of his presentation is that of a detached, outside observer. "I have done my best to step back from those events in which I have participated" (243). In his role as the scrupulous historian, he insists on his efforts to remain objective and therefore independent; indeed, much of his evidence was obtained from formal interviews, together with retrospective archival and citation studies (163–82, 234, 254, 325–41: Appendices D, F, J). Evidently Hull was reflexively self-aware regarding his own possible biases and the limits of his methods:[28] perhaps he knowingly minimized evidence of his own role.

The unresolved themes at issue in the debates described by Hull concerned the choice and criteria of choice of features to use in characterizing a species, on which in turn depends any measure of closeness to features in other species and thence the evaluation of evolutionary relatedness. The issues were contentious because, for some, they called into question supporting evidence for evolution itself. Throughout the fraught episodes he witnessed, Hull appears to have struggled to make sense of what was going on. He hints that the taxonomists were sometimes confused about fundamental matters; disagreements between them and him (242) even caused friction.[29] When, on the rare occasion, Hull gives his own, reserved opinions on the relative merits of the rival taxonomic methods (most clearly on pages 241–42, 273–76, 517–19), this is a philosopher's interpretation of the underlying issues: his undogmatic, detached, and pragmatic stance may directly echo his own emerging general philosophical conclusions.

Hull came to occupy the most unusual position of directly experiencing the very heart, red in tooth and claw, of how science, in general, operates.

"Investigators who study science frequently express shocked disbelief with respect to actual scientific practice," he wrote (495). Hull was, no doubt, himself considerably surprised to discover the extent of the personal and social factors that go into "scientific discovery"—although any scientist could have told him if he had asked. His study reveals all too clearly how reality departs from the popular "scientific" ideal.

This merely heightened the problem before him. "In general, why does science work so well? . . . [This is the sort of question] . . . that philosophers of science dismiss as not being relevant to philosophical analyses of science. If so, then so much the worse for philosophical analyses of science" (302). He refers to his aim as "explaining the marvellous progress made during the past few centuries by successive generations of scientists."[30] Hull is clearly an admirer of science and a believer in its progressive nature: he expresses a sense of awe and puzzlement at what it has achieved. He is offering a solution to the mysteriousness: "My goal in this book is to show that the coincidence between the professional interests of individual scientists to gain credit and the institutional goals of science to increase our knowledge of the empirical world is not in the least 'mysterious' nor the mechanism that produces this coincidence in the least 'hidden'" (357).

In assessing Hull's program, the following points may be noted:

(i) "My work is a combination of history, sociology, philosophy, and evolutionary biology. Now am I sure that my mix is the right mix? No! . . . My work is very sociological. I am not tremendously well trained in sociology; but what I have done is every time I needed to know something, I went to learn it."[31] Because he employs the methods of all the above disciplines, this makes it difficult to judge where his originality lies (Toulmin, Kuhn, and Popper had arrived at similar evolutionary perspectives): but at the very least he has provided invaluable historical and sociological data.

(ii) Hull is trading entirely on what may merely be an analogy—evolutionary selection—derived from biology but transferred over to sociology and conceptual development. Hull energetically rebuts the charge: "The . . . differences between biological and conceptual change turn out to be illusory. At most they are differences in degree, not kind."[32] Even accepting the value of the analogy, it could be considered so general in nature—amounting to a "universal Darwinism"—that it tells you little that is truly explanatory of scientific change.

(iii) Throughout (241–42, 483–92, 508–11), Hull moves seamlessly between social and conceptual systems as realms in which selection explains change. However, his documentary evidence almost entirely addressed social systems: particularly the dynamics of successful scientific groupings and alliances (241–44) and their "influence," for example through leadership and presentational skills (239, 249, 253, 273, 519). There is little attention to the scientific ideas and arguments: Hull rarely quotes from the scientific literature (there are only brief, background overviews in his own words). Hull's account—apart from the where, when, who—revolves around descriptions of participants' opinions about other participants, often in personal and scurrilous terms. Hull is avoiding the *content* of the science (255). As he recognizes, "I do not spend sufficient time in discovering which ideas actually gave rise to other ideas" (207).

(iv) Taxonomy is a descriptive exercise: the schools of taxonomists were arguing only about methods and criteria of description. Moreover, these did not affect the basic material—only the finer details—of evolution theory or of accepted phylogenetic trees. Choice of methods must depend on the effectiveness of their use in context (e.g., in museum curating). Taxonomy is far from being representative of science in general; in laboratory science hypotheses are open to testing and phenomena can be manipulated or effects predicted in ways that do not arise in taxonomy. Alternative taxonomic methods cannot be "tested," for example against any absolute validating standard.

In summary, then, it remains unclear whether Hull as a philosopher made any significant impact on the scientists engaged in the taxonomy debates. Many of the issues had been under scrutiny in the "New Systematics" of the 1940s (102–9). There is stark evidence that two decades after Hull's study, little has changed. According to a recent consideration of the current position: "Unfortunately, the great revolution inspired by Hennig was stopped before its completion—before its positive impacts on taxonomy were fully realized. . . . A pseudoscience based on quantitative manipulations of questionable phenetic data has focused on cladograms at the exclusion of character analysis."[33] The report continues, "The full negative consequences . . . are yet to be generally acknowledged, much less evaluated or appreciated."[34] Taxonomy is not a good example of scientific progress because it has progressed little.

There is perhaps one dominating factor in Hull's approach to scientific change. As against the logical positivist picture of his training, Hull seems to have been struck by the diversity of scientific opinions on any issue at any one time. As he saw it, any one individual scientist cannot represent (or even appreciate) an emerging consensus. "In order to understand science, it is absolutely essential that we do not view scientists as isolated knowing subjects . . . this sort of understanding lacks several characteristics . . . [it] . . . cannot be transgenerationally cumulative the way science is. . . . Individual bias is extremely difficult to eliminate" (22). Conceptual progress can only be defined and achieved at the group, collective level. Hull's outsider's perspective is dominated by the superficial irrationality and unreliability of what scientists report: here he seems to have been heavily, perhaps too heavily, influenced by his experience in his role as a journal editor, overemphasizing transmission of scientific information (155–56).

Hull clearly believed that science progresses towards a greater understanding of reality (458, 462–66, 476). Insofar as Hull's account of ("internalist") conceptual change can be inferred, it is "pragmatic." Regarding the survival of concepts or theories (as in his attitude to essentialism and the theory-ladeness of observations and data), "the proof of the pudding is in the eating" (491). His pragmatic approach shows through in his oft-repeated reference to operationalism.[35] His evolutionism implies an opportunism, complexity, and "randomness" applicable to scientific thinking: he seems to accept—in taxonomy as with the species problem—that there can be no final, ideal method: hence his evident reservations about any forms of "essentialism"[36] (512–13). Given the evolutionary nature and complexity of biological systems, all entities in biology (including human thought and action) are open to change and cannot be delimited by essences.

Hull sums up his critique of conventional philosophy of science as follows. "Logical (positivists) were . . . aware . . . that science . . . is both temporal and social, but these aspects were thought to be irrelevant to a 'philosophical' analysis of science. Philosophers must limit themselves to more modest activities, e.g., to the analysis of rationality as such, the abstract relation between theories and data, and the like. I suspect that one reason for philosophers ignoring science as it actually takes place is that they fear that it will prove to be too complicated and haphazard to allow for any general conclusions. One purpose of this book is to show that this fear is mistaken" (27). He continues: "If philosophic theories about science are to be taken seriously, . . . they cannot remain mere gestures containing nothing but infinitely renewable promissory notes. Some 'hard work' must be done" (29). He means close acquaintance with the scientific, and historical, issues. Hull has recast philosophy in the scientific image, as "the science of science": a late paper is entitled "Studying the Science of Sciences Scientifically."[37]

(RE)DEFINING THE RELATION OF PHILOSOPHY AND SCIENCE

Introducing a collection of appraisals of Hull's philosophical contribution, Griffiths argues: "David Hull has set new standards for close involvement between philosophers of science and the science they study. He has shown us that the philosophical analysis of science can be introduced into the actual process of science as critical and contestable commentary, rather than being produced after the fact and for a separate audience."[38] One suspects that Hull's impact among philosophers of science has been considerable, but primarily through the intangible route of personal influence. His is a moderate version of externalism (of which there are many versions: e.g., Jerome Ravetz,[39] John Ziman[40]). At the same time, he stoutly defends the notion of scientific truth and rationality. His evolutionary model has been absorbed into a wider "evolutionary epistemology,"[41] and alternative perspectives compete; for example, cognitive approaches emphasizing "internal," psychology-based explanations for concept change. A good recent example is Nancy Nersessian's *Creating Scientific Concepts*: Hull is not mentioned.

Among philosophers who have followed Hull most directly, Philip Kitcher is a prominent example, particularly in his 1992 monograph;[42] another is Peter Godfrey-Smith.[43] Textbook treatments of the philosophy of science still revolve around standard positions (of Kuhn, Popper, Lakatos, Feyerabend) defined fifty years ago. It can be argued that no equally well-defined position has emerged since. Was Hull's contribution a possible model for such a position in the future?

The problem facing philosophers of science today is perhaps captured by Ruse as follows. "The biologist *qua* biologist dissects and studies frogs and humans and oak trees and bacteria. But when it comes to questions about what the biologist is doing, then we are in the realm of philosophy. Is this not all somewhat presumptuous? A nonexpert in the field . . . tells its practitioners . . . how they should behave. Sometimes I confess it is presumptuous; unless the philosopher makes some effort to understand the field being studied, grief will not be far behind. . . . A detached general knowledge is not necessarily bad."[44] The assumption is that philosophers' work is closely linked to the scientists' work. But how?

It is not difficult to find evidence of unease regarding this question. "The philosophy of science has been in a state of fermentation and uncertainty in recent years," Godfrey-Smith writes.[45] "All of philosophy is plagued with discussion and anxiety about how philosophical work should be done and what a philosophical theory should try to do. So we . . . have to deal with disagreements about the right *form* for a philosophical theory of science."[46] One senses a distancing of philosophy from science, judging from the following sample of commentary: "It is . . . rare for a philosopher's view of science to be used within a scientific debate to justify one position over another."[47] "Naturalists think that the project of trying to give general philosophical foundations for science is always doomed to fail. They also think that a philosophical foundation is not something science needs in any case."[48] "In a different version of naturalism, there is such a thing as a *philosophical question*, distinct from the kinds of questions asked by scientists."[49] Philosophy "is the discipline that addresses those questions that the sciences cannot (yet, or perhaps ever) answer and the questions about why the sciences cannot answer these questions."[50] Philosophers, it would seem, are well aware that scientists often find them superfluous.

Historically, philosophy and science were coextensive: science only gradually became distinct from philosophy over the last two hundred years. But the possibility arises that they have by now become so divergent as to have become fundamentally detached, even incompatible. Have we perhaps continued to take for granted a historical leftover: an unexamined retention of a vague residual notion assuming their connectedness? The methods and aims of philosophy imply the use of general, abstract concepts covering the widest possible range of evidence and experience to a degree that is inappropriate for science, and especially for biology, where individual variation is of the essence. Compared to philosophy, science is instrumental and aims to limit itself to "answerable" questions.

No doubt the drawing of lines between philosophy and science is a matter of definition and opinion. But our discussion in the context of Hull's work has made it clear that bridging across the disciplines is vastly difficult. As Hull said: "Bringing data to bear on disputes in biology is far from easy. . . . Moving up a level to the philosophy of biology, data are even more difficult to bring to bear on philosophical theses. . . . Certainly, philosophers of biology fill their publications with references to lots and lots of biology. But the bearing that this biology has on genuine philosophical issues is far from clear."[51]

Hull has offered us a perspective that must be considered unmatched before or since: not only was he an observer of science in action, he was also a recorder and interpreter of the historical course of events and a participant with relevant skills and interests complementary to those of the professional scientists. Hull's project represents an outstanding critical test of the proposition that a person trained in one discipline can contribute importantly in another. As we have argued, the outcome is limited, and not only because the project reflects the priorities of its time and Hull's specific circumstances. Why? Hull's attempt to characterize the scientific process revealed some of the difficulties intrinsic to any such exercise if it is to engage with any real aspect of scientific progression: "Historians are committed in principle to detailing all the causal factors that actually influenced the views which the scientists under study held and the decisions that they made. The trouble is that these causal influences are extremely difficult to uncover, and the resulting story tends to be so complicated and detailed that few are willing or able to read it" (115). The requirements, it would seem, are "prohibitive" (72).

In one of his later papers, Hull provided an even more salutary analysis of interdisciplinary collaboration.[52] He writes, "During the past hundred years or so, those scholars studying science have isolated themselves as much as possible from scientists."[53] "The founders of the various branches of Science Studies felt obliged to exclude scientists from their emerging disciplines."[54] "As Science Wars heated up, scientists were led to notice us, and they were not amused. They turned on us with a vengeance."[55] "The chief reaction of outsiders to . . . the works of professional philosophers of science . . . is that they might be very sophisticated and precise, but they have little or nothing to do with science."[56] Hull describes how sociologists, historians, and philosophers of science have adopted a protectionist territoriality with respect to each other. If this insularity is true within science studies, what hope for useful communication with the scientists?

Most academics start from a narrow viewpoint dictated and restricted by the methods and perspectives of the discipline in which they were originally trained. If they aim to engage with the actual work and thought of scientists, nonscientists must be able to ask questions that are relevant, contemporary, and sufficiently informed regarding scientific specialities—be they philosophers, sociologists, historians, anthropologists, or psychologists. Passmore expresses the situation particularly well: "Compared with science, philosophy is a relatively loose tradition. . . . Philosophers . . . are much more inclined than are scientists to ask themselves why they are doing what they are doing. . . . 'Philosophy,' as Wittgenstein once remarked, 'ought to liberate us from the idea that there is a kind of academic doctor who can do things for physicists and other scientists that they are incapable of doing for themselves.' . . . Most of the time, we engage in activities without asking ourselves why we are doing what we are doing. Whether as philosophers, scientists, politicians, teachers, doctors, mechanics or artists, we inherit a complex tradition."[57]

CONCLUSION

Two passages fittingly sum up Hull's claims: "Michael Ruse has accurately diagnosed my general views about the relation between history of science, philosophy of science, and science. We are all engaged in the same activity, only with different though complementary training. Increased interactions between scientists, historians, and philosophers are proving productive for all those concerned, and those of us who concentrate on biology have led the way in this symbiotic relationship."[58] "In general, many of the disanalogies raised by critics of treating conceptual change as a selection process stem from a misunderstanding of biological evolution. Others arise from the inability to entertain alternative perspectives. Old-think retains a powerful grip on all of us. The issue is not whether the new perspective that I am urging is *better* than the traditional perspective but that many people seem incapable of even *understanding* this new alternative."[59]

These quotations pinpoint three central themes: the centrality of Hull's concept of evolutionary selection; his belief that the philosophers of biology have led the way in opening up a potentially new science-philosophy relation; his critical attitude to the narrowness of disciplines. Despite his scepticism about conventional philosophy and his remarkable eclecticism, Hull remained at heart a philosopher. His titles included "A mechanism and metaphysics"[60] and *The Metaphysics of Evolution* (1989). His perspective as a trained philosopher dictated his choice of case study and explains his limited scientific engagement.

How, finally, should we regard the status of Hull's contribution to the philosophy of biology? Is his mechanism offering a new method (based on his naturalistic technique, with its aspiration to scientistic standards of evidence and belief) or is it merely a newly generalized description of what characterizes scientific (and possibly philosophical) practice and knowledge? Hull's bridging of disciplines is only possible because his evolutionary "mechanism" is so very general in nature: too general to deal with the full range of complex issues (scientific, philosophical, cognitive, and social, only some of which have been alluded to above) that would have to go into the mix of any "future philosophy of science." But perhaps Hull's greatest contribution may prove to have been, firstly, to demonstrate—perhaps unwittingly—how it is no longer possible to imagine a monolithic, comprehensive "philosophy of biology" as envisaged fifty years ago, and, secondly, by implication to challenge the role of philosophy against the background of present-day scientific understanding.[61,62]

FURTHER READING

Baker, G. "Alternative mind-styles," in *Philosophy in Britain Today,* S. S. Shanker, editor. London: Croom Helm, 1986, 1–57.

Kuhn, T. S. *The Road since Structure.* Chicago: University of Chicago Press, 2000.

Olby, R. C., editor. *Companion to the History of Modern Science.* London: Routledge, 1990.

Popper, K. R. "How I see philosophy," in *The Owl of Minerva: Philosophers on Philosophy,* C. J. Bontempe and S. J. Odell, editors. London: McGraw-Hill, 1975, 41–55.

Waismann, F. "How I see philosophy," in *Contemporary British Philosophy*, H. D. Lewis, editor. London: Allen and Unwin, 1956, 447–90.

NOTES

Grateful thanks for their input to John Dupré, Katherine Morris, Jerry Ravetz, Alex Rosenberg, Michael Ruse, and Simon Saunders.

1. M. Ruse, editor, *The Oxford Handbook of Philosophy of Biology* (Oxford: Oxford University Press, 2008), 3.

2. M. Ruse, *Philosophy after Darwin* (Princeton: Princeton University Press, 2009), 59. For (often enlightening) personal impressions and assessments of Hull, see *What the Philosophy of Biology Is: Essays Dedicated to David Hull*, M. Ruse, editor (Dordrecht: Kluwer Academic, 1989); Michael Ruse, "David Hull: A memoir," *Biology and Philosophy* 25 (2010), 739–47; Peter Godfrey-Smith, "David Hull," *Biology and Philosophy* 25 (2010), 749–53; David L. Hull, "A mechanism and its metaphysics: An evolutionary account of the social and conceptual development of science," *Biology and Philosophy* 3 (1988), 123–55; Issue dedicated to David Hull, *Biology and Philosophy* 15/3 (2000). For his views on scientist colleagues: David L. Hull, "A career in the glare of public acclaim," *Bioscience* 52 (2002), 837–41; David L. Hull, "Ernst Mayr's influence on the history and philosophy of biology: A personal memoir," *Biology and Philosophy* 9 (1994), 375–86.

3. K. T. Grau, "Force and nature: The department of the history and philosophy of science at Indiana University, 1960–1998," *Isis* 90 (1999), S295–S318.

4. British influences on Hull are evident: Hanson and Stephen Toulmin studied and taught in Oxford and Cambridge; both were much influenced by Wittgenstein, as was Scriven; the Halls had migrated from Cambridge. Toulmin lectured at Indiana in 1959 (Stephen Toulmin, *Foresight and Understanding* [New York: Harper and Rowe, 1963]). Other influences were Koyré and Whitehead.

5. D. L. Hull, *Science as a Process: An Evolutionary Account of the Social and Conceptual Development of Science* (Chicago: University of Chicago Press, 1988), 111. All unassigned page references refer to this text.

6. Hull, *Science*, 111.

7. A comparable early text was Michael Ruse's *Philosophy of Biology* (London: Hutchinson, 1973). "Philosophers of biology used the textbook form as a means for conducting serious arguments with one another" (M. Grene and D. J. Depew, *The Philosophy of Biology* [Cambridge: Cambridge University Press, 2004], 312).

8. D. L. Hull, *Philosophy of Biological Science* (Englewood Cliffs: Prentice-Hall, 1974), 7.

9. J. H. Woodger's *Biological Principles* (London: Kegan Paul, 1929) had covered all the ground still familiar in the 1960s. His influential text was part of the tradition in "theoretical biology" that "philosophy of biology" in part replaces: the transition may reflect the professionalization of the subject by philosophers, originally the province of scientists. "Theoretical biology" is an area (concerned characteristically with modeling and quantification) where the methodological interests of philosophers and scientists can coincide, and where only a limited biological knowledge

is required: "At times, philosophy of science is indistinguishable from theoretical science" (Hull, *Philosophy*, 88).

10. Hull, *Philosophy*, xi.

11. David L. Hull, "What Philosophy of Biology Is Not," *Journal of the History of Biology* 2 (1969), 241–68.

12. Hull, "What Philosophy of Biology Is Not," 268.

13. H. E. Longino, *The Fate of Knowledge* (Princeton: Princeton University Press, 2002), 1–6, 388; Hull, *Science*, 1–6, 388.

14. Hull's 1964 PhD thesis was entitled "The logic of phylogenetic taxonomy" (published in *Systematic Zoology*, 1964). For Hull's papers referred to below, see bibliography in Ruse, *What the Philosophy of Biology Is*, 1989. Early taxonomy papers are: David L. Hull, "Consistency and monophyly," *Systematic Zoology* 13 (1964), 1–11; David L. Hull, "Certainty and circularity in evolutionary taxonomy," *Evolution* 21 (1967), 174–89; David L. Hull, "The ontological status of species as evolutionary units," in *Foundational Problems in Special Sciences*, R. Butts and J. Hintikka, editors (Dordrecht, Holland: D. Reidel, 1977), 91–102. Two 1965 papers were written as a student and published on the urging of Popper (Ruse, "David Hull"), whom Hull had met at Indiana. On Popper's evolutionary epistemology see G. Radnitzky and W. W. Bartley, *Evolutionary Epistemology* (La Salle: Open Court, 1987).

15. On Hull's editorial involvement, see Hull, *Science*, xi, 163–91, 199.

16. W. Bechtel and W. Callebaut, *Taking the Naturalistic Turn* (Chicago: University of Chicago Press, 1993), 206.

17. These were issues (Hull, *Science*, 213–23, 496–508) Hull discussed throughout his career (J. S. Wilkins, *Species: A History of the Idea* [Berkeley: University of California Press, 2009]). He pioneered the concept of "species as individuals" (compare Mayr's "population thinking" and Wynne-Edwards's "group selection" [126–27]). There are at least twenty-two definitions of "species": Hull strove to arbitrate between them. "We are drowning in a sea of species concepts. Hence, scientists are justified in being more monistic than they have in the past. Perhaps more than one species concept is justified, but twenty-two?" (D. Hull, in *Species*, R. A. Wilson, editor [Cambridge, MA: MIT Press, 1999], 44).

18. Leon Croizat also triggered Hull's interest (xi): a heretical figure within science, who took on establishment figures such as Simpson and Mayr (150–54, 167–84, 197–98), Hull in *Rebels, Mavericks, and Heretics in Biology*, O. Harman and M. R. Dietrich, editors (New Haven: Yale University Press, 2008).

19. C. Dupuis, "Willi Hennig's impact on taxonomic thought," *Annual Review of Ecology and System*atics 15 (1984), 1–12.

20. Hull in Wilson, *Species*.

21. Hull reviews critically virtually all the available positions within philosophy of science. See Hull, "Edinburgh program," in *Science* (4, 478–81, 513–14); "Mertonian norms" (310, 314, 371–72, 383–84, 392–93); Lakatos (298–300); logical empiricism (494); Popper (251–53, 342–44); see also Hull, *Biology and Philosophy* 14 (1999), 481–504; Kuhn

(12, 463–64, 492–96); Toulmin (447, 456); Kripke (496–500); Kitcher (500–505); the snares of terminology and the tradition of "linguistic philosophy" (6–9, 12, 294–98, 494, 509).

22. For example, Hull in Ruse, *What*.

23. D. Oldroyd, "David Hull's evolutionary model for the progress and process of science," *Biology and Philosophy* 5 (1990), 473–87.

24. D. L. Hull et al., "A general account of selection: Biology, immunology and behavior," *Behavioral and Brain Science* 24 (2001), 511–73.

25. Ruse, *Oxford*, 27.

26. D. L. Hull, "The professionalization of science studies: Cutting some slack," *Biology and Philosophy* 15 (2000), 61–91; D. L. Hull, "Recent philosophy of biology: A review," *Acta Biotheoretica* 50 (2002), 117–28.

27. The only evidence offered is his intervention at two meetings (Hull, *Science*, 126–27, 190–91, and consequential reports, 187–78). On reactions of scientists to his involvement: "irritation" (509); "at odds with scientists" (240); on their use of Popper's philosophy (252–53); frustration at his failure to guide them on terminology (142). Sokal "complained" (242). Mayr, Williams, and Brooks reacted to his views (410, 509). Some disputed his taxonomic conclusions (242). Some were mildly bemused by Hull's project; "It seemed to me that whatever comment I made, to the effect that 'No, David, that was not how it happened, but rather' . . . caused subsequent revision to move ever further from the historical truth. . . . Years later Colin (Patterson) remarked to me that David's book, even if wrong in particulars, was not too misleading in generalities. That might be true, but I still wonder how that is possible" (G. Nelson, "Colin Patterson [1933–1998]," *The Linnean*, special issue No 2. [London: Linnean Society of London, 2000], 9–23).

28. Hull was self-critical, especially about possible biases and assumptions (Hull, *Science*, 2–3, 6–11, 243, 508–11). "There seems to be something inherently self-contradictory about the externalist penchant for presenting data to show how irrelevant data actually are" (Hull, *Science*, 4).

29. See note 27.

30. D. L. Hull, "A mechanism and its metaphysics: An evolutionary account of the social and conceptual development of science," *Biology and Philosophy* 3 (1988), 123–55, 154.

31. Bechtel and Callebaud, *Taking*, 205.

32. D. L. Hull, *Science and Selection* (Cambridge: Cambridge University Press, 2001), 3.

33. Q. D. Wheeler, editor, *The New Taxonomy* (London: CRC Press, 2008), 2.

34. Wheeler, *The New Taxonomy*, 3.

35. D. L. Hull, "The operational imperative: Sense and nonsense in operationism," *Systematic Zoology* 17 (1968), 438–57; also in Wilson, *Species*. Hull often refers to "operationalism" (342, 485), a version of pragmatism introduced by physicists attempting to objectify a theory-free observational scientific method in terms of the overt procedures involved (paralleling the "verification principle" of logical positivism). "To make matters worse, many early geneticists were taken in by the notion of operational

definitions—the proposition propounded by P. W. Bridgeman in physics and J. B. Watson in psychology" (Hull, *Philosophy*, 17). "I am forced to discuss such traditional philosophical topics as the theory-ladenness of observational terms, operational definitions, and incommensurability. My views on these subjects are unusual ... I distinguish between the sort of piecemeal operationalizing that goes on when scientists attempt to test their ideas and the more global views of epistemologists. The former are extremely important in science; the latter are about as irrelevant as anything can be" (Hull, *Science*, 483).

36. Hull in Wilson, *Species*.

37. We continue to focus on the 1988 study because Hull appears not to have modified or updated his position subsequently.

38. P. E. Griffiths, "David Hull's natural philosophy of science," *Biology and Philosophy* 15 (2000), 301–10, 307.

39. J. R. Ravetz, *Scientific Knowledge and Its Social Problems* (Oxford: Clarendon Press, 1971).

40. J. M. Ziman, *Real Science: What Is It, and What It Means* (Cambridge: Cambridge University Press, 2000).

41. On the origins of "evolutionary epistemology" see Radnitzky and Bartley, *Evolutionary*; and Griffiths, "David Hull's natural philosophy of science." D. L. Hull, "Individuality and selection," *Annual Review of Ecology and Systematics* 11 (1980), 311–32, introduced a new term ("interactors": his alternative to Dawkins's "vehicles" [436]). Hull was wary of sociobiology (e.g., cooperation, competition, altruism, memes) despite the parallels in his own thinking (Hull, *Science*, 20, 223–31, 282, 304); see also D. L. Hull, in *Sociobiology and Human Nature*, M. S. Gregory, A. Silvers, and D. Sutch, editors (London: Jossey-Bass, 1978).

42. P. Kitcher, *The Advancement of Science* (Oxford: Oxford University Press, 1993). Hull encouraged the development of new mathematical modelling methodologies for testing his selection theory (Hull, *World Future* 34 [1992], 67–82).

43. P. Godfrey-Smith, *Darwinian Populations and Natural Selection* (Oxford: Oxford University Press, 2009).

44. M. Ruse, editor, *Philosophy of Biology* (Englewood Cliffs: Prentice Hall, 1989), 1–2.

45. P. Godfrey-Smith, *Theory and Reality* (Chicago: University of Chicago Press, 2003), 1.

46. Godfrey-Smith, *Theory and Reality*, 5.

47. Godfrey-Smith, *Theory and Reality*, 57.

48. Godfrey-Smith, *Theory and Reality*, 150.

49. Godfrey-Smith, *Theory and Reality*, 151.

50. A. Rosenberg and D. W. McShea, *Philosophy of Biology* (New York: Routledge, 2008), 3.

51. Hull in Ruse, *Oxford*, 30.

52. Hull, "The professionalization of science studies."

53. Hull, "The professionalization of science studies," 61.

54. Hull, "The professionalization of science studies," 64.

55. Hull, "The professionalization of science studies," 66.

56. Hull, "The professionalization of science studies," 68.

57. J. Passmore, "Why philosophy of science?," in *Science under Scrutiny*, R. W. Home, editor (Dordrecht: Reidel, 1983), 5–6.

58. Hull in Ruse, *What*, 310–11.

59. Hull, *Science and Selection*, 3.

60. Hull, "A mechanism and its metaphysics."

61. Despite the disciplinary divergences, it is not sufficiently realized that scientists have much to gain by being exposed to the alternatives perspectives of colleagues with different backgrounds. It seems unlikely that they can be taught "how to do science," "how to do it better," or "how great discoveries are made" by anyone: there is no hidden "logic of discovery," and so much of "scientific method" is unconscious "tacit knowledge"; that they cannot define it does not mean that scientists are not the best judges and ultimate repository of the method. However, social and strategic aspects of scientific practice are ever-present challenges in the scientist's working life and usually poorly understood. Moreover, historians are best qualified to reveal the historically determined (and possibly false) assumptions that, unsuspected by scientists, may often set limits on their thinking: historians can remind them of forgotten evidence and alternative theories abandoned on inadequate grounds; T. J. Horder, "The organizer concept and modern embryology: Anglo-American perspectives," *International Journal of Developmental Biology* 45 (2001), 97–132. More generally, see: D. L. Hull, "Central subjects and historical narratives," *History and Theory* 14 (1975), 253–74; D. L. Hull, "In defense of presentism," *History and Theory* 18 (1979), 1–15; Mary P. Winsor, "The practitioner of science: Everyone her own historian," *Journal of the History of Biology* 34 (2001), 229–45. No doubt, moreover, philosophers are important in posing significant questions regarding the morality, ultimate objectives, and limits of science, questions scientists only rarely ask.

62. The open question remains (Passmore, "Why philosophy of science?"): is the role of philosophy in respect of science to provide overview, integration (including with other disciplines), awareness of implications, provocation to think the unthinkable, a defining of ultimate ontological and epistemological limits, the revealing of hidden assumptions or unexplained gaps, or simply conceptual clarification?

V

INSIDER-OUTSIDERS

ALFRED I. TAUBER

ILYA METCHNIKOFF
FROM EVOLUTIONIST TO
IMMUNOLOGIST, AND BACK AGAIN

INTRODUCTION

At Berlin's Circus Renz on August 4, 1890, the Tenth International Congress of Medicine opened to an excited crowd of 7,000 participants.[1] Reports in the area of infectious diseases dominated the headlines. Microbiology had recently developed as a distinct science with the discoveries that various bacterial species caused diverse infectious diseases (gonorrhea, malaria, tuberculosis, cholera, diphtheria, tetanus, and so on), which in turn led to study of the host's response to pathogenic microbial invasion. That science quickly assumed the contours around the problem of host defense, and the fortuitous outcome of survival was designated as *immunity*.

Two theories of host defense divided the Congress. One side, led by Robert Koch (the consummate discoverer of pathogenic microbes), gave protective primacy to so-called humoral factors found in the blood of infected animals. An alternative explanation, offered by Elie (Ilya) Metchnikoff (1845–1916), championed the role of defensive cells in killing bacteria. So on one side of the aisle, humoralists sought to elucidate the chemical basis of the immune reaction by focusing on problems of specificity and the molecular mechanisms of antibody responses.[2] And on the other side, the phagocytic theory described immunity as an ongoing physiological process, in which phagocytes ("eating cells") surveyed the animal's internal environment, ridding the body of senile, damaged, or effete cellular debris and mounting an active host defense response when required, but only as a secondary aspect of their more mundane housekeeping functions.[3]

Controversies about theories are rarely only about the "facts." Metchnikoff, had he been at the meeting, would have felt very much "the outsider." The German microbiologists had rejected him when he sought employment after leaving Russia during the political unrest of the 1880s. Not only was he foreign, but his disciplinary background also apparently had no relevance to the problems of infectious diseases. Instead, in 1888, he found a research home at the newly formed Pasteur Institute, where he continued to promote a theory of immunity that directly opposed the dominant views of Koch and other leading German humoralists. Why Metchnikoff missed the meeting is unclear, but perhaps it is noteworthy, given the centrality of Pasteur and his newly formed institute in the nascent field, that no one from Paris attended the Congress except Waldemar Haffkine,[4] who functioned as a reporter (or "spy") for the Paris scientists.[5] So when Metchnikoff anxiously wrote his wife from Odessa that he could find very little about the congress in the newspapers, she in turn asked Haffkine to send him reports of the proceedings.

Metchnikoff's theory was indeed presented in Berlin, but not by him. Joseph Lister made it the major topic of his address, under the incongruous title "On Antiseptic Surgery" (the choice of the topic was last minute), and Koch, for his part, made a few nasty remarks about the theory. So the phagocyte theory had been discussed in Metchnikoff's absence: his allies wildly applauded him, the German hosts snorted in contempt.[6] The argument was far deeper than assigning putative roles to cells and defensive substances. Beyond the nationalistic rivalry of the time, which might account for some hostility, a deep intellectual and social moat separated Metchnikoff from his detractors. He had not only proclaimed the dominant scientific attitudes of the leading immunologists misconceived, he did so by promoting a theory that threatened the organizing principles of their science. Metchnikoff did little to hide his ambitions: he sought to broaden the horizon of immunology's research from a reductionist program designed to examine the molecular elements of the immune reaction with his own global approach that would encompass immunity in a grand theory of animal physiology.

Metchnikoff, a man of hot temper and celebrated disputes since earliest childhood (referred to as "quicksilver" by his parents and siblings),[7] drew fire from several quarters, ostensibly over disagreements regarding experimental observations. However, the controversy reached deep below the surface of the polemics fired by Metchnikoff's flamboyant style. Because he had presented his theory from the vantage of developmental biology and not from the newly emerging fields of microbiology and immunology, he was regarded as an intruder from a disciplinary point of view, which reflected a different perspective on host defense, one altogether outside the immediate concerns of his competitors.[8] Metchnikoff sought to place immunology within a grand synthesis, in which he saw the immune response as part of the larger Darwinian struggle of species. That orientation would be readily accepted as a context for host defense, but Metchnikoff made his case from extrapolations of *embryological* studies of invertebrates he conducted within an evolutionary construct. For microbiologists and the immunochemists who focused their efforts on defining the mediators of the immune response, Metchnikoff's speculations held little interest. And coupled to these major differences in scientific outlook, Metchnikoff worked as a descriptive biologist, employing experimental methods that were displaced by a new reductive biochemistry that had firmly established itself within the new scientific standards of clinical medicine. In sum, Metchnikoff challenged the prevailing paradigm, both methodologically and theoretically, from a point of view with seemingly little relevance to the immediate preoccupations of his competitors. So on that Berlin platform, the drama of an outsider pushing against powerful forces could not be more clearly enacted.

The pendulum swinging between humoralists and cellularists continued a back-and-forth debate into the first decade of the twentieth century. The argument hinged on which mechanism offered the first line of defense against pathogens. When Metchnikoff observed phagocytes devouring bacteria in experimental animals that survived infection, he concluded, appropriately, that these cells conferred protection, although he did not directly assess what other humoral factors might be involved, and he thus failed to appreciate how both arms of the immune response provided immunity. The humoralists suffered a corresponding myopia, inasmuch as they countered that phagocytes were only scavengers of dead bacteria that had been destroyed by factors in the blood. While the humoralists could not deliver a knockout punch, they clearly dominated the fight. Between the short period of 1888 and 1892, several groups had demonstrated the protective effects of serum factors in animals, and when Emil von Behring successfully treated diptheria with immunized serum, the humoral school of immunology seemed vindicated.[9] These successes drove the young discipline toward the prevailing scientific temper of the times, one characterized by precise quantification and definition of molecular species responsible for physiological effects.[10] And beyond scientific ideology, the approach advocated by the humoralists soon translated into dramatic therapies.[11]

Appropriately, the first Nobel Prize in Medicine or Physiology, awarded to Behring in 1901, celebrated both the promise of immune interventions, as well as the importance of the two scientific disciplines that grounded serum therapy—microbiology and immunology. Given the general excitement that basic discoveries in the laboratory could lead to therapeutic triumphs at the bedside, scientists working in these areas dominated the early Nobel awards: Ronald Ross (1902) for studies of malaria; Koch (1905) for elucidating the etiology of tuberculosis; and Metchnikoff shared the prize in 1908 with Paul Ehrlich, the most important immunochemist of the era,[12] "in recognition of their work on immunity." Note, this last prize acknowledged that both antitoxins (humors) and immune cells were critical components of the

Figure 14.1: Elie Metchnikoff, ca. 1910–1915.
Image from the George Grantham Bain Collection,
Library of Congress (LC-B2-2901-8).

immune reaction. Reviewing the records of the Nobel Committee, it appears that a gesture of conciliation had been offered, where complementary roles for phagocytes and serum factors were recognized.[13] Popular histories leave the Metchnikoff controversy at this point.[14]

Although an uneasy truce was called, throughout the next century the rift between the two traditions reappeared in different guises.[15] Therein lies our story, for Metchnikoff's outsider status would no longer be of significance if his own position had truly been submerged beneath the tide of molecular biology.[16] Two cardinal issues command our attention: the first concerns his establishment of phagocyte biology as a critical aspect of modern immunology. We will suspend further comment on this topic, but Metchnikoff's second cardinal insight points to a fundamental problem with which immunologists still struggle, and which represents a basic conceptual challenge: whereas the humoralists were content with assuming a given identity of the organism, Metchnikoff regarded what we now call "the immune self" as a problem of definition. For him *identity* became an achievement of active, ongoing physiological processes. His "outsider" critique remains unresolved, for defining the identity of the individual organism beyond some loose functional parameters has proven intractable. Simply, the dynamic relations Metchnikoff first described preclude any simple identification of such an *entity*. To this issue we now turn.

METCHNIKOFF, THE EVOLUTIONIST

In the clinical context, immunology began as the study of how a host animal reacts to pathogenic injury and defends itself against the deleterious effects from such microbial insult. This is the historical orientation usually assumed when recounting the history of immunology as a clinical science, a tool of medicine, and, as such, the story typically focuses almost exclusively on the role of immunity as a defender of the infected. And note, the paradigmatic host is the patient, an infected "self," which in turn embeds a major presupposition, namely that there is a definable self that might be defended. Thus, the definition of immunology almost invariably claims that it is the science of differentiating self from nonself. Indeed, the prominence of immunology in the life sciences testifies to the power of its position to define and to protect our very selfhood. And here Metchnikoff appears as the consummate outsider, for he not only denied this basic assumption, he proposed a radical alternative. He did so as a zoologist working in response to problems bequeathed by Darwinism.

As already mentioned, within the early immunology-microbiology community, Metchnikoff, alone, was an embryologist.[17] Intrigued by the potential of defining phylogenetic relationships through the study of the embryology of invertebrates, he believed that a deeper understanding of embryonic anatomic structures and functions might lead to insights about adult anatomy and physiology as traced to these more primitive animals.[18] How these interests eventually centered on immunity is a complex and intriguing story, one that not only illustrates the contributions arising from the periphery of a discipline, but also offers us a glimpse of the theoretical "infrastructure" of modern immunology.

Metchnikoff began his descriptive embryological studies shortly after the publication of Darwin's *On the Origin of Species* (1859). In late autobiographical accounts of his scientific career,[19] it is clear that he saw the development of the phagocytosis theory as a response to Darwin's thesis, and indeed it was. Putting aside the details of how Metchnikoff finally arrived at his mature understanding of evolution, the phagocytosis theory arose from a theoretical dispute with the German evolutionist Ernst Haeckel over the genesis of the hypothetical first complex multicellular organism.[20] Both Haeckel and Metchnikoff used recapitulation of ontogeny to understand phylogenetic development, so by studying the development of embryos, they hoped to reconstruct the evolution of animals.[21] However, argument arose regarding which data were considered pertinent.

Metchnikoff and Haeckel each began by modeling the earliest formation of the embryo. Beginning with a fertilized ovum, cellular division creates a sphere of cells. From that sphere differentiation first arises by the formation of two layers of cells—an ectoderm (outer) and an endoderm (inner). From these two cell layers, all the mature organs develop. Haeckel, extrapolating from *Amphioxus* (lancelet) development, suggested that multicellularity arose from an organism formed by an invagination (formation of a pocket) of a primordial gastrula to form a dual-layered embryo. This so-called *gastrea* was thus analogous to the invaginated gastrulas observed in primitive chordates, whose outer layer of cells moved into the empty spherical inner space as a second primary layer, from which digestive (endodermal) structures would develop. But Metchnikoff discovered another pattern of embryonic layer formation. Observing more primitive animals (sponges, hydroids, and lower medusa), he saw "introgression" as the primordial process. (Instead of an infolding of outer cells to form an inner stratum of cells, he saw individual exterior cells migrating into the interior to form a second layer.) From those findings, Metchnikoff argued that embryonic layers were created from an initially undifferentiated cellular mass that arose from cells migrating from the periphery in a less ordered fashion to fill the inner space of the gastrula sphere.

Beyond disputing the mechanism of forming the inner layer, Metchnikoff's original metazoan (which he called *parenchymella*) represented *intracellular* digestion as the common feature of unicellular organisms and most primitive metazoans. This modality requires "eating" cells to wander within the body of the animal and literally feed those cells that have specialized for other functions (e.g., exterior ciliated locomotive cells). In these simple animals, the phagocyte (the "eating cell") is the body's nutrient system. He criticized Haeckel's *gastrea* as placing extracellular digestion as a primary, original function and thus ignoring the more ancient genealogy of feeding seen in sponges and hydrea. Because Metchnikoff based his theory on animals arising earlier in phylogenetic history than those studied by Haeckel, the Russian could claim the evolutionary priority of introgression and intracellular digestion based on its appearance as a more ancient mechanism of gastrulation. Simply, Metchnikoff bruised Haeckel's claims by citing that the older ancestry revealed a more basic developmental process.

Parenchymella's central cavity cells were capable of digestion and in those organisms developing a gut, this parenchymatic mass further differentiated into two layers: endoderm and mesoderm. The first assumes the specialized digestive function, while the second accounts for the circulatory, respiratory, and locomotive functions. And now we come to Metchnikoff's phagocyte theory, in which he posited that the mesodermal phagocyte retains its original mobility and scavenging abilities for various functions other than nutritive. In a series of classic studies, he tracked the phagocyte's appearance and various functions from the simplest aquatic animals to mammals, and from those observations, he concluded that this cell continued to eat for various purposes in higher animals: repair of damage; removal of cellular debris; and most famously, defense against pathogens and repair of physical injury.[22] In short, the phagocyte became the body's arbiter of "self."[23]

METCHNIKOFF, THE IMMUNOLOGIST

The *gastrea/parenchymella* controversy may have been fought over modeling the first multicellular animal, but for Metchnikoff the issue introduced the beguiling problem of not only how competing cell lines were formed, but how they were integrated into a harmonious whole. He discovered that specialization of function bequeathed a set of problems unique to metazoan organization, namely, specialized cell lines develop and function in competition with one another.[24] He recognized with increasing clarity that evolution must be understood by selective processes that operate on the *interactions* of cell lineages to limit self-replication by any one component in favor of the interests of the organism as a whole.[25] (Malignancy is the example par excellence of uncontrolled growth.)

If metazoan evolution designates specialized cells to fulfill specific functions, what coordinates and integrates these competing cell lines? This question underlies the most fundamental challenge of linking evolutionary theory to developmental biology.[26] Metchnikoff's starting point, unlike Claude Bernard's notion of homeostasis of an idealized equilibrium, was *disharmony*.[27] On this view, cell lines competed for their self-aggrandizement. Thus Metchnikoff pushed the Darwinian struggle between species into the very constitution of the organism itself. In other words, competition in nature—whether in the environment or within the body—depicted the essential state of the organism. And, as in the struggle of species, only *active* processes might account for natural balance, or what Metchnikoff called *harmony*. Health was to be actively sought, and, indeed, harmony became an ideal synthesized from the potential disharmonious assembly of constituents. Metchnikoff dubbed the harmonizing process *physiological inflammation* to capture this integrative, restorative, curative undertaking found in all animals.

Physiological inflammation was a profoundly novel concept. Instead of ancient humors in preset balance, the organism was regarded as composed of cellular components in conflict. Health is thus not given; it is actively pursued. So, rather than marvel at cooperative development, Metchnikoff regarded metazoans as intrinsically "disharmonious" and, correspondingly, "harmony" became an *achievement*. Meanwhile, given the unstable state of these competitive cell lines, he sought the mechanisms by which they achieved a harmonious synthesis. This task he assigned to the phagocyte.

In Metchnikoff's scheme, the phagocyte became the mediator of physiological inflammation, an autonomous activity that ultimately defined host identity. He described the primordial mobile cells filling the gastrula interior as possessing a primitive volatility, essentially free of any commitments to anatomic place or function. In short, they possessed an ancient pluripotential independence that conferred upon them the ability to eat and then feed other cells by their dual capacity to ingest particulate nutrients and move at apparent will through the organism. They thereby became the agents of the collective through their ancient digestive role. And in animals with a gut, phagocytes continued to eat, but now with new regulative functions, namely, in the service of preserving host integrity by engulfing and killing bacteria, congregating around foreign bodies, and servicing wound repair.[28] The schema thus enlarged: in the embryo, Metchnikoff assigned the phagocyte the role of mediating development ("physiological inflammation") and when he followed them into the adult, these "eating cells" not only continued to mediate certain physiological functions (e.g., monitoring and disposing of senile cells), they also became the brokers of *pathological inflammation* as agents of repair and defense.

Within a decade Metchnikoff built an entire theory of immunity based upon the diverse function of these cells.[29] Whether harmonizing ontogenetic development or acting on behalf of host defense, phagocytes were engaged in essentially the same process—clearing the body of dysfunctional elements (endogenous "other") and unwanted external intruders. By retaining its original independence to effect the development of the organism, the phagocyte fulfilled the basic role of molding and then maintaining the structures and cellular relationships in the adult. (A favorite illustration was the metamorphosis of the tadpole, whose tail was "eaten" by phagocytes.[30]) When understood in this historical reconstruction, Metchnikoff's research on immunity is of one piece with his earlier embryological studies, and more to the point, the theoretical structure was the same: the phagocytes were the purveyors of organismal identity. Note, *identity* (self-identification/definition) preceded *integrity* (defense).

Metchnikoff's notion of active response would become a foundation of immune theory. This point is clearly illustrated by comparing his formulation with earlier theories of immunity. For instance, when Louis Pasteur and other early microbiologists studied infection, they believed that animals, which survived, did so because the infecting bacteria outgrew the available nutrients (analogous to the experimental conditions of the test tube). And in those animals that succumbed, their theory asserted that the pathogens were able to effectively mobilize nourishment at the expense of the host. In other words, the animal assumed a passive stance in the face of infection, where survival depended solely on the resources available to both the host and its pathogenic invader.[31] Once immunized animals were shown to possess protective humoral substances stimulated by infection, an active immune response was recognized.[32]

Metchnikoff's theory of immunity, from its inception, was conceived in terms of an active process. Regarding the organism as fundamentally existing in a disharmonious state, the phagocyte seemed to exhibit independent volition in its attempt to repair anatomic damage or establish physiological equilibrium. These views were extrapolated to host defense. In a celebrated experiment performed at Messina in 1882, Metchnikoff introduced rose thorns into transparent larvae and observed ameboid cells surround and eventually digest the foreign substance.[33] In that influential (and romanticized[34]) experiment, he saw the primordial phagocyte perform its ancient function, which in the case of invasion, exhibited its eating function to become a mode of defense: So to explain immunity, Metchnikoff argued that the phagocyte attacked the bacteria, and it did so, according to his theory's tenets, in its basic capacity of correcting disruption or damage, i.e., harmonizing a *disharmonious* state.

Metchnikoff concluded that this cell type asserted a jurisdictional prerogative within the body to carry forth its mission, and this teleological construction drew the ire of his detractors.

Humoralists assigned Metchnikoff's phagocyte theory to an older vitalism, where some mysterious teleology had again resurfaced under the guise of the phagocyte theory. They properly asked, What notion of identity directed the phagocyte? By what mechanisms could it determine selfhood? How, indeed, did it function in the physiological setting when not faced with threatening invaders? Note, the humoralists did not pose the deep dilemmas underlying the phagocyte theory to themselves, for they did not regard the identity problem Metchnikoff raised as directly germane to their own agenda. They confined themselves to elucidating the immunochemistry of the immune response. Simply, his theoretical concerns were outside their purview.

Soon thereafter genetics offered a partial answer to the identity problem, but prior to 1900, the critical contentions of the humoralists were well placed. Characteristically, Metchnikoff embraced the charges and claimed that free will and consciousness had an evolutionary history that might well begin with these observations! Such assertions hardly strengthened his cause, but this response suggests the enthusiasm with which he countered his opponents.[35] And here the great debate commenced.

While Metchnikoff's theory portrayed immunity as a special case of physiological inflammation, a normal process of animal economy, a more subtle message emerged: 1) immunity was an active process with the phagocyte's response seemingly mounted with a sense of independent arbitration, and 2) organismal identity was a *problem* bequeathed from a Darwinian perspective that placed all life in a dynamic evolutionary context. Metchnikoff's overall representation constituted the phagocyte as an *agent*,[36] an actor that is the cause of its own action—that is, self-generated and self-directed behaviors. The portrayal of the phagocyte as autonomous largely derives from the linked features of its capacity to sense its environment and move freely within it, and the various degrees of unpredictability and meaningfulness that characterize this behavior. On this view, the phagocyte, as an agent, becomes a metaphorical "self," a primordial microcosmic expression of what later immunologists would extend into an epistemology of biological identity.[37] However, while placing the identity function at the nexus of immunology's concern, Metchnikoff failed to provide the necessary preconditions for those who hoped to demonstrate the molecular reactions that conferred protection of such an *entity*, and thus he failed to directly engage his detractors. With that failure, not only was he marginalized, but his displacement reflected how the main currents of immunology of his era were moving away from his dynamic biology.

CONCLUSION

Immunology followed the track established by the humoralists for the next half-century, and only after World War II did the matters central to Metchnikoff's theory again emerge as central concerns.[38] From our vantage, we can appreciate that genetic definitions of self, while necessary, are insufficient to explain immune behavior. Consequently, supplementary approaches resonating with Metchnikoff's global perspective have gained attention.[39] So while some would assign Metchnikoff to the wine cellar of history, to be pulled out on occasion and celebrated as an old hero, this placement distorts his pivotal contributions and obscures how his insights still have relevance to immunology's current research priorities.

Today, now that the basic molecular mechanisms of the immune reaction have been elucidated, we are still left with the confounding questions of how the immune system is organized and regulated. This large-scale problem drives immunology's prevailing theory back to Metchnikoff's original concerns, which he firmly placed within a Darwinian construct: evolution had presented its students with a problem, What is identity? Metchnikoff's partial answer: developmental processes continue throughout the life of the organism, which constantly evolves as it faces environmental challenges. To be sure, each organism begins with a genetic blueprint, but beyond offering biological capacity, organismal identity is achieved in an ongoing process of dialectical encounters with the environment.[40] The process is *dialectical* because encounters induce change as the organism adjusts and accommodates itself to its habitat, and as a consequence, host integrity is constantly scrutinized. Finally replacing Aristotelian notions of essentialism, this dynamic vision of biology has bequeathed to us persistent dilemmas of defining individuality and placing such individuals within the complex ecology of a world inhabited with others.[41]

For detractors of the phagocyte theory, the more narrow spectrum of understanding immune defensive mechanisms dominated the murkier questions concerning organismic identity. But Metchnikoff recognized that *identity* must be prior to *integrity*, and thereby he offered a theory that drew upon the rich questions evolutionary science proposes. No longer a given, immunity, in its broadest understanding, becomes a crucial problem for biology.

By recognizing the perplexing significance of how the character of individuality is maintained under conditions of constant change, he firmly placed within the conceptual foundations of immunology a scientific enigma still unresolved and whose profound implications are again being assessed.[42] Much of the subsequent history of immunology may be traced to the attempts at establishing a definition and experimental basis that fulfills such an identity function. By the mid-twentieth century, the mechanisms of autoimmunity challenged earlier notions of immune selfhood, and with the complexity of immune tolerance beginning to unfold, the erosion of the simple self–non-self dichotomy raised deep theoretical issues. What is the self, and even, Is there an immune self?[43] These questions, pointing at the very heart of the discipline, remain unresolved.

Perhaps only an outsider, one aloof from the immediate interests of the dominant scientific community, could ask such unorthodox questions. Certainly, Metchnikoff dramatically fulfills the criteria, with a firm footing in another discipline; a deep understanding of a broad scientific problem; a staunch personality that withstood the hostile reactions of competitors; and last, but most strikingly, the creative vision to perceive connections between one set of problems and another. He saw where others did not that the assumption of organismal identity was exactly that, an *assumption*. From the perspective of evolutionary biology, fixed identity does not exist. Evolution presents the dynamics not only of the species, but the constant challenge to individual organisms of adapting and responding to environmental challenges. This *ethos* of change dominated Metchnikoff's own understanding of biology, and this perspective, adopted at a time when Darwinism was still in transition and not fully appreciated, placed him outside the community of early immunologists, who framed their research very differently. Their "boundary conditions" were set with the intent on defining mechanisms of the immune response; Metchnikoff wanted to understand the biology of evolution and development. The immune reaction occupied only a portion of his panoramic vision. So in the end, we might well ask: *who*, in fact, was outside, and, more, saliently, what constitutes the core *inside*?

FURTHER READING

Mazumdar, Pauline M. H. *Species and Specificity: An Interpretation of the History of Immunology*. Cambridge: Cambridge University Press, 1995.

Metchnikoff, Elie. *Lectures on the Comparative Pathology of Inflammation*, translated by F. A. Starling and E. H. Starling. London: Kegan, Paula, Trench, Trubner, 1893, reprinted by New York: Dover, 1968.

Moulin, Ann-Marie and Alberto Cambrosio, editors. *Singular Selves: Historical Issues and Contemporary Debates in Immunology*. Amsterdam: Elsevier, 2001.

Silverstein, Arthur M. *A History of Immunology,* second edition. Amsterdam: Elsevier, 2009.

Tauber, Alfred I. *The Immune Self: Theory or Metaphor?* New York: Cambridge University Press, 1994.

Tauber, Alfred I. and Leon Chernyak. *Metchnikoff and the Origins of Immunology: From Metaphor to Theory*. New York: Oxford University Press, 1991.

NOTES

1. Pauline M. H. Mazumdar, "Immunity in 1890," *Journal of the History of Medicine and Allied Sciences* 27 (1972), 312–24.

2. Pauline M. H. Mazumdar, *Species and Specificity: An Interpretation of the History of Immunology* (Cambridge: Cambridge University Press, 1995); Arthur M. Silverstein, *A History of Immunology,* second edition (Amsterdam: Elsevier, 2009).

3. Elie Metchnikoff, *Lectures on the Comparative Pathology of Inflammation*, translated by F. A. Starling and E. H. Starling (London: Kegan, Paula, Trench, Trubner, 1893, reprinted by New York: Dover, 1968); Elie Metchnikoff, *Immunity in Infective Disease*, translated by F. G. Binnie (Cambridge: Cambridge University Press, 1905); Alfred I. Tauber and Leon Chernyak, *Metchnikoff and the Origins of Immunology: From Metaphor to Theory* (New York: Oxford University Press, 1991); Alfred I. Tauber, "The Birth of Immunology: III. The Fate of the Phagocytosis Theory," *Cellular Immunology* 139 (1992), 505–30; Alfred I. Tauber, "Metchnikoff and the Phagocytosis Theory," *Nature Reviews: Molecular Cell Biology* 4 (2003), 897–901.

4. Haffkine (1860–1930), another Russian émigré, later developed successful vaccines for cholera, which he tested in India in 1893, and an anti-plague vaccine in 1896.

5. Luba Vikhanski, personal communication.

6. The nationalistic lines were not so clearly drawn, for the studies grounding Behring's own work were performed at the Pasteur Institute by Emil Roux and Alexandre Yersin, who had demonstrated in 1888 that filtrates of diphtheria cultures (thus free of bacteria) contained a toxic substance (called *toxin*) that reproduced the pathology of diphtheria infection caused by the intact bacteria. Parallel studies performed by the American, George Nuttal, showed an active immune response and finally put to rest the passive immune model (Silverstein, *A History of Immunology*). And not surprisingly, later competing humoralists ignored the lines established by national borders (Eileen Crist and Alfred I. Tauber, "Debating Humoral Immunity and Epistemology: The Rivalry of the Immunochemists Jules Bordet and Paul Ehrlich," *Journal of the History of Biology* 30 [1997], 321–56).

7. Olga Metchnikoff, *The Life of Elie Metchnikoff 1845–1916*, translated by E. R. Lankester (London: Constable and Boston: Houghton Mifflin, 1921).

8. Tauber and Chernyak, *Metchnikoff and the Origins of Immunology*; Helena D. Gourko, Donald I. Williamson, and Alfred I. Tauber, editors, *The Evolutionary Biology Papers of Elie Metchnikoff* (Dordrecht: Kluwer Academic Publishers, 2000).

9. Behring decisively shifted the balance of the argument in favor of the humoralists with three seminal reports: 1) animals immunized with tetanus (with Shibasaburo Kitasato) or diptheria toxin generated a humoral factor that neutralized bacterial exotoxin (1890); 2) protection was afforded by passive transfer of immune serum (with Erich Wernicke, 1892); and 3) such immune serum cured diptheria infection (1891). Indeed, the first successful treatment of this childhood scourge made Behring a celebrity (Derek S. Linton, *Emil von Behring: Infectious Disease, Immunology, Serum Therapy* [Philadelphia: American Philosophical Society, 2005]).

10. Mazumdar, *Species and Specificity*.

11. The cellular school did not achieve such spectacular results, and the difference was telling. Almoth Wright advocated measurement of immunity by a so-called "phagocyte index", and he championed therapies to augment the phagocyte's role in host defense, but those efforts were soon eclipsed by further developments in serum therapy and had little lasting influence (Tauber, "The Birth of Immunology;" Silverstein, *A History of Immunology*).

12. Ehrlich was the first to accurately measure the potency of immune serum and thus offer a quantitative dimension to immunology, and in addition he presented the first theory of antibody reactions and their physiological origin (Silverstein, *A History of Immunology*), which proved to be highly prescient as confirmed by studies in the mid-twentieth century (Scott H. Podolsky and Alfred I. Tauber, *The Generation of Diversity: Clonal Selection Theory and the Rise of Molecular Immunology* [Cambridge: Harvard University Press, 1997]). He also pioneered chemotherapy, developing early antimicrobial and anti-cancer drugs.

13. Tauber, "The Birth of Immunology;" Alfred I. Tauber, *The Immune Self: Theory or Metaphor?* (New York: Cambridge University Press, 1994).

14. For example, Paul de Kruif, *The Microbe Hunters* (San Diego: Harcourt, Brace, and Jovanovich, 1954).

15. Tauber, *The Immune Self*; Alfred I. Tauber, "The Elusive Self: A Case of Category Errors," *Perspectives in Biology and Medicine* 42 (1999), 459–74.

16. Alfred I. Tauber, "The Molecularization of Immunology," in Sahotra Sarkar, editor, *The Philosophy and History of Molecular Biology: New Perspectives* (Dordrecht: Kluwer Academic Publishers, 1996), 125–69.

17. Tauber and Chernyak, *Metchnikoff and the Origins of Immunology*.

18. Gourko, Williamson, and Tauber, *The Evolutionary Biology Papers of Elie Metchnikoff*.

19. Metchnikoff, *The Life of Elie Metchnikoff 1845–1916*.

20. Tauber and Chernyak, *Metchnikoff and the Origins of Immunology*; Gourko, Williamson, and Tauber, *The Evolutionary Biology Papers of Elie Metchnikoff*; Tauber, "Metchnikoff and the Phagocytosis Theory."

21. Stephen J. Gould, *Ontogeny and Phylogeny* (Cambridge: Harvard University Press, 1977).

22. Metchnikoff, *Lectures on the Comparative Pathology of Inflammation*.

23. The term self and its congeners were introduced by Macfarlane Burnet in 1949 and became the central motif of immunology only in the 1970s (Tauber, *The Immune Self*).

24. Leo Buss, *The Evolution of Individuality* (Princeton: Princeton University Press, 1987).

25. A single-celled protist must simultaneously express specialized modes of locomotion and feeding, and yet retain the capacity for cell division (Buss, *The Evolution of Individuality*). Metazoans have no such constraint, having taken the strategy of differentiating germ cells from somatic functions. At issue is the simultaneous need of the organism to move through fluid with cilia or flagella and to divide using a mitotic spindle. Unless a cell possesses multiple microtubule organizing centers capable of performing both tasks, functional range will be constrained. In certain protist groups, cell division and locomotion can occur simultaneously; in others they cannot. While many protist taxa overcome the ciliation constraint to division, those protists giving rise to metazoans did not. Metazoans inherited the constraint limiting simultaneous mitosis and ciliation. Leaving ciliated cells on the surface for locomotion and feeding, the movement and subsequent proliferation of germ cells from the surface into the center of the blastula sphere is gastrulation—a metazoan solution to the requirement of simultaneous movement and reproduction.

26. Scott F. Gilbert and David Epel, *Ecological Developmental Biology. Integrating Epigenetics, Medicine, and Evolution* (Sunderland, MA: Sinauer Associates, 2009).

27. Tauber and Chernyak, *Metchnikoff and the Origins of Immunology*; Gourko, Williamson, and Tauber, *The Evolutionary Biology Papers of Elie Metchnikoff*; Tauber, "Metchnikoff and the Phagocytosis Theory."

28. This orientation extended Bernard's earlier conception of the organism as an autonomous entity. By radically changing the inside-outside topology so that the organism's interior becomes the determining context of function, Bernard effectively isolated the organism from its environment, and immunology became one of its defining sciences. Bernard furnished biology with a new concept of the organism, one that would have wider ramifications than the establishment of physiology and biochemistry. Obviously, interchange with the environment was a necessary requirement for life, but Bernard emphasized how boundaries provided the crucial metabolic limits required for normal physiological function. With his concept of the *milieu interieur*, the body was envisioned as a demarcated, interdependent yet autonomous entity characterized by "corporeal atomism" (Ed Cohen, "Figuring Immunity: Towards the Genealogy of a Metaphor," in *Singular Selves: Historical Issues and Contemporary Debates in Immunology*, Ann-Marie Moulin and Alberto Cambrosio, editors [Amsterdam: Elsevier, 2001], 179–201, 190). This orientation established the theoretical grounding for the development of later life sciences, i.e., infectious diseases, genetics, neurosciences, and immunology. However, as important as Bernard's concept proved to be, his construction also obfuscated certain aspects of biology's complexity. Most importantly, the ecological consciousness that emerged in the twentieth century set up a conceptual struggle to promote a contextualist

approach to complex biological environments populated by multiple species, against a biology dominated by the centrality of the autonomous organism, which in turn has had an impact on contemporary immunology as well (Alfred I. Tauber, "The Immune System and Its Ecology," *Philosophy of Science*, 75 [2008], 224–45).

29. Metchnikoff, *Lectures on the Comparative Pathology of Inflammation*.

30. Metchnikoff, *Lectures on the Comparative Pathology of Inflammation*.

31. Tauber and Chernyak, *Metchnikoff and the Origins of Immunology*.

32. Silverstein, *A History of Immunology*.

33. Metchnikoff, *The Life of Elie Metchnikoff*.

34. Tauber and Chernyak, *Metchnikoff and the Origins of Immunology*, 107–11.

35. Metchnikoff, *Lectures on the Comparative Pathology of Inflammation*, 192–93.

36. Crist and Tauber, "Debating Humoral Immunity and Epistemology."

37. Tauber, *The Immune Self*; Alfred I. Tauber, "Moving Beyond the Immune Self?" *Seminars in Immunology* 12 (2000), 241–48; Alfred I. Tauber, "The Biological Notion of Self and Nonself," *Stanford Encyclopedia of Science*, 2009, http://plato.stanford.edu /entries/biology-self/.

38. Podolsky and Tauber, *The Generation of Diversity*.

39. Tauber, "The Biological Notion of Self and Nonself."

40. Richard Levins and Richard C. Lewontin, *The Dialectical Biologist* (Cambridge: Harvard University Press, 1985).

41. Tauber, "The Immune System."

42. Tauber, "Moving Beyond the Immune Self?"; Tauber, "The Biological Notion of Self and Nonself."

43. Tauber, *The Immune Self*, 1999; Tauber, "Moving Beyond the Immune Self?"

FRANÇOIS JACOB
TINKERING WITH ORGANISMS AND MODELS

INTRODUCTION

It might seem absurd to call François Jacob an "outsider."[1] He received a Nobel Prize in Physiology or Medicine at the age of forty-five (with André Lwoff and Jacques Monod), for a discovery still considered to be one of the most elegant[2] of the classical era of molecular biology: with Monod, he elaborated the operon model, the first molecular model explaining how the activity (expression) of genes can be controlled. It opened the way to the study of higher organisms and their development. It was a result as important as the discovery of the double helix structure of DNA for the expansion of molecular biology.

From three different points of view, however, Jacob was an outsider. He initiated his work on bacteria and bacteriophages at the age of thirty, with a biological background limited to what a future medical student acquired during his training and his unsuccessful activity in antibiotic production after the war.[3] The reason for this delay is that he left France in June 1940 to join the French Free Army, organized by General de Gaulle in London, and fought in Africa for four years. He was severely wounded in Normandy in August 1944 and spent a full year in the hospital. War experience was important for many scientists, but for Jacob it was totally disconnected from his future research activity.

When, at the end of the 1960s, Jacob turned to the study of mouse development, he was not the only molecular biologist abandoning bacteria and bacteriophages to study more complex organisms and their development. But he made the biggest leap, entering a field where embryologists and geneticists had already accumulated a huge amount of data and elaborated plenty of models. Finding a place in this complex field was not obvious. The success of the extension of the molecular vision to higher organisms was not immediate.

But Jacob is also an outsider because of the attention he paid to the place of molecular biology in the larger history of biology, and to the philosophical consequences of the recent transformations in biology. His work *The*

Figure 15.1: François Jacob.

Logic of Life,[4] lauded by the philosopher Michel Foucault, was a deeply original presentation of the history of biology. *The Possible and the Actual*,[5] which stemmed from a series of conferences, was probably Jacob's most original contribution to the philosophy of science and simultaneously to the theory of evolution.

Most of this article will be devoted to the transition that Jacob made from the study of bacteria and bacteriophages to work on mice and their embryogenesis. It corresponds exactly to the middle of his academic career: twenty years devoted to microorganisms, and twenty years to mammals. I will examine the reasons for this transition, but also the difficulties that were encountered and the long-term success of the approach opened up by Jacob.

To fully understand this shift from one organism to another, it is necessary to reconsider Jacob's belated and difficult entry into biological research, the characteristics of the then-young field of molecular biology, and the nature of the research pursued at the Pasteur Institute as well as in a limited number of places around the world. But it is also necessary to take into account Jacob's third "shift"—his sudden interest in the history and later the philosophy of science that was initiated simultaneously with his move to the study of higher organisms. The vision of science expressed by Jacob in three successive books[6] explains his scientific redeployment, as well as the difficulties he experienced. There is a deep unity in Jacob's conception of science behind the apparent discontinuities in his academic career, a conception progressively elaborated through his earlier war and research experience.

NOT PREDESTINED FOR SCIENTIFIC RESEARCH

Under the influence of an admired uncle who was a doctor, Jacob's initial plan was to become a surgeon. His mother's illness—she was diagnosed with cancer after he started his medical studies—further motivated him in medicine. It was not the possibility of using medical knowledge to explain physiological and pathological observations that attracted Jacob, but the chance it offered to fight actively against disease and death. This explains why, at the end of the war, when his wounds denied him entry to the world of surgery, he never considered the possibility of being a practitioner, and completed his medical studies only to have the diploma. Jacob has described the nearly miraculous path that led him to the laboratory of André Lwoff, who opened its door.[7] Jacob admits that he himself would never have welcomed to his own laboratory a student such as he was at that time! And Lwoff confirmed that, while he did not regret his decision, twenty years later he still found it difficult to understand why he'd taken him.[8] Lwoff worked on lysogeny, the complex relation between certain bacteriophages, viruses of bacteria, and their bacterial hosts; temperate bacteriophages may remain silent within bacteria, or abruptly start to reproduce within them, which leads to the lysis of bacteria. In 1949, Lwoff had found a way to induce bacteriophages to multiply within bacteria, which opened the way to an easier experimental study of lysogeny, and put Lwoff in an excellent mood: two complementary reasons to say "yes" to Jacob!

Jacob's entry into microbiology and molecular biology was perhaps smoother and more foreseeable than the "legendary" version of events might suggest. The first step was the work Jacob did at the end of the war on antibiotic production in the recently created French centre for penicillin. Although not scientifically original and lacking any serious prospect of production, this work allowed him to discover the world of research and the pleasure of designing experiments and of progressively overcoming obstacles. It also offered an opportunity to explore the field of microbiology and to become acquainted with its leading figures, as well as participate in his first international scientific congress.

Jacob progressively realized that something was emerging at the boundary between microbiology, genetics and physics, what would later be called "molecular biology." He found out which few French laboratories were engaged in this revolution. And the successful career of his cousin Herbert Marcovich showed Jacob that his lack of familiarity with biochemistry and all aspects of microbiology was not a barrier to him making a contribution to research in this field. Jacob has emphasized in his autobiography how few people were working in the new field. This allowed him to know everything that was happening by reading just a few articles and by meeting and discussing with their authors.

This account does, though, mask two difficulties that Jacob had to face and overcome. The first was the highly abstract nature of the genetic observations made on microorganisms at that time. Since its origin, genetics had been an "abstract" discipline, but the situation was worse in bacterial genetics, where the not-yet-explained complexity of the mechanisms of genetic exchange delivered plenty of puzzling results. The second difficulty was the transdisciplinary nature of the new field. Researchers working in the field had to be able to grasp observations made by crystallographers, and more generally the results obtained by the use of physical techniques, the statistical methods necessary to interpret them, and the chemical and biological knowledge required to design experiments and analyze their results.

There is something exceptional in the work of Jacob in the fifteen years that separated his entry into André Lwoff's laboratory and his Nobel Prize: his capacity to replace abstract representations by mechanistic models, and to do so alone when researchers accustomed to a particular representation were far less likely to abandon it. During his first four years at the Pasteur Institute, Jacob replaced the mysterious and complex relation between a lysogenic bacterium and its virus by a series of precise results and well-formulated questions on the number of silent phages present in the lysogenic bacterium, their position, their relations with other phages, and so on. The same is true for the work done later with Elie Wollman on conjugation: this complex genetic exchange was replaced by a simple scheme in which a donor bacterium progressively introduces into a recipient one its circular DNA molecule opened at one precise position. In the case of the operon model, Jacob was able with Monod to transform the vague notion of an inhibitor controlling gene activity into a precise scheme in which the product of a regulatory gene directly interacts with DNA to block the transcription of the downstream genes.[9]

FROM BACTERIA TO THE MOUSE EMBRYO

Jacob was not unique among the founders of molecular biology in abandoning bacteria and bacteriophages at the beginning of the 1960s to move toward the study of more complex organisms. Such a general movement was the direct consequence of the construction of the operon model.[10] Ever since the 1930s, development and differentiation had been considered by geneticists as a problem of gene regulation. Although discovered in microorganisms obviously lacking any form of development, the operon model might easily be adapted to explain gene regulation in higher organisms. This is what Monod and Jacob did as early as 1961, the same year they published the operon model, at the conclusion of the famous Cold Spring Harbor meeting.[11] Jacob chose to study the development of the mouse, a mammal, whereas Gunther Stent adopted the leech; Sydney Brenner the nematode, a small worm; and Seymour Benzer, the fruit fly *Drosophila*.

In various texts written at the time of his decision or later,[12] Jacob explained the reasons for his choice. Many years were necessary to favor one of two opposing strategies: to choose a simple system, easily accessible with the tools developed in the study of bacteria, with the risk that mechanisms specific to the development of complex organisms would not be discovered in this way, or to turn immediately to the study of complex organisms such as mammals, with the risk of being unable to overcome the experimental difficulties associated with their study, the tools of genetic engineering having only been developed in the mid-1970s.

For Jacob, three criteria were essential in his choice: the need to select a system in which genetic and biochemical approaches could be combined—this blend had been the recipe which allowed the construction of the operon model; the selected system had to be of value for the Pasteur Institute, at which Jacob intended to pursue his scientific career; and the system had to be "pleasant" to manipulate. Jacob did not appreciate nematodes. But he was interested by the embryonal carcinoma (EC) cell lines studied by Boris Ephrussi at Gif-sur-Yvette, to the south of Paris. These murine cell lines, derived from spontaneous tumors of the gonads called teratocarcinomas, generated tumors when injected into a recipient animal. Depending upon the culture conditions, they could remain undifferentiated in cell cultures, or differentiate into different tissues. I remember, as a PhD student working in a laboratory close to Jacob's, his fascination while he was observing cells differentiated in vitro into heart muscle cells "beating" regularly on the culture plates. In addition to being pleasant to observe, these cell lines provided biochemists with an amount of material that could not be obtained by directly working on the embryo. These in vitro systems provided the opportunity to isolate and characterize the proteins involved in the process of differentiation. In addition, EC cells could be reintroduced into an early embryo, a blastocyst, lose their oncogenic potential, and participate in the formation of a new organism. This demonstrated that EC cells had much in common with early embryonic cells and gave much physiological significance to the results that might be obtained by using such an experimental system.

Two additional reasons probably played a part in Jacob's decision. The first was the use of a model that had not already been selected and developed before by another molecular biologist; that is, a direct competitor sharing the same techniques and models. The second was the connection between differentiation and development on the one hand, and cancer exhibited by teratocarcinoma on the other. Throughout his scientific career, Jacob paid attention to cancer and its mechanisms. He had already tried to cast some light on it through the observations he made on lysogeny and with the operon model: the rapid formation of a tumor had similarities with the sudden induction of a silent prophage; cancer might also be seen as a dysregulation of the mechanisms controlling cell division.

Jacob's final choice is also explicable in terms of his own conviction that evolved mechanisms operate during the development of higher organisms. He did not share Monod's opinion that the negative model of regulation corresponding to the operon was the only possible mechanism of regulation. In fact, he proposed a positive regulation in the model that he elaborated with Sydney Brenner and François Cuzin in 1963 to explain the control of cell division, the replicon model.[13] This does not mean that Jacob considered that the mechanisms operating in higher organisms were of a completely different nature, but rather that they were an extension and recombination of mechanisms operating in simpler organisms.

This movement in a new direction would not have been possible without the help of Jacob's collaborators: not only those who allowed him to discover the new field, the way to cultivate cells and manipulate the embryos, but also those who, like Hubert Condamine and Charles Babinet, had taken part in the work on bacteria and had spent months or even years in embryology laboratories learning the techniques and concepts required to study early mammalian development before introducing them into Jacob's laboratory.

Was Jacob's decision the right one? The answer is clearly yes in the long term. The embryological development of the mouse has been extensively studied in recent years, thanks in particular to the knockout technology developed at the end of the 1980s that allowed researchers to inactivate at will any gene or to replace it by a modified copy. This technique was based on the capacity of embryonic stem (ES) cells to integrate into the early embryo and to participate in its development, a property shared by the EC cells Jacob studied. Mice have become the model favored by biologists in their studies of the mechanisms of human diseases and how to combat them.[14] The work done on EC cells, the efforts made to orient their differentiation in one or another direction, anticipated the numerous efforts made today in regenerative medicine using ES cells.

But the work was not immediately successful, and the obstacles took a long time to overcome.[15] The strategy Jacob adopted was to concentrate his efforts on the first developmental steps and on the proteins present in the cell membranes. There was good experimental evidence that in mammals the relative position of the cells within the embryo and the contacts they establish with neighboring cells determine the fate of these cells. In addition, immunological techniques had recently been devised to characterize the proteins present in cell membranes and even to check for their functions. For Jacob and other molecular biologists, the main players in the control of cell differentiation were regulatory proteins controlling gene expression. But these proteins were supposed to be present in small amounts and therefore inaccessible by traditional biochemical methods, in contrast to the proteins present in cell membranes. Characterizing membrane proteins was clearly seen as a first step toward the characterization of the regulatory genes and the proteins that controlled their expression.

The experimental approach Jacob preferred in 1970 was reasonable, but it was rapidly rendered obsolete by two events. The first was the development of genetic engineering techniques in the mid-1970s that accelerated, but also redirected, experimental efforts. It became easier to go from genes to proteins, rather than laboriously purifying proteins in order to have access to the genes. The second was the discovery that the genes controlling development, the "master" developmental genes, had been conserved during evolution. To isolate them, it was no longer necessary to purify the proteins they encode. The simplest way was to directly isolate the encoding genes by taking advantage of their resemblance to the developmental genes present in well-studied organisms such as *Drosophila*.

In his first years of work, Jacob's work on the early development of mice was misdirected by a hypothesis that proved inaccurate and was the cause of numerous disillusions.[16] A protein named the F9 antigen was rapidly described by Jacob's lab as playing a major role in early mammalian development. It was involved in a process of "compaction" of the embryo that immediately preceded the first differentiation event. It was not restricted to the mouse embryo, but conserved in all mammalian embryos. Through the use of sophisticated immunological techniques, it was demonstrated by Karen Artzt in Jacob's laboratory that this protein was encoded by the *T*-complex of the mouse, a gene complex in which many mutations affecting mouse early development had been localized. An appealing model was elaborated by Artzt and Dorothea Bennett: the *T*-complex was structurally related to the major histocompatibility complex and controlled the early development of mammals through the successive synthesis of proteins present in the cell membrane. Jacob did not participate directly in the development of this model, but he received it quite favorably. Unfortunately, the immunological observations on which the model was based were wrong as a result of a careless interpretation of insufficiently controlled data. It was later discovered that the *T*-complex, which had been studied for fifty years by the most eminent geneticists, did not exist. It was an illusion, resulting from a lack of genetic recombination in the chromosome on which the so-called complex was located. This episode shows how difficult it is for an outsider to check the experiments and the hypotheses in a new field with which he is not familiar.

While the work on EC cells and their in vitro differentiation can retrospectively be seen as an anticipation of the work on human ES cells and regenerative medicine, the relation between them is far from direct. Even if it was obvious for Jacob and his collaborators that EC cells were equivalent to early embryonic cells, it was not his lab that directly tested this hypothesis by putting early embryos into culture and showing that the cells obtained in this way, the ES cells, were similar to EC cells. In addition, it would be historically wrong to consider that the characterization of ES cells in mice naturally opened the way to the production and use of ES cells in humans. Mouse ES cells were seen as a tool to explore mammalian development, a view supported by their use in knockout experiments. Human ES cells were obtained with the idea of using them to regenerate tissues. Nearly twenty years separated the production of human ES cells from that of mouse ES cells. Such a long interval explains why a certain form of "culture"—technical tricks and tacit knowledge obtained from the study of mouse EC and ES cells—was not transmitted to the laboratories working on human ES cells.[17] Whence some recent "surprises" for people working on human ES cells, such as the discovery that ES cells can be cancerous!

Teratocarcinoma was not a good model for cancer in the 1970s. A new vision of cancer emerged in those years in which the formation of a tumor is the result of the accumulation of mutations in a group of genes called the oncogenes. Observations made with EC cells, in which a cancerous cell could revert to a normal phenotype when surrounded by "normal" cells, did not fit the new theory and were swept under the carpet. The observation made by one of Jacob's collaborators that teratocarcinoma cells and the foetus produce the same factor to inhibit the immune response of the host against, respectively, the tumor and the embryo, was not reproduced.

Jacob was well accepted by the specialists of mammalian development: he was an invited speaker in many of the scientific meetings organized during these years. The biggest challenge for him was not intellectual: to understand the early development of mammals was no more difficult than to unravel the complex system of lysogeny, and the nature of the molecular and mechanistic explanations he looked for was similar. The strength of molecular explanations was that they could be used for most—if not all—biological phenomena. It was to Jacob's credit to be one of the first to import into embryology the explanatory power of molecular descriptions. What was probably the most difficult for him personally was the abrupt change in the rhythm of experiments. Whereas in microbiology the results of an experiment were obtained within a few hours, in the new domain in which Jacob had entered experiments had to be planned months in advance, and the same length of time was often necessary to obtain results. The nature of the game to which Jacob often compared scientific research had dramatically changed.

Being an outsider clearly helped Jacob to see the potential of the mouse and the importance of cellular systems in studying differentiation and development. But it prevented him from being sufficiently cautious in the interpretation of the first observations he made. The transition from bacteria to mice was also for Jacob a transition between a direct involvement in the realization of experiments, and a new, more distant role in organizing the experiments done by his collaborators. The price to pay for long-term success was a painful pathway littered with errors and disillusions.

THE PLACE OF HISTORY AND PHILOSOPHY IN JACOB'S WORK

It is traditional for scientists to consider the philosophical and ethical consequences of the discoveries in which they participated. Jacob's final book, *Of Flies, Mice, and Men*, belongs to this tradition, as does Monod's *Chance and Necessity*, written in 1970.[18] But Jacob's first nonscientific book, *The Logic of Life*, a history of biology, is different and original. The only possible comparison would be with *The Growth of Biological Thought: Diversity, Evolution and Inheritance*, published by Ernst Mayr in 1982.[19] But the comparison goes no further. When Jacob wrote his book, he was, unlike Ernst Mayr, on the verge of exploring a new field of biology and fully engaged in research. Whereas Ernst Mayr's intention was to demonstrate the importance of the "modern synthesis" (i.e., his own contribution to biology), *The Logic of Life* is not the mere acknowledgement of recent results in molecular biology. It is a complete historical description of the development of the life sciences since the Renaissance. Deeply influenced by Michel Foucault, Jacob minimizes the role of the scientific actors in the production of biological knowledge. The "game of the possible" that scientists play is limited by the available techniques and the "episteme" of their era. It is not William Harvey who is responsible for the discovery that the heart is a pump, but rather the time in which Harvey lived that made the vision of the heart as a pump possible.[20]

Many biologists are quoted in *The Logic of Life*. But they are not mentioned (or only indirectly) for the discoveries they made, but rather because their writings are useful in grasping the epistemic context in which these discoveries were possible. Nevertheless, the historical vision of Jacob, at odds with the traditional "Whiggish" history written by scientists, does not exclude the possibility of progress of scientific knowledge, in what Stéphane Schmitt has rightly called a "Hegelian" view.[21] The four successive levels of observation that have prevailed in life sciences since their beginnings—from organism to molecules through organs and genes—is not the replacement of old paradigms by new incommensurable ones. It has more to do with the progressive deployment of the intimate organization of the living world.

Jacob's second book, *The Possible and the Actual*, is highly different. It belongs to what is today called the philosophy of biology. Jacob proposes a new interpretation of Darwinian theory, one that gives tinkering (i.e., recombination of preexisting elements), the major role in evolution. The notion of tinkering had already been used by Darwin in his treatise on orchids.

But Jacob gave it a wider extension, showing how this tinkering process is particularly obvious at the macromolecular level.

Whereas this new vision of evolution has become part and parcel of present evolutionary thought, *The Logic of Life* remains a solitary gem in the history of science: a new way to write history that thus far has not been followed by others.

Jacob is also the author of an autobiography, *The Statue Within*. Many famous scientists write their autobiography. But Jacob's has two unusual characteristics. The first is that science occupies only a limited place, one third of the book; the rest of the work covers his youth and his war experience, which for Jacob was the most significant part of his life. The second is his peculiar style, the quality and originality of which allowed him to be acknowledged as a writer, and to be inducted into the prestigious French Academy (of literature). This is unusual. In the French tradition a deep separation exists between the studies of humanities and those of scientific matters. To be a writer is not the attribute of scientists, who are often accused of using an impoverished language. Buffon was an exception, and it is probably not an accident therefore that he is the most cited author in *The Logic of Life*.[22] In literature too, Jacob was an outsider!

CONCLUSION:
ONE OR MANY OUTSIDERS IN THE SAME PERSON?

Was Jacob an outsider in many different ways, or can we find similar characteristics in the different disciplinary switches he made during his scientific career?

Jacob has repeatedly argued that scientific knowledge is not important per se, but as a process. Well-established knowledge is cold and boring; it has no flavor. Only the construction of science is of value. Such a sharp differentiation between what Jacob has called the "night" science, the obscure phase during which science is elaborated, and the "day" science, science as it is taught, obviously favors movements both within and between disciplines.

Such a vision of science represents, in addition, a depreciation of scientific knowledge, a partial negation of the cumulative process in science. It is clearly linked with the historical vision of science presented in *The Logic of Life*.[23] The models and theories elaborated by scientists are determined by the methodologies they have at hand, and even more by the time at which they work. Categories of thought are limited. Often, new theories and models are nothing more than the recombination of elements present in previous models.

Jacob has been criticized for having overemphasized the metaphor of the genetic program. He introduced the latter with Monod in the conclusion

of the *Journal of Molecular Biology*[24] article in which they described the operon model and distinguished regulatory instructions from structural ones.
Ernst Mayr simultaneously introduced the notion of program.[25] In 1970, in
The Logic of Life,[26] Jacob explicitly compared the genetic program with the
program of a computer, and asserted the existence in the genome of a genetic program of embryological development, but also of ageing. Many philosophers have underlined the weaknesses of this comparison: within an
organism, one can neither distinguish the program and the machine (i.e.,
the software and the hardware), nor the data from the instructions of the
program. These criticisms are well founded. But in the sixties and seventies
the comparison between computers and organisms nevertheless helped to
assimilate the new molecular vision of life. For Jacob, using metaphors is
also a part of the tinkering game. They are useful for testing new combinations. They should be abandoned only when their heuristic value vanishes.

Many mechanisms specific to eukaryotic cells (i.e., cells with a nucleus)
and multicellular organisms have been discovered during the last four decades (gene splicing, editing, epigenetic modifications of chromatin, the
huge regulatory role devolved to non-coding RNAs, etc.). These mechanisms
could not have been anticipated from the studies done on microorganisms.
It does not mean that the hypothesis that mechanisms operating in eukary-
otes and multicellular organisms are similar to those operating in bacteria
was not historically useful, and it paved the way for the discovery of the new
mechanisms.

Resorting to conceptual tools (models, metaphors, kinds of explanation)
by scientists is no different from the tinkering action of evolution, as shown
in *The Possible and the Actual*.[27] Recombination of preexisting elements is a
process shared by the world and by human knowledge, and also represents
a limit to this knowledge.[28] Although the development of modern science,
with the necessary focusing on questions of limited amplitude, has generated
different disciplines and subdisciplines insulated one from the other, scientific activity remains the same whatever the questions asked. It consists in
creating novelty by recombining previously existing conceptual material.[29]
Creating myths or new scientific knowledge is the same activity, the only
difference being that scientific theories and models must withstand their
confrontation with reality.

Often, outsiders are the best positioned to introduce new combinations. Not only can they import these new combinations from other fields of
research, other disciplines, but they are also less conscious of the obstacles,
both mental and social, that limit this activity of recombination in a particular discipline. They are more prone to "transgressions," because they are

less familiar with the "rules of the game" dominant within the disciplines they have recently entered.

In a sense, every creator in art, literature, or science is an outsider, trying to find beyond his or her own field of competence recipes to recombine pre-existing parts of models in a different way. But in some sense also, none is an outsider, since the tinkering process permeates both all human activities and the natural world.

FURTHER READING

Jacob, François. *The Statue Within: An Autobiography*. New York: Basic Books, 1988.

Judson, Horace F. *The Eighth Day of Creation: Makers of the Revolution in Biology.*
   Cold Spring Harbor: Cold Spring Harbor Laboratory Press, 1996.

Morange, Michel. *A History of Molecular Biology*. Cambridge: Harvard University
   Press, 1998.

NOTES

1. Horace F. Judson, *The Eighth Day of Creation: Makers of the Revolution in Biology* (Cold Spring Harbor: Cold Spring Harbor Laboratory Press, 1996), 343–432; Michel Morange, *A History of Molecular Biology* (Cambridge: Harvard University Press, 1998), 150–163.

2. Gunther S. Stent, "That Was the Molecular Biology That Was," *Science* 160 (1968), 390–395.

3. François Jacob, *The Statue Within: An Autobiography* (New York: Basic Books, 1988).

4. François Jacob, *The Logic of Life* (Princeton: Princeton University Press, 1973).

5. François Jacob, *The Possible and the Actual* (Seattle: University of Washington Press, 1982).

6. François Jacob, *The Logic of Life*, *The Possible and the Actual*, *Of Flies, Mice, and Men* (Cambridge: Harvard University Press, 1998).

7. François Jacob, *The Statue Within*.

8. André Lwoff, "The Prophage and I," in *Phage and the Origins of Molecular Biology*, John Cairns, Gunther S. Stent and James D. Watson, editors (Cold Spring Harbor: Cold Spring Harbor Laboratory Press, 1992), 88–99.

9. Thomas D. Brock, *The Emergence of Bacterial Genetics* (Cold Spring Harbor: Cold Spring Harbor Laboratory Press, 1990).

10. Michel Morange, "The Operon Model and Its Legacy," *Journal of Biosciences* 30 (2005), 313–316.

11. Jacques Monod and François Jacob, "General Conclusions: Teleonomic Mechanisms in Cellular Metabolism, Growth and Differentiation," *Cold Spring Harbor Symposium on Quantitative Biology* 26 (1961), 389–401.

12. François Jacob, "Biologie moléculaire: la prochaine étape," *Atomes* 271 (1969), 748–750; François Jacob, *Of Flies, Mice, and Men*, 47–64.

13. François Jacob, Sydney Brenner and François Cuzin, "On the Regulation of DNA Replication in Bacteria," *Cold Spring Harbor Symposium on Quantitative Biology* 28 (1963), 329–348.

14. David Malakoff, "The Rise of the Mouse, Biomedicine's Model Mammal," *Science* 288 (2000), 248–253.

15. Michel Morange, "Introduction," in *Travaux scientifiques de François Jacob*, Nadine Peyrieras and Michel Morange, editors (Paris: Editions Odile Jacob, 2002), 7–66.

16. Michel Morange, "François Jacob's Lab in the Seventies: The *T*-complex and the Mouse Developmental Genetic Program," *History and Philosophy of the Life Sciences* 22 (2000), 397–411.

17. Michel Morange, "Twenty-Five Years Ago: The Production of Mouse Embryonic Stem Cells," *Journal of Biosciences* 31 (2006), 537–541.

18. Jacques Monod, *Chance and Necessity* (London: Collins, 1972).

19. Ernst Mayr, *The Growth of Biological Thought: Diversity, Evolution and Inheritance* (Cambridge: Belknap Press of Harvard University Press, 1982).

20. François Jacob, *The Logic of Life*, 34–35.

21. Stéphane Schmitt, "François Jacob: une nouvelle vision de l'histoire des sciences," in *Une nouvelle description du vivant: François Jacob, André Lwoff et Jacques Monod*, Claude Debru, Michel Morange and Frédéric Worms, editors (Paris: Editions Rue d'Ulm, 2012), 45–53.

22. François Jacob, *The Logic of Life*.

23. François Jacob, *The Logic of Life*.

24. François Jacob and Jacques Monod, "Genetic Regulatory Mechanisms in the Synthesis of Proteins," *Journal of Molecular Biology* 3 (1961), 318–356.

25. Ernst Mayr, "Cause and Effect in Biology," *Science* 134 (1961), 1501–1506.

26. François Jacob, *The Logic of Life*, 274.

27. François Jacob, *The Possible and the Actual*.

28. Michel Morange, "Introduction."

29. François Jacob, *Of Flies, Mice, and Men*, 125–145.

VI OUTSIDERS FROM INFORMATICS

EHUD LAMM

16

THEORETICIANS AS PROFESSIONAL OUTSIDERS
THE MODELING STRATEGIES OF JOHN VON NEUMANN AND NORBERT WIENER

INTRODUCTION

Somewhat ironically for a discipline known for its austerity, the folklore of mathematics has more than its fair share of anecdotes and myths about heroes, mavericks, and eccentrics. Typically, one is introduced to these characters in the course of becoming a mathematician, via anecdotes and tall tales that are, so to speak, passed from father to son. Two twentieth-century mathematicians that are the subject of often-repeated anecdotes are John von Neumann (figure 16.1) and Norbert Wiener (figure 16.2). Both are well known as significant mathematicians and both worked at influential centers of learning (von Neumann eventually residing at the Institute for Advanced Study at Princeton, Wiener at MIT). However, the two are typically portrayed in very different terms. Anecdotes portray von Neumann as a "mathematician's mathematician"—the one who is able to outsmart other mathematicians. Wiener is typically portrayed as the absent-minded professor. The role von Neumann played in the history of computing is well known, as are his contributions to systems biology. Wiener's contribution is often downplayed, and the cybernetic research program he is best known for is portrayed as being ultimately a failure.

Both von Neumann and Wiener were outsiders to biology. Both were inspired by biology, and both proposed models and generalizations that proved inspirational for biologists. Around the same time in the 1940s that von Neumann developed the notion of *self reproducing automata*, Wiener suggested an explication of teleology using the notion of *negative feedback*. These efforts were similar in spirit. Both von Neumann and Wiener used mathematical ideas to attack foundational issues in biology, and the concepts they articulated had lasting effect. But there were significant differences as well. Von Neumann presented a how-possibly model, which sparked interest from mathematicians and computer scientists, while Wiener collaborated more directly with biologists, and his proposal influenced the philosophy of biology. The two cases illustrate different strategies by which mathematicians, the "professional outsiders" of science, can choose to guide

Figure 16.1: John von Neumann.
Courtesy of Los Alamos National Laboratory Archives.

their engagement with biological questions and with the biological community, and illustrate different kinds of generalizations that mathematization can contribute to biology. The different strategies employed by von Neumann and Wiener and the types of models they constructed may have affected the fate of von Neumann's and Wiener's ideas—as well as the repu-

Figure 16.2: Norbert Weiner.
Photograph by SPL/ Photo Researchers, Inc.

tation, in biology, of von Neumann and Wiener themselves. The history of their reputation as forefathers of systems biology is an interesting example of how new disciplines construct their past.

Our two distinguished suitors, overbearing and brash as mathematicians are wont to be when discussing mathematical ideas, were pursuing in this case a rather reluctant, bashful, bride-to-be. Nine years before Wiener and his co-authors John Bigelow and Arturu Rosenblueth published their paper about teleology, E. B. Wilson articulated the reserved attitude of biologists toward uninvited theoreticians.[1] Wilson's remarks at the Cold Spring Harbor Symposia on Quantitative Biology in 1934 were ostensibly about the "mathematics of growth," but it is impossible to fail to notice their tone and true scope. Wilson suggested orienting the discussion around five axioms, or "platitudes" as he called them. The first two are probably enough to get his point across. Axiom 1 states that "science need not be mathematical," and if that's not bad enough, axiom 2 solidifies the reserved attitude toward mathe-matization by stating that "simply because a subject is mathematical it need not therefore be scientific." Our two protagonists, renowned and accom-plished mathematicians however they clearly were, had a lot of courting to do. Still, Wilson seemed to leave an opening for the two prospective suitors. Despite his otherwise disparaging remarks, he concluded by noting, "One must not fail to mention, as contrasted with empirical curve plotting analy-ses, the attempts at fundamental rational analysis." Mathematics, it turns out, is not all of a piece. Fundamental rational analysis was precisely what Wiener and von Neumann purported to do, but entering the world of biology, as we shall see, each suitor would adopt his unique approach to courtship.

BEHAVIOR, PURPOSE, AND TELEOLOGY

Wiener (1894–1964) and von Neumann (1903–1957) are probably the most well-known American mathematicians of the mid-twentieth century. In the mid-1940s to mid-1950s, they were driving forces behind the Macy Conferences, one of the most celebrated multidisciplinary series in recent scientific history. The first meeting was held in 1946. The annual meetings, entitled "Confer-ence on Circular, Causal and Feedback Mechanisms in Biological and Social Systems," were by-invitation-only events and were chaired by the neurophys-iologist Warren McCulloch. Among the participants were the psychiatrist and cybernetician William Ross Ashby, the anthropologists Gregory Bate-son and Margaret Mead, the sociologist Paul Lazarsfeld, and the ecologist G. E. Hutchinson. Partly as a result of their shared interest in computing machines, both Wiener and von Neumann pursued related questions about the organization and functioning of the brain and the analysis of behavior and social behavior. Their perspective was that of the then cutting-edge sciences of computing automata and information theory. Wiener's work on target-tracking machines for the Air Force led him to think about feedback mechanisms, specifically negative feedback. This became a central organiz-ing notion in his conception of *cybernetics*. Experience with the behavior of actual target-tracking mechanisms led to a conjecture about intentional-ity and purpose-driven behavior, which Wiener then tried to generalize by arguing that negative feedback is the defining characteristic of purpose-ful behavior. Von Neumann, in turn, grew increasingly frustrated with at-tempts to understand the brain. Trying to understand the brain using the techniques of neurology was like trying to understand the ENIAC computer "with no instrument . . . smaller than about 2 feet across its critical organs, with no methods of intervention more delicate than playing with a fire hose."[2] He argued that this hopeless task be replaced by the attempt to arrive at a complete and full understanding of less-than-cellular organisms, namely viruses and bacteriophages. Their fundamental property is that they self-reproduce, and von Neumann devoted a lot of energy to a formal analysis of the question of self-reproduction. Von Neumann and Weiner worked in the same milieu, had similar interests, and even corresponded, and yet they arrived at two very different questions—the nature of purpose, and the necessary conditions for self-reproduction—and would approach the two questions in remarkably different ways.

Both Wiener and von Neumann were early starters, and began their intellectual journey being home-schooled. Wiener started his academic studies at the tender age of eleven and referred to himself in later life as an ex-prodigy. He studied philosophy, the field in which he obtained his PhD, and biology, where he preferred theorizing to anatomical work, before becoming a mathematician. Around the time Wilson expressed the skeptical view about the role of mathematics in biology, Wiener began attending an interdisciplinary seminar group at Harvard Medical School and developed an interest in physiology. What better background for the kind of work we are discussing? Von Neumann was and remained a true outsider—a mathematician, first and last, who contributed to many scientific fields, from quantum physics to economics.

Wiener and his junior colleague, the electrical engineer Julian Bigelow, developed their ideas about negative feedback and purpose while working on the problem of predicting the location of enemy aircraft during WWII.[3] While working on this problem, they noticed that systems governed by negative feedback may fall prey to ever more powerful oscillations, finally losing track of the target. They wondered if similar phenomena are found in human pathology, since this would suggest that it too is governed by negative feedback. They approached Wiener's longtime friend Arturo Rosenblueth, a physiologist then at Walter Cannon's lab at Harvard, who told them that exactly this phenomenon is found in patients suffering from intention tremors. These patients exhibit oscillatory behavior with ever-wider oscillations around the target they aim for. The chain connecting intentionality and feedback was being closed. The idea emerged from the interaction of Wiener (a mathematician by self-determination and institutional affiliation), Bigelow (an engineer), and Rosenblueth (a physiologist). This was an interdisciplinary group through and through.

The programmatic paper that resulted from this work, "Behavior, Purpose and Teleology," authored by Rosenblueth, Wiener, and Bigelow (referred to henceforth as RWB), was published in January 1943 in *Philosophy of Science*. Several things about RWB's article are worth noting. The authors stress that their interest lies in the "behavioristic study of natural events," which is concerned with a black-box analysis of the behavior of systems. This they contrast with functional analysis that is concerned with the internal organization of systems. The tension between these two approaches is endemic in biology in general, and was particularly painful in the context of studying animal learning and behavior in the heyday of behaviorism. RWB used the first paragraphs to make sure their commitments were known to the reader, and were unwavering about the idea that behaviorist analysis is applicable to machines and to living organisms alike, though organisms and machines may be radically different when it comes to functional analysis. The paper then delves into a series of distinctions that are summarized in the single, and not visually stimulating, figure in the paper (figure 16.3).

According to RWB, purposeful behavior is behavior aimed at fulfilling a particular goal, such as picking up a glass of water from the table. Attaining the goal, or failing irrevocably, may be an immediate result of the action taken by the organism, as happens when a frog strikes at a fly. Alternatively, the behavior of the system may be continuously guided by input from the environment, leading the system to correct its behavior, a mechanism referred to as *negative feedback*. Negative feedback was used by the target-tracking systems Wiener and his colleagues studied as part of the war effort. There they observed that undamped negative feedback quickly leads

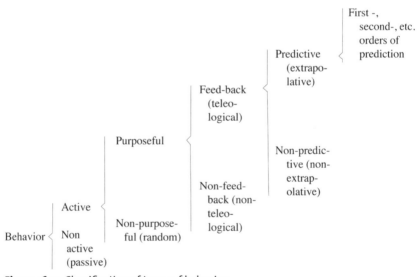

Figure 16.3: Classification of types of behavior.
From *Behavior, Purpose and Teleology* by Rosenblueth, Wiener, and Bigelow (1943).

to oscillatory behavior, which results from overcorrection. This observation, which is immediately apparent to anyone who tries to build a system that relies on negative feedback, led Wiener and his colleagues to raise a startling suggestion:

> This picture of the consequences of undamped feed-back is strikingly similar to that seen during the performance of a voluntary act by a cerebellar patient. At rest the subject exhibits no obvious motor disturbance. If he is asked to carry a glass of water from a table to his mouth, however, the hand carrying the glass will execute a series of oscillatory motions of increasing amplitude. . . . The analogy with the behavior of a machine with undamped feed-back is so vivid that we venture to suggest that *the main function of the cerebellum is the control of the feed-back nervous mechanisms involved in purposeful motor activity*.[4] [italics added]

In a sense, Wiener and Bigelow used the target-tracking system and its formal analysis as a model, albeit one that was found serendipitously, and appealed to it in asking concrete questions about the human nervous system. The model in this case is not a representation of the target system, the human brain, but rather an example system that exhibits properties that are of interest. If similar behavior were found in patients, the model could then provide a tentative hypothesis about mechanisms that can bring it about. The model provides a *how-possibly* account of the behavior in question.

How does the promise for black-box modeling sit with the article's focus on feedback? Wiener and his colleagues defined negative feedback as referring to behavior that is controlled by the margin of error of the system relative to some specific goal. The term feedback is also commonly used to refer to the way components of a system interact with one another, thereby creating "feedback loops." RWB were not interested in this kind of functional analysis. Their black-box model of intention tremors referred to properties of behavior, not directly to neuronal mechanisms.[5]

The discussion of intention tremors is scientifically interesting, but the bulk of the article is devoted to establishing the set of distinctions that appear in figure 16.3. The authors acknowledged that this is merely one way to classify behaviors. Their main justification for their particular conceptual scheme was that it highlights the importance of the notions of purpose and teleology, which they defined as "purpose controlled by feed-back":

> Teleology has been interpreted in the past to imply purpose and the vague concept of a "final cause" has been often added. This concept of final causes has led to the opposition of teleology to determinism. . . .

purposefulness, as defined here, is quite independent of causality, initial or final. Teleology has been discredited chiefly because it was defined to imply a cause subsequent in time to a given effect. When this aspect of teleology was dismissed, however, the associated recognition of the importance of purpose was also unfortunately discarded. Since we consider purposefulness a concept necessary for the understanding of certain modes of behavior we suggest that a teleological study is useful if it avoids problems of causality and concerns itself merely with an investigation of purpose. . . . causality implies a one-way, relatively irreversible functional relationship, whereas teleology is concerned with behavior, not with functional relationships.[6]

SELF-REPRODUCING AUTOMATA

In 1947, a year after the first Macy Conference, Wiener published in the *Atlantic Monthly* a letter written in December 1946 in which he advocated against cooperating scientifically with the military. After Hiroshima and Nagasaki, Wiener wrote, the scientist knows that if he works with the military he will end up putting unlimited powers in the hands of those whom "he is least inclined to trust." At the same time Wiener was taking this stand, von Neumann was getting more and more involved with the workings of the recently formed Atomic Energy Commission (AEC). Already deeply involved in strategic thinking in the Navy and Air Force, and sitting on numerous governmental committees, von Neumann was eventually appointed a member of the AEC in 1955. Throughout this time, von Neumann's and Wiener's scientific interests continued to overlap.[7]

At around the time Wiener published his thoughts on teleology, John von Neumann became actively interested in the brain in the wake of work by Warren McCulloch and Walter Pitts; his groundbreaking book on social behavior, *The Theory of Games and Economic Behavior*, coauthored with Oskar Morgenstern, was published in 1944.[8] After several years thinking about the problem of self-reproduction, von Neumann discussed his thoughts on the subject in September 1948 at the Hixon Symposium on Cerebral Mechanisms and Behavior.[9] Von Neumann began his talk by asking his audience for forbearance, emphasizing that he was an outsider to the fields to which the conference was dedicated. His goal was to give the audience of psychologists and biologists a picture of the mathematical approach to their problems, and to "prepare you for the experiences that you will encounter when you come into closer contact with mathematicians."

Living organisms are more complicated and subtle than automata, von Neumann argued, but each can provide lessons applicable to the other. He distinguished between the study of the elementary units from which organisms are composed, and the study of how the organization of these components leads to the functioning of the whole. Those with the background of the mathematician or logician, von Neumann explained to his audience, will be attracted to questions of the second kind. Like RWB, who distinguished between behaviorist and functional analysis, von Neumann was interested in high-level behavior, namely self-reproduction. In contrast to them, however, he was concerned with functional organization. Instead of black-boxing the system as a whole, his approach was to black-box the components by axiomatizing their behavior. Essentially, his goal was to consider the functional organization of systems composed of idealized components. Starting with the work of McCulloch and Pitts to which von Neumann referred, this type of idealization has been typical in the study of artificial neural networks by computer scientists. Analysis of the kind von Neumann proposed can support generalizations that are not otherwise easy to make, as McCulloch's work demonstrated. It is not, however, as he himself noted, a very effective way to determine if the idealization provides a good representation of reality, of the sort presumably sought by biologists. It is also not obvious that when models of this sort exhibit behavior that is similar to that of the modeled system they in fact help explain it. This may depend on whether the model provides necessary or sufficient conditions, and on the extent of idealization involved in defining the components. It may also depend on whether the behavior of the model is simple enough for us to understand. If the model is capable of self-organization and learning, abilities that were later introduced to artificial neural network models, the problem is exacerbated. In the discussion of the applicability of von Neumann's model to the real world following the talk, McCulloch observed that while his own results proved that neural networks can compute any computable number, in Turing's sense, they did not explain how the nervous system achieved any particular result. Other participants of the Macy conferences had similar reservations.

While the idea of self-reproduction seems incredible, and some might even have thought it to involve a self-contradiction, with objects creating something as complex as they are themselves, von Neumann's solution to the problem of self-reproduction was remarkably simple. It is based on two operations: 1) constructing an object according to a list of instructions, and 2) copying a list of instructions as is:

> The general constructive automaton A produces only X when a complete description of X is furnished it, and on any reasonable view of what

constitutes complexity, this description of X is as complex as X itself. The general copying automaton B produces two copies of $\varphi(X)$ [the instructions which represents X], but the juxtaposition of two copies of the same thing is in no sense of higher order than the thing itself. . . . Now we can do the following thing. We can add a certain amount of control equipment C to the automaton $A + B$. The automaton C dominates both A and B, actuating them alternately according to the following pattern. The control C will first cause B to make two copies of $\varphi(X)$. The control C will next cause A to construct X at the price of destroying one copy of $\varphi(X)$. Finally, the control C will tie X and the remaining copy of $\varphi(X)$ together and cut them loose from the complex $(A + B + C)$. At the end the entity $X + \varphi(X)$ has been produced. Now choose the aggregate $(A + B + C)$ for X. The automaton $(A + B + C) + \varphi(A + B + C)$ will produce $(A + B + C) + \varphi(A + B + C)$. Hence auto-reproduction has taken place.[10]

This procedure is trivial for anyone who is computer-literate to understand; yet it was a remarkable theoretical result in 1948. What, however, does it tell us about biology? It is often observed that von Neumann's explanation, which involves treating the genetic material both as instructions and as data that is copied as-is, is analogous to the reproduction of cells, since DNA, the analogue of the instruction list, is passively replicated. Von Neumann compared the construction instructions that direct the automaton to genes, noting that genes probably do not constitute instructions fully specifying the construction of the objects their presence stimulates. He warned that genes are probably only general pointers or cues that affect development, a warning that, alas, did not curtail the "genetic program" metaphor that became dominant in years to come.

Von Neumann further noted that his model explained how mutations that do not affect self-replication are possible. If the instruction list specifies not only the self-replicating automaton but also an additional structure, this structure will also be replicated. "Mutations" in the additional structure will be copied indefinitely, since they do not affect self-replication. This could be thought of as an explanation of nonlethal mutations. Back in 1922, the geneticist H. J. Muller observed that genetic material retains the ability to reproduce even after an unlimited number of mutations have occurred. He considered this special property to be a crucial difficulty for theories that ground the origin of life in autocatalysis. Muller initially considered the possibility that genetic replication involves the help of external machinery (in the protoplasm) that acts as a general purpose copier or "mimeograph," akin to automaton B in von Neumann's model. Eventually he came to dismiss this solution as far as the origin of life was concerned, because an early division of labor did not make evolutionary sense.[11] Von Neumann's highly abstract existence proof does not help answer Muller's evolutionary conundrum.

Unsatisfied with a purely formal proof, von Neumann developed a series of models that tried to put flesh on the abstract notion of construction. He eventually came up with five models, the most famous of which is the cellular automaton model.[12] In this model, construction activities are modeled explicitly, yet the model abstracts away unessential properties of motion in space, energetic considerations, and so on. Cellular automata are comprised of a homogenous grid of cells, each of which is in one of a finite number of states. Time proceeds in discrete steps; the state of a cell at any given step is a function of the states of its immediate neighbors in the previous time step. Von Neumann sketched a cellular automaton consisting of cells with twenty-nine states in which self-reproducing ensembles of cells could be embedded (see figure 16.4).

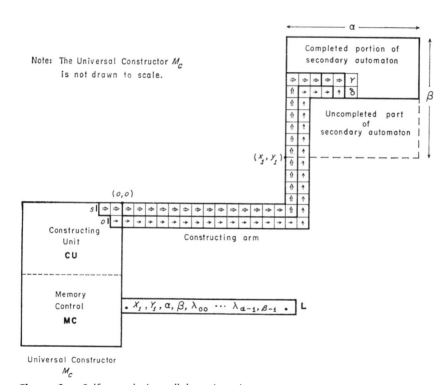

Figure 16.4: Self-reproducing cellular automaton.
From *Theory of Self-Reproducing Automata*, A. Burks, ed. (1966).

Although more concrete than the formal proof quoted above, the cellular automaton is no truer to biological detail. However, it is this model that is most closely associated with von Neumann's work on self-reproducing automata, and figure 16.4 has become iconic. Von Neumann's work on self-reproducing automata is often given as an example of the essence of artificial life research.[13] Since von Neumann's work, cellular automata have become a standard modeling approach, used heavily in theoretical biology and physics.[14] In addition to these uses, there has been continual research on formal models of self-replication.[15] This theoretical work is, however, largely divorced from the empirical study of self-reproduction by mainstream biologists.

■

In 1955, the soon-to-be Nobel laureate geneticist Joshua Lederberg exchanged several letters with von Neumann. Lederberg began this remarkable correspondence by asking von Neumann what his work indicated concerning "the minimal information required for 'self-reproduction.'"[16] Lederberg was concerned with the notion of self-reproduction as it applied to intracellular particles such as genes, noting that their reproduction depended on an appropriate surrounding cell. He was thus enthusiastic about von Neumann's black-boxing of the components of the system, allowing him to focus on the functional organization of the system, only the whole of which is self-reproducing. In this way, the issue with self-reproducing genes is seemingly avoided, and the mathematical model could provide insight.[17] But Lederberg was searching for a model that would help identify the minimal biological structures that underlie reproduction. He hoped for criteria indicating how intracellular components correspond to the elements of von Neumann's model, but noted that he would be surprised if von Neumann's conceptual model was intended as a structural representation of the biological system.[18] Like Muller, Lederberg was concerned with the evolution of self-reproducing systems from simple autocatalytic processes, and envisaged *chemical models* of self-reproducing systems.

On the notion of information, the original topic raised by Lederberg, the illustrious mathematician and the illustrious geneticist had difficulty finding common ground even after exchanging long and detailed letters. Von Neumann emphasized the independence of a self-reproducing organism embedded in the cellular automaton grid from the definition of the cellular automaton itself. The former is simply an arbitrary collection of cells in specific states, while the latter is essentially the definition of the function determining the transition between states. Von Neumann stressed that the information content of the organism is not contained in the definition of the transition function. Lederberg, in turn, could not regard the cellular automaton definition, independent of any particular self-reproducing organism embedded in it, or the universal constructor, as mere material resources that do not contain information.[19] Noting that they were talking at cross purposes, Lederberg highlighted the two issues that concerned him most: how can autocatalytic molecules be combined so that they can store an arbitrary amount of information, and how does organismal complexity come about? Both questions were not answered by von Neumann's model.

Inspired by von Neumann's formal proof, the British geneticist Lionel Penrose built a series of mechanical models of self-reproduction, which he published in 1958 in, of all places, the *Annals of Human Genetics*.[20] Penrose designed wooden tiles that could hook together in one of two configurations. Shaking a series of unhooked tiles arranged on a horizontal track did not cause the tiles to hook up—unless one hooked-up pair, which he called a "seed," was introduced to the chain, in which case the shaking caused other tiles to hook up in pairs having the same configuration as the seed (see figure 16.5). This model

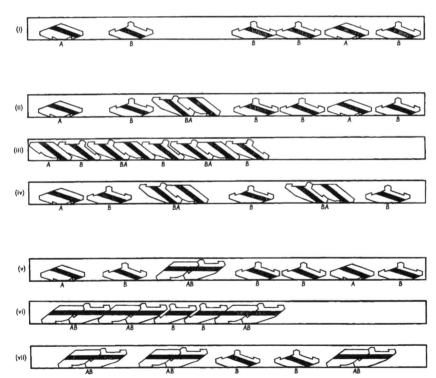

Figure 16.5: Self-replicating chain with units of two kinds.
From "Mechanics of Self-Reproduction," L. S. Penrose (1958).

showed that reproduction could be achieved by very simple mechanisms—if the notion of reproduction is indeed an appropriate description for what happens in the model.

Penrose elaborated this simple model, designing tiles that could propagate increasingly complicated seeds and in a way addressing one of the two issues that concerned Lederberg. The final, Rube Goldberg-esque tile he called the S-unit (figure 16.6). Each component of the S-unit provides the model with a specific capability. For example, pendulums are used to count the number of units that together make up one replicating organism, and the wedges are used to control the order in which units are assembled.

Watching Penrose's ingenious tiles on film is mesmerizing. But as Penrose acknowledged, while there were some similarities to DNA, they were not conclusive. He made some preliminary suggestions about the function of various chemical components of the DNA molecule by comparing them to the elements of the S-unit, but thought the further speculation was not worthwhile.

Von Neumann's model interested and inspired biologists. It was not able to answer the kinds of questions they had, significantly those that dealt with the physical and chemical aspects of self-reproduction and questions about the evolution of the machinery involved in self-reproduction. Penrose's

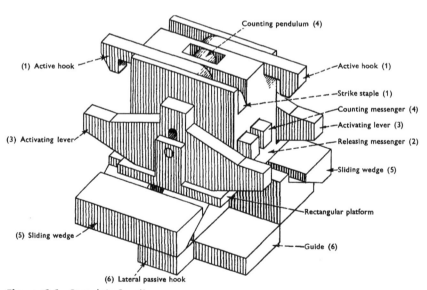

Figure 16.6: Complete S-unit.
From "Mechanics of Self-Reproduction," L. S. Penrose (1958).

physical models, which are closer in some respects to the phenomena, also fell short. Like the RWB article, what these models could do was to help clarify and pinpoint the phenomenon in question. Von Neumann demonstrated that the notion of self-reproduction does not involve a logical contradiction, he mitigated the implications of arguments based on considerations of complexity, and he opened the way to a discussion about the minimal requirements for self-reproduction. His models did not represent phenomena; they carved out of the biological mélange one question amenable to formal study. Whether it was appropriate to study this question independently from thinking about the development of the organism as a whole or of the evolution of genetic systems remained open questions.

TELEOLOGY

Wiener's article about teleology led to a flurry of responses. Many valid, supposedly fatal criticisms were raised. For example, groping in the dark for matches that are not there cannot be considered purposeful behavior, if purpose is understood as behavior aimed at achieving a desired relation with an existing aspect of the environment.[21] More fundamentally, it was argued that attributing purpose purely by observing behavior, while ignoring intention, simply misses the point. The simplistic identification of purpose with negative feedback was rejected.[22] Further philosophical reflections clarified tremendously that aspect of teleology that RWB tried to capture, and influenced thinking on *biological function*, on the notion of a *genetic program*, and on teleology in evolution. Often described as flawed, the article remains a classic treatment of the notion of teleology. A fundamental goal of the article was to encourage conceptualizing both living and artificial systems as goal-directed systems, controlled by feedback. These notions are now commonplace.

In a 1954 Princeton lecture devoted to the role of mathematics in science and society, von Neumann also reflected on the question of teleology and the opposition between causal determinism and teleological laws, which apply to a whole process "viewed as a unity."[23] He used the example of mechanics to argue that mathematical transformations can show that the two supposedly contradictory explanations are in certain cases formally equivalent. Two formulations of mechanical laws, the Newtonian or causal formulation, and the teleological principle of least action, were shown to be mathematically equivalent. According to the first formulation, motion is determined by causal laws applied to the state of the object at each time-point. According to the second, the trajectory of objects is such that a certain formally defined quantity is minimized when the trajectory is considered as a whole. Teleology, von Neumann acknowledged, may be important when thinking about biology, but only mathematical reasoning can tell us when the distinction is in fact meaningful.

Both Wiener and von Neumann suggested ways to diffuse the problem of teleology that besets biology. Wiener, seemingly more modestly, restricted his "solution" to the behavioral level, leaving aside the question of causality and determinism. Von Neumann, who, playing the role of the mathematician, emphasized that only by doing math can the question be sensibly addressed, seems more hubristic. On the other hand, Wiener redefined words and concepts to suit his perspective—while RWB acknowledged that their conceptual scheme is one among many, the article did not endorse pluralism and suggested that conceptual housecleaning was in order. While seemingly very different, Wiener and von Neumann's reflections on teleology are not mutually exclusive. Indeed, a teleological description of behavior of the sort suggested by Wiener and his colleagues can be deterministic and causal, in the sense used by von Neumann, and the formal equivalence highlighted by von Neumann in no way prohibits teleological behavioral descriptions—in fact, it legitimizes them. What remains, however, is a striking difference in rhetoric and emphasis between the two men.[24]

CONCLUSION

So how did the two suitors fare? Writing in 1951, the geneticist Theodosius Dobzhansky reflected widely held sentiments about the role of theory in biology when he wrote,

> experience has shown that, at least in biology, generalisation and integration can best be made by scientists who are also fact-gatherers, rather than by specialists in biological speculation.[25]

Quoting this negative sentiment, the cyberneticist Michael Apter offered a rebuttal culminating in a quote from von Neumann's 1948 lecture on self-reproduction, in which he elaborated on the distinction between studying the elements of a system and attempts to study how elements, defined by stipulation, constitute an integrated system.[26] Von Neumann argued that in spite of the limitations of this approach, it is "important and difficult."[27] The goals of this *systems biology*, given its obvious limitations, are to study the larger "organisms" that "can be built up from these elements, their structure, their functioning, the connections between the elements, and the general theoretical regularities that may be detectable in the complex syntheses of the organisms in question."

As Claude Shannon put it in a 1958 review of von Neumann's contributions to automata theory, and specifically self-reproducing automata:

If reality is copied too closely in the model we have to deal with all of the complexity of nature, much of which is not particularly relevant to the self-reproducing question. However, by simplifying too much, the structure becomes so abstract and simplified that the problem is almost trivial and the solution is un-impressive with regard to solving the philosophical point that is involved. In one place, after a lengthy discussion of the difficulties of formulating the problem satisfactorily, von Neumann remarks: "I do not want to be seriously bothered with the objection that (a) everybody knows that automata can reproduce themselves (b) everybody knows that they cannot."[28]

The empirically minded biological retort to this view was articulated bluntly by the neurophysiologist John Eccles in his review of the published record of the Hixon Symposium. Wiener and Von Neumann sought to bring mathematical abstraction to biological questions. Eccles's objections apply to both:

> It seems to the reviewer that the development of neurophysiology is likely to be impeded rather than aided by superficial analogies with automata. Despite all its grandiose claims cybernetics has contributed nothing to neurophysiology except the confusion of some neurophysiologists. . . . One further criticism concerns the section on the reproduction of automata. One may doubt if von Neumann expects us seriously to accept this logical game which is but a mere caricature of reproduction, for it involves the tacit assumption of a supervising genius who not only designs automata and has blue-prints of them, but also initially inserts instructions into them so that in principle they would go through the motions of a reproductive cycle![29]

In contrast, Warren McCulloch, who presided over the Macy cybernetics conferences, looked to mathematics for a theory "so general that the creations of God and men must exemplify it," acknowledging that these necessary conditions could not determine what neural mechanisms are to be found in humans. Robots, the quintessential how-possibly models, then suggest specific hypotheses about the human brain that can be tested experimentally. The very generality of math, McCulloch told a reserved psychologist, meant that the influence of mathematicians should be welcomed rather than feared.[30]

But even he sounded downtrodden in the concluding comments he prepared for the tenth and final conference in 1953. After noting the diversity of the research fields of the participants, he wrote,

Our most notable agreement is that we have learned to know one another a bit better, and to fight fair in our shirt sleeves . . . our consensus has never been unanimous. . . . In our own eyes we stand convicted of gross ignorance and worse, theoretical incompetence.[31]

What role did the theoreticians play in all of this? Wiener took a specific biological phenomenon, intention tremors, and generalized. Von Neumann did the opposite: he took a general biological category, "reproduction," and developed a concrete, though formal, and hence general, model. While in some respects these look to be exactly the same type of work—a model of teleology (as negative feedback), and a model of reproduction (as self-reproducing automata)—the endeavors are in some respects mirror images. Von Neumann, the ultimate outsider, worked by himself and developed a formal but concrete model, seemingly unconcerned in this work with being faithful to biological knowledge. Most of this work was sketched out in talks and presentations. Wiener, who had biological training himself, speculated about neuroanatomy, worked closely with collaborators, thinking with and "like" engineers and physiologists, and published in a philosophy of science journal. Von Neumann developed a series of models that illustrated how self-reproduction is possible. Wiener argued for a particular and stringent definition of a central notion in biology. It also seems at first as if von Neumann inspired a significant body of work that explicitly traces itself back to his work on self-reproducing automata, and is portrayed as a father figure of artificial life research, while Wiener's analysis of purposeful behavior, though influential for a short period, can safely be categorized as flawed and without lasting effect.

On reflection, the differences in the ways our two suitors went about trying to woo their coy muse are rather less clear-cut, as is the true impact of their different approaches to the study of biological questions. Wiener and Von Neumann were corresponding and collaborating about these issues since 1944. Together they pushed forward what became the Macy Conferences, which brought together many people interested in these ideas. They both were evangelizing, and evangelizing together—even if the two had their differences and a level of mutual dislike. Each employed various infiltration tactics and invested time engaging with biologists. I discussed only two infiltration attempts, Wiener's conceptual analysis and von Neumann's models, works in which mathematics clearly played very different roles. However, the influence of both can be found in contemporary systems biology. Wiener's more engaged approach and his emphasis on concrete phenomena such as intention tremors, which seemed to end in failure, appears in retrospect to be as relevant if not more characteristic of contemporary modes of interdisciplinarity in biology than the idealized modeling and somewhat less collaborative approach of von Neumann. Likewise, the two styles of systems biology that I characterized earlier as "behaviorist" (Wiener) and "functional" (von Neumann) turn out not to be mutually exclusive, and both are included in the modeling toolkit of contemporary practitioners. The story of two grand strategies that "invading" mathematicians can employ is too neat. The real story illustrates the wide range of tactics mathematicians routinely deploy when engaging with biology. Von Neumann's and Wiener's different personalities and styles may have affected how their contributions to biology are portrayed in the lore of systems biology and artificial life, but not their lasting impact as two founding fathers of contemporary systems biology.

FURTHER READING

Heims, Steve J. *John Von Neumann and Norbert Wiener: From Mathematics to the Technologies of Life and Death*. Cambridge, MA: MIT Press, 1980.

Israel, Giorgio and Ana Millán Gasca. *The World as a Mathematical Game: John von Neumann and Twentieth Century Science*, translated by Ian McGilvay. Basel/ Boston: Birkhäuser Verlag, 2009.

Kay, Lily E. *Who Wrote the Book of Life?* Stanford, CA: Stanford University Press, 2000.

Keller, Evelyn Fox. *Making Sense of Life: Explaining Biological Development with Models, Metaphors, and Machines*. Cambridge, MA: Harvard University Press, 2003.

Wiener, Norbert. *I Am a Mathematician: The Later Life of a Prodigy*. Cambridge, MA: MIT Press, 1964.

NOTES

1. For a discussion of this paper and its significance, see Evelyn Fox Keller, *Making Sense of Life: Explaining Biological Development with Models, Metaphors, and Machines* (Harvard University Press, 2003), 84–87.

2. Letter to Wiener, November 29, 1946, McCulloch Papers, American Philosophical Society.

3. Norbert Wiener, *I Am a Mathematician* (New York: Doubleday, 1956), 252–54.

4. Arturo Rosenblueth, Norbert Wiener and Julian Bigelow, "Behavior, Purpose and Teleology," *Philosophy of Science* 10 (1943): 20.

5. For a discussion of whether feedback can be defined solely by reference to external behavior see William C. Wimsatt, "Some Problems with the Concept of 'Feedback,'" *Proceedings of the Biennial Meeting of the Philosophy of Science Association* 1970 (1970): 241–56.

6. Rosenblueth, Wiener and Bigelow, "Behavior, Purpose and Teleology," 23–24.

7. See Steve J. Heims, *John Von Neumann and Norbert Wiener: From Mathematics to the Technologies of Life and Death* (Cambridge, MA: MIT Press, 1980).

8. Steve J. Heims, "Gregory Bateson and the Mathematicians: From Interdisciplinary Interaction to Societal Functions," *Journal of the History of the Behavioral Sciences* 13 (1977): 141–59.

9. "The General and Logical Theory of Automata," read at the Hixon Symposium in September 1948; published in 1951. John von Neumann, *Collected Works*, edited by A. H. Taub (New York: Macmillan, 1961–1963), vol. V, 288–328. Von Neumann's publications on self-reproduction are surveyed by Burks in the preface to John von Neumann, *Theory of Self-Reproducing Automata*, edited and completed by Arthur W. Burks (Urbana: University of Illinois Press, 1966).

10. John von Neumann, "Theory and Organization of Complicated Automata," in von Neumann, *Theory of Self-Reproducing Automata*, 85. Original lecture delivered at the University of Illinois in December 1949. Note that automaton A is assumed to be a *universal constructor*, able to construct any machine described in its input.

11. Muller recounts the trajectory of his thoughts in Hermann J. Muller, "The Gene Material as the Initiator and the Organizing Basis of Life," *The American Naturalist* 100 (1996): 493–517.

12. See von Neumann, *Self-Reproducing Automata*.

13. Christoper G. Langton, "Artificial Life," in *The Philosophy of Artificial Life*, Margaret A. Boden, editor (New York: Oxford University Press, 1996).

14. See Palash Sarkar, "A Brief History of Cellular Automata," *ACM Computing Surveys* 32 (2000): 80–107; G. Bard Ermentrout and Leah Edelstein-Keshet, "Cellular Automata Approaches to Biological Modeling," *Journal of Theoretical Biology* 160 (1993): 97–133.

15. Moshe Sipper, "Fifty Years of Research on Self-Replication: An Overview," *Artificial Life* 4 (1998): 237–57.

16. Lederberg to von Neumann, March 10, 1955, Joshua Lederberg Papers, National Library of Medicine. For more on this correspondence and how the notion of information invaded biology see Lily E. Kay, *Who Wrote the Book of Life?* (Stanford, CA: Stanford University Press, 2000).

17. Lederberg to von Neumann, April 3, 1955, Joshua Lederberg Papers, National Library of Medicine.

18. Lederberg to von Neumann, April 3, 1955, September 3, 1955.

19. Von Neumann to Lederberg, August 8, 1955; Lederberg to von Neumann, August 8, 1955; von Neumann to Lederberg, August 15, 1955, Joshua Lederberg Papers, National Library of Medicine.

20. Lionel. S. Penrose, "Mechanics of Self-Reproduction," *Annals of Human Genetics* 23 (1958): 59–72. The idea was first proposed in Lionel S. Penrose and Roger Penrose, "A Self-Reproducing Analogue," *Nature* 179 (1957): 1183.

21. Richard Taylor, "Purposeful and Non-Purposeful Behavior: A Rejoinder," *Philosophy of Science* 17 (1950): 327–32.

22. See Kay, *Book of Life*, chapter 3; Israel Scheffler, "Thoughts on Teleology," *British Journal for the Philosophy of Science* 9 (1959): 265–84; Larry Wright, "The Case against Teleological Reductionism," *British Journal for the Philosophy of Science* 19

(1968): 211–23; Larry Wright, "Explanation and Teleology," *Philosophy of Science* 39 (1972): 204–18.

23. John von Neumann, "The Role of Mathematics in the Sciences and in Society," address at the Fourth Conference of the Association of Princeton Graduate Alumni, June 1954 (*Collected Works*, vol. VI).

24. Arturo Rosenblueth and Norbert Wiener, "Purposeful and Non-Purposeful Behavior," *Philosophy of Science* 17 (1950): 318–26, makes stronger metaphysical claims.

25. Theodosius Dobzhansky, "Mendelian Populations and Their Evolution," *The American Naturalist* 84 (1950): 401–18.

26. Michael J. Apter, *Cybernetics and Development* (New York: Pergamon, 1966), 23.

27. Von Neumann, "General and Logical Theory of Automata," 290.

28. Claude E. Shannon, "Von Neumann's Contributions to Automata Theory," *Bulletin of the American Mathematical Society* 64 (1958): 123–29.

29. John C. Eccles, untitled review, *The British Journal for the Philosophy of Science* 4 (1954): 345–47.

30. Warren S. McCulloch to Hans-Lukas Teuber, December 10, 1947, McCulloch Papers, American Philosophical Society.

31. Warren S. McCulloch, "Summary of Points of Agreement Reached in the Previous Nine Conferences on Cybernetics," in *Cybernetics: Circular Causal and Feedback Mechanisms in Biological and Social Systems: Transactions of the Tenth Conference on Cybernetics*, Heinz Foerester, editor (Princeton, NJ: Macy Foundation, 1953).

OREN HARMAN

ON THE IMPORTANCE
OF THE PARVENU
THE AMAZING CASE OF
GEORGE PRICE IN EVOLUTIONARY
BIOLOGY

INTRODUCTION

The barbed wire around Tolmers Square did nothing to take away from the ambience. On the contrary, it only added to its somewhat fantastic aspect. There, on a bench in the southwest corner in the summer of 1974, Sylvia Stevens, a young print artist of twenty-four, sat listening to George Price, a gaunt, sickly man. Price was actually just fifty-one, but he looked much older. His hair was scraggly, his teeth were decaying, his fingernails covered in grime. Hunched over himself, George sat cross-legged and, eyes burning, told the most amazing stories. He had worked in the Manhattan Project as a chemist, he said, and later built a teaching machine for the use of the Harvard behaviorist, B. F. Skinner. He'd invented computer-aided design, was snagged by IBM, lived among musicians in Greenwich Village, and had written stories about how to beat the Soviets for *Life* magazine. He'd befriended Senator Hubert Humphrey, corresponded with Nobel laureates H. J. Muller and Sir John Eccles, and with the writers Upton Sinclair and Aldous Huxley. And, at Bell Labs, Price told Sylvia, he had known the great Claude Shannon, father of information theory. In fact, he had written an equation that he hoped would do for biology what Shannon had done for communications. He'd also recently figured out, together with Professor John Maynard Smith, how to apply the theory of games to the evolution of animal behavior, a feat he was especially proud of.

Tolmers was a bohemian cooperative fighting against the encroachment of real estate speculators in London, and had drawn to it artists and activists. George, it seemed, was just an itinerant homeless man who had chanced to be living among the squatters. But what stories these were, Sylvia thought. How incredible it would be if they were true.[1]

Who was George Price, this unlikely man who in a few months would put an end to his life, leaving a suicide note addressed to Sylvia? And what equation was he talking about? What theory of games?[2]

In this essay, Price will be presented as a radical outsider who ventured into biology and ended up making seminal contributions. In particular, I will focus on two contributions—what became known as "the Price equation," a covariance relationship referring to selection processes, informed by Price's interest in Shannon and information, and the concept of the evolutionary stable strategy (ESS) informed by Price's infatuation with game theory and the Cold War. While the former has recently reemerged as an important actor in the levels-of-selection debate, one that invites a multilevel selection approach, the latter almost immediately spurred a revolution in the way biologists study and ask questions about the evolution of behavior. Both were accomplished by a man who died virtually unknown in a squat, after having converted to evangelical Christianity and establishing a "hot line to Jesus."[3] More importantly, though, for the purposes of this collection, they were accomplished by a man who had never studied biology. George Price was, if not quite the classic, then a very unique and unusual, parvenu.

"MAY GO HAYWIRE BUT WILL NEVER BE HUMDRUM"

It began in Scarsdale, New York, where Price was born on October 16, 1922. Losing his father at the age of four, Price was brought up in New York City by a struggling mother who fought off creditors for the stage-lighting company her husband left behind, just barely providing for little George and his older brother Edison. The young Price soon distinguished himself as unusually bright at the Stuyvesant School for boys, where, one former classmate remembered, he was "leagues above everyone." In the class yearbook, the *Indicator*, Price wrote that he hoped to become a "research physicist" and promptly graduated second out of a class of 708. But he was a strange bird, almost too bright, unsettling. The Harvard interviewers who met with him for all of forty-five minutes got it right: "May go haywire but will never be humdrum," they concluded, accepting him to the freshman class of 1940.[4]

Leaving Harvard the next year, Price ultimately graduated in chemistry from the University of Chicago, where he was awarded the Eli Lilly Fellowship to continue for a doctorate. It was the fall of 1943, and the slotted enzyme chemistry project he had been working on soon morphed into one of detecting trace elements of uranium in human bodies: made to sign declarations of secrecy, Price was told that he was now a part of the Manhattan Project. It was here that he met Julia Madigan, a devout Roman Catholic who objected to the use of nuclear weapons. A rabid atheist himself and an atomic hawk, Price baffled his friends when he announced that she would be his wife. Before long he had completed his PhD, and the unlikely newlyweds settled in Cambridge, where Price assumed a teaching job in the Chemistry Department back at Harvard.

Figure 17.1: George Price in his University College London office, 1973. Courtesy of the Price family.

But academia, Price soon found, was not his port of call. He was intellectually restless and searching for a way out. Then, in the summer of 1948, he read the first part of a general theory using mathematics to quantify information, published in the *Bell System Technical Journal*. The paper, "A Mathematical Theory of Communication" by Claude E. Shannon, immediately caught his attention.[5] It was elegant, precise, and got straight to the essence of the matter. This was how science should be done, he thought, and where it was going. And so, with an offer doubling his Harvard salary, Price and his growing family moved to Morristown, New Jersey, where he joined Bell Labs. Contracted to work on long-term basic science in the Chemistry Division, Price nevertheless decided to pursue his own interests. He was grinding and mapping germanium surfaces to calculate the relative importance of surface treatment and bulk properties in determining transistor characteristics, short-term applied electrical engineering work that had future Bell Nobel laureates John Bardeen and William Shockley intrigued.[6] This kind of tinkering Price had never formally studied, but rather picked up as a boy helping his mother run the Display Stage Lighting Company. Those who knew him were not all that surprised: George was always prone to strike out his own path.

Still, he would not stay at Bell for long. The marriage was falling apart, and a desperate move to Minnesota was hatched to try to save it. There, under his old Manhattan Project chief, Samuel Schwartz, George would return to the kind of work he had done at Chicago, and, moving to a quaint two-story cottage in St. Paul with a front and back lawn for his two baby daughters, perhaps provide an environment more conducive to family peace. Schwartz had discovered that by localizing the fluorescence of intravenously administered porphyrin in tumor tissues, surgeons could get an idea of the metastatic spread of a cancer, an implication Merck and Co. had great interest in, for obvious reasons. However, animal tissues showed weak fluorescence unless stained with dyes called fluorochromes, which made it difficult to recognize cell types and to observe the exact location of the porphyrin. Price figured out how to allow for combining a phase contrast image of contrasting color and smoothly controllable intensity with the fluorescence image using a beam splitter of his own making.[7] Schwartz was over the moon, saying Price was the only person he knew who deserved to be called "a man of genius."[8] But if the science was going great, no such luck befell the marriage. Soon, it had broken up acrimoniously.

Following a bout of polio, alone, depressed, and restless once more, Price began to strike out. In 1955, still entirely anonymous, he penned a scathing attack on the then-popular claims for parapsychology. The resulting article was published by *Science* and triggered heated reactions from the maligned ESP experts as well as Aldous Huxley and Upton Sinclair.[9] Encouraged by this success, Price wrote a piece for *Fortune* magazine on "How to Speed up Invention" based on a design machine of his own making that could manipulate three-dimensional objects on a computer screen.[10] Challenged by the experts, he produced a 70-page proof of concept, replete with the necessary code, for computer-aided design (CAD). Worried about the Soviet threat, he sold a story to *Life* magazine and struck up a correspondence on the matter with the Minnesota Senator Hubert Humphrey.[11] Leaving St. Paul and more or less abandoning his family, he moved to Greenwich Village and stamped "writer" on his passport. When IBM's director of research saw his "proof of concept" and offered him a handsome job in R&D, Price insisted on going it alone, consulting a fancy Fifth Avenue lawyer with intentions to develop the idea himself instead. Down and out on drugs in his Village apartment, he missed the deadline for application for a patent, and, with a new director of research now at the helm, joined the company as a rather low-rung technical worker. To relieve his boredom, he wrote stories on everything from "How to Hatch an Egghead" to "The Physics of Bowling" to "The Real Threat from Red China"[12] and, using the journalist's prerogative, met men of stature. Claude Shannon remembered him from Bell Labs, the Nobel laureate geneticist H. J. Muller found him amusing if somewhat strange, and B. F. Skinner took a real liking to him when, following a visit to his Harvard conditioning lab, Price designed for him a "teaching machine" using an acoustic phonograph with spring motor plus a paper disc.[13]

He was shooting in all directions, desperate to make some kind of breakthrough. To Harper & Brothers he now sold a proposal for a book on how to win the Cold War. To *Science* he sent a paper on the hymen that was rejected; one of the readers called it "a crotchety, verbose diatribe" by an amateur that had "no place in a scientific journal."[14] To the Nobel laureate John Eccles he sent a "discovery of major importance" on the "Structure and Function in the Invaginated Synapses of Retinal Receptor Cells"; Eccles wrote back politely that it was "remarkable" but that he himself could not follow it beyond his rather "mediocre attainments."[15] And at IBM, unbeknownst to and uncommissioned by his bosses, he worked independently on developing *sui generis* methods for finding nearly optimal solutions to multiple-activity, multiple-factor situations—a computer programming conundrum—which he promptly sent to the future Nobel laureate economist Paul Samuelson. (Samuelson replied: "I am sure that much of what you are doing would interest economists, particularly if it could be related to earlier work."[16])

Finally, in the winter of 1966, Price underwent an operation to remove a thyroid tumor, which got rid of the cancer but left his arm and shoulder partially paralyzed. Falling into a deep depression, he decided to leave everything behind. He was on his way to England, where, he wrote to a friend, he hoped he might have better luck making an important discovery. An outsider all his life, he was about to enter the world of evolutionary biology and finally leave his mark.

THE PRICE EQUATION

It was the end of 1967, and George Price was forty-five years old. Besides an undergraduate course in physiology, he'd never studied biology in any formal way. But he had now become infatuated with the evolution of family and fatherhood, a problem he began thinking about in America, but which, as he devoured the popular books of Robert Ardrey and Desmond Morris, now loomed even larger in his mind.[17] Whether or not this was a psychological mechanism to turn unruly personal circumstances into a controllable scientific project, Price was determined to crack the mystery.

He started with basics: food allocation in the Pleistocene. Having worked on optimality models at IBM, it was clear to him that the way to approach the problem would be to consider a number of alternative strategies and observe which of them complied more directly with observed behavior. One optimal solution for our ancestors on the African plains could have been complete sharing alongside promiscuous, noncompetitive mating and cooperative rearing of the young with little or no recognition of paternity. An alternative would be allocation of hunting spoils between males, who would then distribute their share to females and children individually. Such a system would favor monogamy, since if a man kept the food to himself, he would scarcely leave any progeny. On the other hand, if a man provided generously to many women and kin, changing partners every so often, most of the time he'd be feeding children of other males, thereby contributing to the decrease of frequency of his own genes in the population. Since most human societies looked more like the second model, it stood to reason that family had developed under selection pressures related to food distribution at a time in human evolution when hunting by all-male bands became important. Exalted "fatherhood" was no more than a genetically optimal solution to the challenge of securing and distributing daily grub.

But this, as in Ardrey and Morris, was conjecture; to really make any headway, Price knew—"to protect against biasing effects of emotional prejudice"— mathematical tools for evolutionary inferences were necessary.[18] Already he had written to Bill Hamilton, whose paper on altruism and inclusive fitness Price had read in the late-opening public library at Holborn. Price was impressed. Here was a mathematical treatment of a genetic behavioral trait, cast in evolutionary terms. But something seemed too limiting about Hamilton's equation: was it really the case that altruism could only evolve between genetically related individuals? Was kin the sine qua non of kindness? Seeking a broader treatment, Price set out to re-derive Hamilton's math. After all, perhaps shared genes for altruism would suffice, and relatedness would be one way, but not the only way, to share them. Hamilton's rule might just be an instance of a wider phenomenon.

He remembered Shannon. "The recent development of various methods of modulation such as PCM and PPM which exchange bandwidth for signal-to-noise ratio," the Bell giant had opened the 1948 paper Price so admired, "has intensified the interest in a general theory of communication."[19] And indeed, Shannon had laid out for the first time a mathematical framework for measuring the information sent across noisy communication channels. Here was a generalized extension of what had previously been seen as an engineering problem to a generic process of transmitting information under conditions of uncertainty. Shannon wrote:

> The fundamental problem of communication is that of reproducing at one point either exactly or approximately a message selected at another point. Frequently the messages have meaning; that is, they refer to or are correlated according to some system with certain physical or conceptual entities. These semantic aspects of communication are irrelevant to the engineering problem. The significant aspect is that the actual message is one selected from a set of possible messages.[20]

Shannon's system presupposed a source and a transmitter, a channel and a receiver and a destination. It used the bit as its primary unit of information, and entropy, a term borrowed from thermodynamics, to quantify the amount of uncertainty associated with knowing the value of a bit of information. In essence, Shannon wanted to know how bandwidth and noise affect the rate at which information can be transmitted over an analog channel. This was about selection, just as he had explained: a source transmitting to a receiver and the question of the difference in information between the two. As Price turned to the genetics of evolutionary traits, it seemed to him that this was pretty much a similar story.

After all, even though "information theory" dealt with the possibility of preserving information and evolutionary theory with the possibility of it changing, the problem in both was of tracking the change in a character (rate of information, frequency of a genetic trait) over time from one point (source, parent generation) to the next (receiver, child generation). The way to do that for information, Shannon had argued, was to describe the maximum possible efficiency of error-correcting methods versus levels of noise interference and data corruption.[21] The way to do that for behavioral traits, it seemed to Price, could be represented thus:

$$\Delta \bar{z} = \frac{Cov(w,z)}{\bar{w}},$$

where the change in the frequency of a trait, z, is equal to the covariance of that same trait with fitness, w, divided by the mean fitness of the group. This was a mathematical tautology, but Price could already see that it was a more abstract approach than using coefficients of relatedness à la Hamilton. The spread of a trait like altruism, for example, could be tracked by statistical covariance of the character with fitness rather than via calculations of the pathways of relatedness, which meant that, in principle and under the right conditions, altruism might depend on association rather than family. This was rather encouraging.

But as with the entropy associated with information, there was also the matter of how loyally the given trait was passed from one generation to the next. If the trait was genetic, it would be important to know whether the gene had undergone a mutation or whether it biased the system to pass itself on more frequently than would be expected. To take this into account, Price added a further term:

$$\Delta \bar{z} = \frac{Cov(w,z)}{\bar{w}} + \frac{E(w\Delta z)}{\bar{w}},$$

where $E(w\Delta z)$ is a measure of the extent to which the trait, z, will be passed on faithfully. There was no doubt Shannon had played an influential role. Price even wrote to him to inquire whether Shannon thought the covariance equation could be interpreted in terms of channel capacity.[22]

The equation partitioned trait change in evolution into selection (Cov) and transmission (E). This was valuable enough,[23] except that the new term did much more. With a bit of algebraic shuffling, it showed that selection could work at two levels simultaneously; in fact, it could even partition them to see how much each contributed to the overall change. Instead of defining the two terms of the equation as the selection and transmission terms, corresponding to the individual and the genes in the sperm and egg respectively, they could be bumped up one notch in the rung and redefined as relating to the individual and the group.[24]

What this meant was that selection could be viewed as working on multiple levels of the biological hierarchy simultaneously, a result that undercut decades of debate.[25] No longer was it mandatory to argue for an exclusive

Figure 17.2:
George Price with his wife Julia and two daughters, Annamarie and Kathleen, Morristown, New Jersey, 1949.
Courtesy of the Price family.

level at which selection works; it was now an empirical matter in each and every case at what level selection operated most strongly. Without entirely meaning to, Price had written an equation that could be used to solve a conundrum going all the way back to Darwin. And he had done it with no biological training whatever. Staring at his own creation, unbelieving, he thought of it as a "miracle."[26] And, sending it off to *Science* in the fall of 1969, in capitol letters so that no one could miss it, he stated in a cover letter that this was his attempt at a general "Mathematical Theory of Selection" analogous to what Claude Shannon had done with the theory of communication. Fittingly, there were no references, not even one. This was biology outside the box.

EVOLUTIONARILY STABLE STRATEGY

Very much contrary to Price's expectations, *Science* rejected the paper, as did *Nature*, until Hamilton himself, bowled over by the equation, intervened and, using his clout as a respected modeler, pulled a fast one over its editors that led to publication.[27] By this time Price had already shown his result to Hamilton's former PhD supervisor, the biostatistician C. A. B. Smith and secured a research grant and honorary position at the genetics department at University College London. (Indeed, Price had literally walked off the street in September 1968 into the department to show the equation to any biology professor he might meet, an outlandish move that led within the hour to his being offered—a completely unknown stranger—keys to his own office in the department). Now he was already working on a second paper, influenced like the first one by preoccupations very far afield from biology. If Price's selection mathematics had been spurred to life via his meeting with Shannon and "information theory," his next project would be born of an infatuation with the Cold War.

"I have just been reading your very interesting paper on 'The genetical evolution of social behavior," Price had written to Hamilton after encountering his paper in Holborn Public, "and would very much like to have reprints if you still have any to spare." In connection to his interest in the evolution of family, Price inquired whether Hamilton knew of any evidence for genes being able to somehow "recognize" copies of themselves in other bodies. Perhaps in humans, where cultural evolution worked alongside biological, "some interesting effects could, in theory, occur."[28] Here was Price trying to break the narrow, and to him depressing, bounds of kin selection.

"So far I haven't arrived at any clear idea even as to what sort of 'game' the genes are expected to be playing when operating together," a polite Hamilton replied to this perfect stranger. "Something like socialism (or is it racialism = can't tell), admittedly, seems indicated, but I have only vague ideas as to the mechanisms by which biological and cultural evolution interact." Hamilton seemed to be implying that if a feeling of repugnance from the narrowness of altruism confined to kin does exist in humans, then it too, doubtless, has been selected for: the broadening of altruism beyond the family, after all, would produce the salutary effect of increasing genetic diversity. More importantly, it would help safeguard the group against competing rivals. Kindness was always functional. "I am sure that prisoner's dilemma situations," Hamilton ended his letter, "are common and important in biological evolution."[29]

■

But what did Hamilton mean by "game"? Price was not sure. It all sounded at once familiar and vague. Then, in the Senate House Library, he came across a paper by one G. Stonehouse regarding deer antlers. Whereas for

Darwin these cumbersome structures seemed to be maladaptive results of sexual selection, Stonehouse argued that antlers were adaptive solutions to the demands of thermoregulation (males grow them, he claimed, because they are larger and need to dissipate more heat).[30] But Price, the outsider, wasn't convinced. Skin flaps and large ears seemed like much better solutions if thermoregulation was indeed a problem. Besides, reading up on the topic, he discovered that Roe deer were in velvet in spring and without velvet in summer, whereas lowland tropical species would have to dissipate sweat year-round. A better solution was called for. Suddenly, it hit him: games!

As with Shannon, Price was harking back to his nonbiological past. With Julia, at the Manhattan Project, he would argue into the night about the merits of building an atomic bomb. Precisely then, in 1944, John von Neumann and Oscar Morgenstern had published their *Theory of Games and Economic Behavior*, which detailed the logic of détente based on the threat of deterrence. Increasingly, Price was becoming worried. If America didn't understand that the Soviets were a real threat and acted accordingly, she'd be doomed. To exercise true deterrence, she'd need to overhaul her values, placing a higher tab on liberty than luxury and concerning herself more about "who wins the Nobel Prize in physics than about who wins the World Series." She'd have to double her defense budget and build more weapons of mass destruction. Price's "Arguing the Case for Being Panicky" appeared in *Life* magazine in 1957 and presented, as the editors announced in the headline, "a strongly worded and provocative warning of impending national disaster."[31] Indeed, Price was so worried, he decided to write a book about how America could win the Cold War. *No Easy Way* was eventually contracted with Doubleday, though never delivered. Still, there was an entire section planned on the logic of détente based on the threat of deterrence.[32]

Hamilton's clue now suddenly revealed itself: The evolution of antlers followed a classic von Neumann game. Working out the logic, Price began to write a paper. If a group of deer varied in both fighting ability, E signifying greater fighting ability than e and ability to deescalate combat indicated by D versus d, then games could be set up between contestants to determine how De, DE, de, and dE would fare. Attaching probabilities to injury, survival, and victory, Price discovered that taking into account four or five rutting seasons per generation over a hundred generations, limited combat strategies would evolve. One obvious way to achieve such a result would be to grow ornate antlers. Locking heads with antlers in ritualized combat instead of mortally puncturing each other with sharpened horns actually made evolutionary sense. The logic of one-off games applied repeatedly made this clear.

But was the logic strong enough to explain the evolution of behavior? After all, a malicious deer always stood to triumph by breaking the rules. Here is where iterated games came in, and a notion that would soon be dubbed the evolutionarily stable strategy (ESS). "A sufficient condition for a genetic strategy to be stable against evolutionary perturbations," Price wrote,

> is that no better strategy exists that is possible for the species without taking a major step in intelligence or physical endowment. Hence a fighting strategy can be tested for stability by introducing perturbations in the form of animals with deviant behavior, and determining whether selection will automatically act against such animals.[33]

Soon, a number of rules emerged from the games: 1) An animal should avoid battle with a stronger animal, 2) an animal should be aggressive against a weaker animal, and 3) when fighting an equal opponent, an animal should try an occasional "probe"—an escalation meant to judge an adversary's reaction. Most important of all was the principle of "getting even," a strategy that would in later years be named "Tit-for-Tat."[34] This notion hinged on a fundamental game theoretic principle—that one's best strategy always depends on the strategy of the opponent, a principle which, as Price now wrote, characterized human "two-person game" conflict "at all levels from kindergarten to nations." An evolutionary arms race had been put in place between signals for strength (and "wildness" and "unpredictability"), and the ability to judge their "honesty." The more elaborate the cognitive apparatus the species had evolved, the more would this principle apply—most markedly, doubtless, in humans; but the greater the variation in antlers in a population of deer, the greater the chance fighting might be avoided, since a glance from afar would suffice to preclude much of the combat.

Like the covariance equation, this was not yet a solution to the problem that had most occupied Price—the evolution of family—but, he thought, it was inching him closer. Crucially, it was an original application of game theory to the evolution of behavior. True, the Austrian ethologist Konrad Lorenz had considered ritualized combat before Price, but he put it down to group selection, and like other proponents of this idea, offered no well-delineated mechanism.[35] And true, the mathematical population geneticist Richard Lewontin, having read Luce and Raiffa's textbook on game theory, had already set out in 1960 to model a game between animals and nature, though one that could only measure the fitness of a species against a changing environment rather than genes of animals interacting between themselves *and* the environment. Modeling these more complex interactions seemed to Lewontin to entail simplifications that furthered one from reality, and

Lewontin wasn't interested in "interesting but not true," so let it go.[36] Hamilton had applied game theoretic thinking and the notion of an "unbeatable strategy" to a recent paper on extraordinary sex ratios—this was the origin of his clue to Price—but he didn't deal per se with behavior.[37] The notion of an evolutionarily stable strategy, though Price had not yet defined it formally, was new. From the otherwise economic and international relations–applied theory of games, the Cold War (with a little help from the clue from Hamilton) had made its way via Price into biology.

The story of how John Maynard Smith became involved and helped to introduce the ESS to the world has been told elsewhere.[38] Just as with Hamilton and the covariance paper, it is doubtful whether without Maynard Smith's intervention Price would have been successful in publishing his ideas alone: "Antlers, Intraspecific Combat, and Altruism" of 1968 was an unwieldy long draft that had been rejected by *Nature* and shelved, Price having failed to successfully complete the computer simulations. "The Logic of Animal Conflict," on the other hand, co-authored by both men and published in *Nature* in 1973, almost immediately became a classic. Finally, an ESS was formally defined: it was a behavioral strategy such that, if the majority of the population adopts it, there is no "mutant" of higher reproductive fitness. Now, "Dove," "Hawk," "Bully," "Retaliator," and "Prober-Retaliator" were introduced and set against each other to simulate evolved strategies that might match observed behavior.[39]

Predictions became possible. For example, the simulations showed that the optimal strategy against "Hawk" is "Dove," or immediate retreat. This seemed to suggest that anyone who would simulate wild, uncontrollable rage would be at an advantage, as his opponents would yield to the threat. So a pretend "pseudo-Hawk" should evolve, leading in turn to the selection of a second type with the ability to "call a bluff." To beat out this type, a true maniacal craziness would then evolve, one that could not be easily counterfeited. Elephants "on musth" seemed to fit the prediction perfectly: as they raged uncontrollably through villages, often with fatal results, a brown tarlike fluid secreted by the temporal glands ran down their faces. Locals believed that such beasts were invaded by wild spirits, but this was just nature's way of signaling that here was no charade, no fakery. Thanks to George Price, with the crucial help of insider friends, the union of game theory and evolution had been cemented. This was very real and was here to stay in biology. Price had come entirely from outside, but he did so, like the elephants, with no guile.

LEGACY OF AN OUTSIDER

It is impossible to imagine evolutionary theory today without the theory of games. Popularized by Richard Dawkins in *The Selfish Gene*, the ESS soon became central to the toolbox of the evolutionary biologist. It turned out that Maynard Smith and Price made a mistake in their 1973 paper: "Retaliator," after all, was not an ESS.[40] But this mattered very little. Before long, game theory had been incorporated into undergraduate textbooks and continues to be used to solve evolutionary mysteries from seed dispersal and dimorphism to parental care, division of labor, and the evolution of cooperation.[41] Indeed, evolutionary game theory has been so successful in biology that it has been reincorporated in its mother discipline, economics, and is regularly used far afield from biology by sociologists, psychologists, anthropologists, and philosophers.

The legacy of the "Price equation" is not as unequivocal. Undoubtedly, it has played a central role in the reawakening of a valuable multilevel selection approach to evolution, and provides a crystal-clear way of thinking about selection processes, as a number of enchanted theorists have argued.[42] It has also been put to use in attacking problems as diverse as reproductive value, evolutionary epidemiology, genetic programs, human cultural evolution, rogue genes, meiotic drive, biodiversity, and ecology.[43] But there are other biologists who find its dynamic insufficiency and abstract generality prohibitive in applying the equation to any real evolutionary scenarios. Still others argue that the very claim of dynamic insufficiency betrays a deep misunderstanding: since the equation is an identity, not a model, it can neither be dynamically sufficient or insufficient. On this line of thought, a number of theorists claim that it is logically impossible to derive any theoretical predictions from the equation since it contains no modeling assumptions, and that those who use it are confusing identity with causality and probability theory with statistics.[44] Whether the Price equation is valuably invaluable or invaluably valuable remains a point of argument.[45]

Still, George Price's place in the history of biology has been reclaimed, which brings us full circle to the issue of the parvenu. Clearly, Price represents an outsider in more than one sense: always on the outskirts, never quite fitting in, working alone and following personal hunches, Price moved from chemistry to engineering to cancer research to writing to computer programming to evolutionary genetics with unusual alacrity. Long before his arrival in England and splash unto the biological scene, Price was an intellectual scavenger in search of a breakthrough. An autodidact and cocksure, he approached problems from first principles, usually ignoring the work accomplished by "insiders" in any given field.

But what was it about his "outsiderness" that eventually made his enduring contributions to biology possible? In the case of the equation, there is
no doubt that Price's search for a highly abstract general law of selection
played a role. In this respect, he was similar to R. A. Fisher, another kind of
outsider, with whom he shared an abstract theoretical cast of mind.[46] Here
a penchant for breaking a problem into its component parts, simplifying
and using as little biological assumptions as possible served Price well. The
"Price equation," after all, is totally devoid of mechanism. Few pure biologists would deign to think in this way. That Price had been impressed by
Shannon's work in information theory makes this orthogonal outsider approach all the more clear.

Introducing game theory into the study of the evolution of behavior similarly betrays Price's formal approach. But here a political preoccupation—
Price's infatuation with Cold War dynamics—seems to have played the
more direct and decisive role. And being an outsider may have made it easier
for Price to think more associatively than an insider trained in the field: Not
having been "indoctrinated" into evolutionary genetics or biology may in
some sense have rendered associative thinking more probable, since association becomes useful precisely when there is a lack of direct local knowledge. Yes, Hamilton had provided the all-important clue, but it was from
the fire of Price's previous preoccupations that the crucial metaphor was
stoked. In principle, and also in practice, this kind of associative thinking
is possible for "insiders" as well, but it may be an advantage enjoyed more
easily by "outsiders," whether born by necessity or otherwise.

That Price needed the help of the insiders Hamilton and Maynard Smith
to publish is an interesting point for reflection; however novel and path-
breaking his interventions were, scaling the walls of publication proved
difficult. Indeed, though the difficulty in both cases was not for the same
reasons (the covariance paper was complete but misunderstood, whereas
the animal conflict paper was incomplete and needed more work), the fact
that these interventions were so novel and pathbreaking may have been the
cause for their difficult birth. In this case, Hamilton and Maynard Smith's
"insider" status proved crucial and serves as a salutary reminder that disciplinary walls are real things, protected and strengthened by such things as
name recognition, pedigree, conventional notation, and method, and more
generally by what can be called an "insider culture." Outsiders often need
help from insiders to get in.

Ultimately, while Price was revolutionizing evolutionary biology with formal treatments of selection and behavior, he was also undergoing a radical
evolution in his personal life, becoming an evangelical Christian and then

a vagabond. We can only guess precisely what was on his mind when he put an end to his life on January 6, 1975, in a cold London squat. But that this unsteady and confused outsider gave us the stability of the evolutionary game and the clarity of the covariance equation is another sure instance of history's sense of irony.

FURTHER READING

Harman, Oren. *The Price of Altruism: George Price and the Search for the Origins of Kindness*. New York: W. W. Norton, 2010.

Okasha, Samir. *Evolution and the Levels of Selection*. Oxford: Oxford University Press, 2006.

Price, George R. "Covariance and Selection," *Nature* 227, 1970, pp. 520–521.

Smith, J. Maynard and G. R. Price, "The Logic of Animal Conflict," *Nature* 246, 1973, pp. 15–18.

Smith, John Maynard. *Evolution and the Theory of Games*. Cambridge: Cambridge University Press, 1982.

NOTES

1. Communication with Sylvia Stevens, March 18, 2008. On the story of Tolmers Square, see Nick Wates, *The Battle for Tolmers Square* (London: Routledge and Kegan Paul, 1976).

2. See Oren Harman, *The Price of Altruism: George Price and the Search for the Origins of Kindness* (New York: W. W. Norton, 2010) for a full biography.

3. "Jesus Hot Line," *Sennet*, January 15, 1975.

4. Interview with Richard A. Bader, May 7, 2008; " 'Til We Meet Again," *Indicator*, June 1940, 102; "Harvard College Freshman Scholarship Personal Interview Report," May 24, 1940, Harvard University Archive.

5. Claude E. Shannon, "A Mathematical Theory of Communication," *Bell System Technical Journal* 27, July 1948, pp. 379–423.

6. George Price, "Transistor Work Already Started That Would Be Simple and Valuable to Finish," draft, June 27, 1949, George Price Papers.

7. G. R. Price, "Some Relationships of Porphyrins, Tumors, and Ionizing Radiations," *University of Minnesota Medial Bulletin* 27, 1955, pp. 7–13.

8. Samuel Schwartz letter to Dr. Avram Goldstein, January 25, 1978, Box 33, Samuel Schwartz Papers.

9. George Price, "Science and the Supernatural," *Science* 122, 1955, pp. 359–367.

10. George Price, "How to Speed Up Invention," *Fortune*, November 1956, pp. 150–153 and 218–228.

11. George Price, "Arguing the Case for Being Panicky," *Life*, November 18, 1957, pp. 125–128; George Price–Hubert Humphrey correspondence, George Price Papers.

12. George Price articles in *THINK* and *Popular Science Monthly*, March 1959–November 1960.

13. George Price, "The Teaching Machine," *THINK*, March 1959, pp. 10–14.

14. Graham DuShane (editor) letter to George Price, July 19, 1962, George Price Papers.

15. John C. Eccles letter to George Price, January 20, 1965, BL:KPX1_3.2, George Price Collection, British Library.

16. George Price–Paul Samuelson correspondence, George Price Papers. Letter quoted is from January 13, 1966.

17. George Price letter to Kathleen Price, April 11, 1968, George Price Papers. In particular, Price had enjoyed Morris's *The Naked Ape* and Ardrey's *Genesis Flood*.

18. George Price, "Supplementary Details of Intended Research," draft proposal to the Science Research Council, BL:KPX_5.4, George Price Collection, British Library.

19. Shannon, "A Mathematical Theory of Communication," p. 379.

20. Ibid.

21. For a good explanation of Shannon's theorem, see David J. C. MacKay, *Information Theory, Inference, and Learning Algorithms* (Cambridge: Cambridge University Press, 2003).

22. George Price letter to Claude Shannon, October 16, 1969, George Price Papers. No reply was either written or survives.

23. See S. A. Frank and M. Slatkin, "The Distribution of Allelic Effects Under Mutation and Selection," *Genetics Research*, 55, 1990, pp. 111–117.

24. For a technical exposition of the levels of selection debates, see Samir Okasha, *Evolution and the Levels of Selection* (Oxford: Oxford University Press, 2006). On Price specifically see S. A. Frank, "George Price's Contributions to Evolutionary Genetics," *Journal of Theoretical Biology* 175, 1995, pp. 373–388.

25. On these historical debates see Harman, *The Price of Altruism*; Elliott Sober and David Sloan Wilson, *Unto Others: The Evolution and Psychology of Unselfish Behavior* (Cambridge, MA: Harvard University Press, 1998); and Mark E. Borrello, *Evolutionary Restraints: The Contentious History of Group Selection* (Chicago, University of Chicago Press, 2010).

26. This is how Price described the equation to family and friends, as well as to Bill Hamilton. See William Hamilton, *Narrow Roads of Gene Land*, Part I (Oxford: W. H. Freeman/Spektrum, 1996), p. 173.

27. J. H. Morris (editor) letter to Bill Hamilton, April 10, 1970, BL:KPX1_10.6.2, Bill Hamilton Collection, British Library; George R. Price "Selection and Covariance," *Nature* 227, 1930, pp. 520–521.

28. George Price letter to Bill Hamilton, March 5, 1968, George Price Papers. The paper Price refers to is W. D. Hamilton, "The Genetical Evolution of Social Behavior," *Journal of Theoretical Biology*, 7, 1964, pp. 1–16.

29. Bill Hamilton letter to George Price, March 26, 1968, BL:KPX1_4.5.5, William Hamilton Collection, British Library.

30. G. Stonehouse, "Thermoregulatory Function of Growing Antlers," *Nature* 218, 1968, pp. 870–872.

31. Price, "Arguing the Case for Being Panicky," p. 128.

32. George R. Price, *No Easy Way*, draft, George Price Papers.

33. George Price, "Antlers, Intraspecific Combat, and Altruism," unpublished draft, 32 pages, George Price Papers.

34. See Robert Axelrod, *The Evolution of Cooperation* (Harmondsworth: Penguin, 1984).

35. Konrad Lorenz, *On Aggression* (London: Methuen, 1966).

36. He did publish one paper, however; R. C. Lewontin, "Evolution and the Theory of Games," *Journal of Theoretical Biology* 1, 1961, pp. 382–403; R. D. Luce and H. Raiffa, *Games and Decisions: Introduction and Critical Survey* (New York: Wiley, 1958); Interview with Richard Lewontin, December 31, 2007.

37. W. D. Hamilton, "Extraordinary Sex Ratios," *Science* 156, 1967, pp. 477–488.

38. See Oren Harman, "Birth of the First ESS: George Price, John Maynard Smith, and the Discovery of the Lost 'Antlers' Paper," *Journal of Experimental Zoology B* (*Molecular and Evolutionary Development*) 316, 1, 2011, pp. 1–9.

39. J. Maynard Smith and G. R. Price, "The Logic of Animal Conflict," *Nature* 246, 1973, pp. 15–18.

40. See J. S. Gale and L. J. Evans, "Logic of Animal Conflict," *Nature* 254, 1975, pp. 463–464.

41. Maynard Smith's *Evolution and the Theory of Games* (Cambridge: Cambridge University Press, 1982) remains a classic. See also the more recent Martin A. Nowak with Roger Highfield, *SuperCooperators: Altruism, Evolution, and Why We Need Each Other to Survive* (New York: Free Press, 2011) for an example of a career in mathematical biology informed by the theory of games.

42. See, for example, S. A. Frank, "George Price's Contributions"; Samir Okasha, "Why Won't the Group Selection Controversy Go Away?," *British Journal of the Philosophy of Science* 52, 2001, pp. 25–50 (though Okasha here also points to certain limitations); Alan Grafen, "Developments of the Price Equation and Natural Selection Under Uncertainty," *Proceedings of the Royal Society London B* 267, 2000, pp. 1223–1227 and "The First Formal Link Between the Price Equation and an Optimization Program," *Journal of Theoretical Biology* 217, 2002, pp. 75–91; Benjamin Kerr and Peter Godfrey-Smith, "Generalization of the Price Equation for Evolutionary Change," *Evolution* 63, 2009, pp. 531–536.

43. Examples include William B. Langdon, "Evolution of GP Populations: Price's Selection and Covariance Theorem," in *Genetic Programming and Data Structures* (Norwell, MA: Kluwer Academic Press, 1998), pp. 167–208; J. W. Fox, "Using the Price Equation to Partition the Effects of Biodiversity Loss on Ecosystem Function," *Ecology* 87, 2006, pp. 2687–2696; T. Day, "Insights from Price's Equation into Evolutionary Epidemiology," *DIMACS Series in Discrete Mathematics and Theoretical Computer Science* 71, 2006, pp. 23–43; and Stephen C. Stearns, "Are We Stalled Part Way Through a Major Evolutionary Transition from Individual to Group?," *Evolution* 61, 2007, pp. 2275–2280.

44. Matthijs van Veelen, "On the Use of the Price Equation," *Journal of Theoretical Biology* 237, 2005, pp. 412–426; and Matthijs van Veelen, Julián Garcia, Maurice W. Sabelis, Martijn Egas, "Call for a Return to Rigor in Models," *Nature* 467, 2010, p. 661.

45. The usual experience of people with the Price equation is that they find it difficult to decide whether there is something very deep about it that they usually don't understand, or whether it is actually rather useless. My own thinking on the matter has changed somewhat since writing *The Price of Altruism*, and I am more skeptical now about the usefulness of the equation. I thank Matthjis van Veelen for discussions on the matter.

46. Price was very interested in Fisher, and ended up providing the best explanation of his often misunderstood "fundamental theorem." See G. R. Price, "Fisher's 'Fundamental Theorem' Made Clear," *Annals of Human Genetics* 36, 1972, pp. 129–140.

LUIS CAMPOS

OUTSIDERS AND IN-LAWS
DREW ENDY AND THE CASE
OF SYNTHETIC BIOLOGY

18

INTRODUCTION

On the banks of the Charles River on a beautiful sunny day in June 2004, MIT hosted what was billed as the "First International Conference on Synthetic Biology," soon to be known as "Synthetic Biology 1.0." Intended to bring "together, for the first time, researchers who are working to . . . design and build biological parts, devices, and integrated biological systems," as the call for papers stated, the conference also immediately drew international attention from the scientific press: *Nature* sent a total of six representatives, including the editor-in-chief. Although the conference organizers expected only 150 participants to attend and booked a room that could hold twice that, over 500 people expressed interested in attending with six weeks' notice. As an interdisciplinary scientific meeting—a heterogeneous group putatively gathering together for the first time, and intentionally proclaiming itself to be doing so—the first synthetic biology meeting was almost by definition a group of outsiders. Many participants felt they were coming from distinctly different disciplinary approaches, and more than a few seemed concerned that their research might not be "proper" synthetic biology.

Why couldn't engineers hold themselves to similar standards of success in biological systems? asked the conference organizer that day in June. Couldn't the biological world be tamed in the same sorts of ways and with the same sorts of tools with which engineers had already experienced such great success in taming the nonliving world? Another speaker described repressible promoters as being "functionally identical" to transistors, while others described the biological and genetic equivalents of logic gates, toggle switches, and oscillators. To many who had been trained in more classical modes of biology, such talk of "biological circuits" and "design constraints" seemed utterly foreign; to others, this was entrancing. One member of the audience, seeing a genetic circuit diagrammed on the screen, whispered to his neighbor: "it looks cool to see it look like a . . . an integrated circuit!" A unique synthesis of engineering with biology was underway, and the event

Figure 18.1: Drew Endy.
Photograph by Sam Ogden/Photo Researchers, Inc.

would later be described as "the first conference of its type, anywhere."
Behind this gathering of outsiders was the young, soft-spoken, and charismatic assistant professor, Drew Endy (see figure 18.1).

AN OUTSIDER INAUGURATES A FIELD:
DREW ENDY AND THE ENGINEERING IDEAL IN BIOLOGY

As the mild-mannered, witty host and one of the principal conveners of
the first conference, Endy had something of the quality of a visionary and
prophet—a true believer in the power of what he was calling "synthetic biology" to change the world for the better. Endy was (and remains today) one of
the foremost advocates of this newest introduction of an engineering ideal

into biology. He has on various occasions repeated his call for an "open source biology" based on "tools of mass construction," with the ultimate goal of "rebuilding the living world." Complaining at the 1.0 meeting that what had been called genetic engineering "doesn't look or feel like any form of engineering," Endy shared a vision of biology as a kind of black-boxable technology, rather than as something immutable and natural. Would-be synthetic biologists at the 1.0 meeting similarly dismissed genetic engineering as mere "genetics . . . nothing more than selective breeding." Doing away with inherent inefficiencies of natural systems was a key first step in emphasizing the centrality of design and in constructing biological systems with desired functionality.

Such a vision for biology was not one ordinarily held by biologists. Indeed, Endy noted early on that "many of the folks coming into synthetic biology are not biologists," counting himself among that number.[1] Born in 1970 and raised in Pennsylvania, Endy described himself as having long been fascinated by "Legos, Lincoln Logs, this and that." He received a D in high school biology, however, ostensibly for failing to commit to memory the Latin names of 200 insects. ("That didn't go so well.") Matriculating at Lehigh University ("next to Bethlehem Steel") to train as a civil engineer, he earned his bachelors (1992) and master's of science (1994) degrees in civil and environmental engineering, respectively, and worked for a time on problems of wastewater treatment. He also stumbled into a molecular genetics course after another engineering course he intended to take, on structural dynamics, was canceled. These inadvertent exposures to biology led Endy to Dartmouth College, where he received his PhD in biochemical engineering (1998) under "an engineer who knew something about DNA." Although the narrative would seem to be a steady path from the physical and toward the biological, Endy has insisted that "that's not it . . . the simpler way to say it is, I like to build stuff, and biology is the best technology we have for making stuff—trees, people, computing devices, food, chemicals, you name it. I somehow found my way to biology."[2]

Endy was fundamentally interested in biology "not as a science, but as a technology platform."[3] As he would later put it: "I come from the clan of the opposable thumbs. I like to make stuff."[4] He soon realized, however, that the computer models he had been using to explain the changing "architecture of the natural genetic systems" were insufficient, and that he would have to get his hands wet with lab work. He began to learn the basics of molecular biology and trained in microbiology and genetics while completing a postdoc in "one of the last bacterial virus labs in the country" at the University of Texas–Austin, with Ian Molineux. Endy's growing familiarity with mapping and cloning DNA prepared him for a move in the late 1990s to Berkeley, California, where he would work with Roger Brent and Sydney Brenner at the newly established, not-for-profit Molecular Sciences Institute (MSI), "where our mission was to go do the next generation of biology, whatever makes sense." Endy's attempt at being an insider was not the easiest, and he concluded that his modeled predictions and his attempts to "refactor" (or systematically redesign) the T7 phage with more than 600 simultaneous mutations were, by and large, failures: "I would want one behavior, and when I went to make the change, exactly the opposite would happen." Ever the engineer, he sought to conduct a "failure analysis" and learned that his modeling tools simply "weren't good enough to support purposeful determinative changes that result in the behavior that I expect." Endy ultimately concluded that "evolution is not selecting for designs of natural biological systems that we can understand, the things we inherit from the living world have not been selected for ease of understanding, let alone ease of manipulation. It's not part of evolution's objective function."[5]

To be able to successfully model, much less ultimately engineer biological systems, a choice was called for. Endy decided that he could either "go back and understand a whole bunch more about the science of the organism in order to model it better"—the insider approach, and what he called "a fine and valid traditional path" at the heart of the institute's approach—or he could play the outsider and start from scratch. "I thought, screw it," Endy recalled. The complications of biology were of little interest. "Let's build new biological systems—systems that are easier to understand because we made them that way," he concluded.[6] Such unusual ideas were not yet commonplace at the time, and yet this was the moment that Endy later characterized as his "transition into what is now called synthetic biology."[7]

While at the MSI, Endy had ongoing conversations with Brent and with fellow outsider Robert Carlson about the nature of a new approach to biological engineering they tentatively called "open source biology," in direct reference to the open source software movement.[8] This dovetailed with Endy's goal of developing a framework that would permit the free exchange of standardized biological parts, and which, again in analogy to the realm of software engineering, could eventually contribute to the development of a thriving industry based on "foundational technologies." The name did not catch on, however, and an alternative name for the new field—"intentional biology"—similarly faced catcalls of derision, failing to go over well with traditional biologists who resented the implication that their biology was somehow "unintentional."[9]

Endy had first tried to drum up funding for his new approach in October 1999 by coauthoring a report on "A Standard Parts List for Biological Circuitry" for the ultimate insider agency, the military's Defense Advanced Research Projects Agency (DARPA).[10] The agency declined to fund the proposal, and Endy later acknowledged that his vision was outside the norm: "It's a radical research agenda. The idea of standard biological componentry is either dismissed as a research question because [people think] it's irrelevant or dismissed as a research question because [they believe] it's impossible."[11] Endy's idea that "we should rebuild the natural biological systems we most care about and domesticate their genomes" remained clearly foreign to most biologists.[12] But Endy persisted: "I started broadcasting that idea in the 1990s," he recalled, and "the only person who returned a coherent signal intellectually was Tom Knight, in the electrical engineering department at MIT."[13]

THE OFFICE OF BIOLOGICAL DISENCHANTMENT

Endy had found another "outsider" colleague with a similar vision. Knight, a computer scientist by training, had first met Endy about five years after Knight himself began working on questions in biology. The two found that they shared a common vision for an engineered biology based on standardized biological parts, which Knight had begun to theorize in 2001. As Endy joined MIT as a Fellow in Biology and Biological Engineering in 2002 in a relatively new Department of Biological Engineering (founded in 1998), "synthetic biology" came to find one of its first homes at the Computer Science and Artificial Intelligence Laboratory (CSAIL), where Knight, a senior research scientist, wrote up a white paper on the idea of a standardized biological part in 2003.[14]

Endy became a faculty member a year later, but even so faced repeated challenges in communicating his engineering vision of life to more traditionally oriented researchers in biology:

> I run into a curious problem, a reaction I get from fellow researchers, biologists, physicists when talking about designing biological systems. They say, "Biology is not like that." And what that means usually is there is something magical, there is something enchanted, about life—and certainly there is from an experiential perspective—which doesn't apply to biology. But I think we need to disenchant the molecular details of biology without destroying its magic. So that we begin to navigate a path where the idea of design is accessible, but a sense of awe is maintained.[15]

Knight, famous for the utterance "the genetic code is 3.6 billion years old—it's time for a rewrite"—shared Endy's frustrations.[16] Few biologists, however, seemed taken with the vision of biological engineering with parts: "They are not excited. Nor should they be. It's a different agenda."[17] It was a familiar agenda at MIT, however, an institution with a long tradition of hacking. Endy repeatedly noted that he wanted to engineer biology along the lines of the electronics, software, and Internet industries—indeed, the 1.0 conference included an address by one of the creators of the Internet. But Endy found biology "much cooler than electronics or computers. . . . It's the stuff of life."[18] As the open-source ethos conceived of the genetic code as a code like any other that could be hacked, what had once been an increasingly moribund metaphor—the genetic code—was soon resurrected into a second life. When a 2004 issue of *2600: The Hacker Quarterly* ran an unsigned manifesto by a mysterious "Prof. L" calling for the hacking of life, some even suspected Endy as the author (he wasn't). Endy's commitment to an outsider-oriented open source ethos for synthetic biology took him down increasingly unusual paths for one busy working with genes, and he was soon invited to address other hacker constituencies, including the world-famous Chaos Computer Club of Berlin. Some of Endy's own students even began to wonder about the possibility of a "do-it-yourself" (DIY) biology, hacking biology outside the walls of academic laboratories.

Although initially programmatic, Endy's call for an open source approach to standard biological parts turned increasingly pragmatic as he viewed open source parts as the best way of permitting the "uncoupling of concept and manufacture" from each other—two modes he felt were tightly allied in most biological research to date. With a future toolkit of "parts" at hand, Endy argued that researchers would not only be able to share their creations more readily, but would in the first instance be more free to think conceptually about possible interrelations and inventions without having to worry about inventing the parts themselves. "When we arrive at the future with a first generation of parts that can work together," he said, "we'll have the parts open and free, and people will be able to build what they want."[19]

Countering possible critiques that there was no such thing as a standard biological part and that biological functions always depend on environment—critiques that came even from his own friends and colleagues—Endy injected a quick dash of humor into his presentation at the 1.0 meeting: naysayers and critics could send their complaints to the "Office of Biological Disenchantment, MIT 68-580, 77 Massachusetts Ave., 02139." (This was Endy's office.) Even more importantly, Endy encouraged fellow true believers in the community to send him reports of their failures, so that these could be organized into "data sets" and analyzed to greater purpose.

REFACTORING BIOLOGY AND THE ETHOS OF COOL

The hacking ethos of the outsider was not universally shared, however, and conference participants at the 1.0 meeting spent more than a few moments clarifying among themselves whether "hacking" was to be considered an unadulterated good or not (those bred in MIT's culture generally seemed to claim that it was, while some others felt that "hacking" conveyed problematic mischievousness or even criminality). In fact, the successful social application of engineering principles and terminology to biology would take some effort. At the 1.0 meeting, the newness of this interface between "biology" and "engineering" was most evident in the polite confusion that at times reigned in the audience, demonstrating the difficulties of establishing a common discourse and a common identity—of demarcating an "inside" and an "outside" to the field. A certain insecurity of method and of presence infected speakers and audience alike, with much apologizing for one's particular disciplinary interest or methodological approach. Most participants at this meeting were obviously not quite sure what synthetic biology was and said so repeatedly, a situation that began to change already by the 2.0 meeting. And yet, for several years afterwards, it still remained markedly unclear just who really counted as a "synthetic biologist" as other competing "schools" of synthetic biology began to emerge. "Synthetic biology" was arguably at first more a like a collection of symptoms than a coherent new discipline.

Endy, however, had an unequivocal vision of what "synthetic biology" meant: a better-engineered biology based on clear design principles and functional modular components. His two-pronged approach thus involved both "refactoring" (or simplifying the genetic make-up from its natural state of confusion to a state of clear design) and the establishment of "foundational technologies," including the setting of standards and the creation of standardized biological parts. Endy took familiar conceptual tools from engineering—"synthesis, standardization, and abstraction"—and actively applied these to biology, claiming that these were as important as the more familiar tools of recombinant DNA, PCR (polymerase chain reaction), and sequencing were to molecular biologists. In his opening salvo at the 1.0 conference, Endy had applied these ideas of "functional composition" to break down basic biological functions in ways that might be encoded as genetic material. He also promoted a novel nomenclature of an "abstraction hierarchy" with tripartite "levels of complexity": "parts," made of protein or DNA, that could be combined, given numbers, and "black-boxed" to make "devices" at the next level up, which he defined as "combinations of one or more parts that encode human-defined functions." These devices, in turn, were to be designed to be able to "hook up" one to another. "Systems," then, ultimately, were "combinations of one or more devices that encode human-defined functions." The end goal of his enterprise was nothing less than the reliable synthesis of biological systems that, as he would later write in *Nature*, "behave as expected."[20]

Endy's vision as he presented it at 1.0 was entrancing, and his outline of a new synthetic—and not just analytic—biology was as electrifying as it was foreign. "Will the substrate of biology permit this?" Endy asked the room provocatively, referring to the doubts of his biologist friends. He had no such doubts himself. Ever the resourceful outsider, he began to refactor life. To bring about his vision of standard interchangeable black-boxed biological parts, Endy envisioned the creation of a system of "BioBricks," or "short pieces of DNA that constitute or encode functional genetic elements," such as promoter sequences, terminator sequences, repressors, ribosome-binding sites, and reporter genes.[21] Endy envisioned that, like Legos, these BioBricks could then be arranged to form genetic circuits. "In other experiments," Endy said, "BioBricks have been combined into devices that function as logic gates and perform simple Boolean operations, such as AND, OR, NOT, NAND, and NOR."

Another effect of standardization was a transformation of a classic approach that used "model organisms" to study biological systems. By boiling all organisms down to what were taken to be their most fundamental and important characteristics—a set of basic parts of the simplest organisms that encode the fundamental basis of life, and from which all higher complexities could be constructed—synthetic biology seemed to promise to "vanish" the model organism itself. Rather than deal with all the particular idiosyncrasies of particular model organisms, Endy's vision of synthetic biology instead had as its goal the invention of "parts" that would presumably work across organisms and their systems in any number of theoretically possible and engineerable ways. This was the promise of an ultimate outsider biology—biology outside the confines of the species or the organism. It was a vision of biology outside the comfort zone of many biologists, as well.

Employing the ultimate outsiders in a university research environment—undergraduates—Endy and colleagues set about the wholesale synthesis of a synthetic biology community, first through a winter-term class and then through a "jamboree" known as the International Genetically Engineered Machines competition (iGEM), which came to function in part as the major source of new BioBrick parts. Over the years, as attendance doubled nearly every year—from four teams in 2004 to 165 teams and over a thousand participants by 2011—the number of parts grew accordingly. Together, Knight and Endy created the Registry of Standard Biological Parts (originally http://

parts.mit.edu, now available at http://partsregistry.org), which is both an online registry to keep track of these parts and a real-world deep-freeze repository. By mid-2006, there were already some 167 basic BioBricks and 421 parts in the Registry, and in short order iGEM organizers were regularly shipping multiple tens of thousands of parts per year.[22] Assorted iGEM paraphernalia—T-shirts, buttons, and stickers encouraging participants to "Share your science!"—not only marked synthetic biology as cool, they also helped to convey the field's purportedly free and open spirit. The dream of a synthetic biology based on engineering principles of design and simple, interchangeable, interrelatable parts was beginning to be realized, all through the ethos of what an undergraduate might consider "fun."

Claiming for itself a certain aura of revolutionary "outsiderness" and deliberately seeking to distinguish itself from institutionalized and academic biology, synthetic biology's ethos of cool was pervasive, spread rapidly, and worked well in attracting the younger generation into synthetic biology. And Endy himself, time and again, suggested that synthetic biology was novel and revolutionary. He attempted to convey the field's youthful appeal and outsider edge even beyond the iGEM competition, most notably through the production of a comic book called *Adventures in Synthetic Biology*. The comic, illustrated by Chuck Wadey of *Spiderman* fame, with its boy-protagonist "Dude" (named after a character in the 1998 cult classic film "The Big Lebowski") was even eventually published in *Nature* alongside Endy's visionary manifesto for the future of synthetic biology.[23]

Consciously or unconsciously, Endy himself also seemed to embody this ethos of cool, whether from his interests in skateboarding to kite surfing in Kahului. As one reporter noted, "Endy emits a sense of barely contained energy; he'll speak in fluid, page-length arguments while absentmindedly picking up and twirling a small end table, or tapping his foot so animatedly that his sandal flies off."[24] When not playing the set of conga drums in his office or making cut-glass mosaics of his beloved T7 phage at his lab bench, Endy envisioned himself outside of the laboratory either "surfing . . . [or] writing poetry and surfing."

Within a few short years, Endy was profiled in both *Esquire* (as one of the 75 "best and brightest" and "most influential people" of the twenty-first century) and the quirky science-feature magazine *Seed*, among others. Even years later, as Endy began to downplay the "revolutionary" angle of synthetic biology for a variety of reasons, "cool" talk continued to permeate his vision of synthetic biology as he worked tirelessly behind the scenes to help run the iGEM competition, to champion the "flagship" conference series, and to establish the nonprofit BioBricks Foundation.[25]

But outsider status is relative, and Endy himself was emerging as a foundational figure within the growing constellation of synthetic biology. Viewed from outside of the community of synthetic biologists, Endy's central role in envisioning and developing the emerging field defined him as the ultimate *insider*. This has been most evident in the response his vision received from some European scientists, from some North American civil society organizations, and from the complicated role he has taken negotiating the future of synthetic biology vis-à-vis the economic and security interests of the U.S. federal government.

A EUROPEAN BACKLASH

When the Swiss Academy of Sciences had *Adventures in Synthetic Biology* translated into German, one observer noted, the reaction was "not very positive." In the United States, the narrative told of the boy-protagonist (whose forgetting to insert a regulator in his bacterium led to its filling up with gas and uncontrollably exploding) would be "seen as progress, you've done something, learned something, now try again." But the idea of people "manufacturing life and making mistakes was like the horror vision of Europeans," according to Markus Schmidt, director of the EU-funded Synbiosafe Project. Indeed, some of the most striking early responses to Endy's vision came from researchers gathered at the European Science Foundation's first European Conference on Synthetic Biology, held in November 2007 in Sant Feliu de Guíxols, Spain. A Spanish investigator, Victor de Lorenzo, spoke publicly and humorously there, attempting to put the differences in "very very extreme terms . . . even at the risk of being a little politically incorrect . . . to make a cartoon" of what he saw as the differences between the "American" views of synthetic biology promoted by Endy and a broader set of "European" views.

De Lorenzo shared the sense of many European biologists, who were confounded that Endy had claimed a revolutionary status for his field. Many of the techniques that synthetic biology depended on—DNA synthesis and PCR, among others—were the results of decades of work by many researchers around the world: "You cannot ignore that," de Lorenzo insisted. Even the proposed new special jargon of the field, like Endy's proposed polymerases-per-second (PoPS) unit of measurement, was alienating: "I worked for twenty years in transcription and to this day I have no idea what PoPs means," de Lorenzo said. Endy was here understood less as an "outsider" overturning established traditions of biology than as a brash American using the extravagant and alienating rhetoric of "revolutionary" discovery to reappropriate knowledge and shared techniques from a global commons. Indeed, Endy's vision was viewed by another Swiss researcher not as some kind of cool revolutionary breakthrough, but rather as "typically American salesmanship."

The emerging "European" view at this conference held that the claim that engineers were "refounding" biology was problematic: "I think that in Europe we see synthetic biology as an empowering of existing interfaces. There's something new but there's also something of tradition," de Lorenzo said. Participants proposed that perhaps there was a need for an explicitly "European" approach to synthetic biology, which would be based instead on the coordination of a broad array of existing areas of research under one umbrella—a different meaning of "synthesis" that would leave fewer traditional biologists out in the cold. For de Lorenzo, synthetic biology was "about the improvement of existing interfaces, not foundational technologies and other things. Europeans like to have roots." (A Spanish student later corroborated this view, saying: "Europeans are more skeptical of [synthetic biology], being how we, in Europe, we are used to having a history.") Whether this next relabeling of synthetic biology was a European community-building exercise or something else, it was clear that the standardized parts–based view of synthetic biology was not the only definition Europeans would countenance.

POLICING THE BOUNDARIES

"That's not synthetic biology," Endy noted quietly a few months later, upon hearing of the European discussion and their efforts to include a wide variety of approaches not necessarily having anything to do with standardized parts under the new label for the field. But endless debates about the definition of "synthetic biology" and other assorted niceties were of little interest to Endy. Although he would publicly often remain pluralistic about various meanings that emerged, in defining and ultimately in defending a particular view of synthetic biology himself he was actively defining an inside and an outside. In fact, soon after, Endy would help to create the BioBricks Foundation, in large measure initially to trademark the term "BioBrick" as a means of protecting the "integrity" of his vision of interchangeable parts when the perceived integrity of the moniker was threatened as many different standards for interchangeable parts began to emerge, and as incompatible submissions were being added to the Registry

The policing of the boundaries of synthetic biology was not only metaphorical or legal, however—it was also often quite literal, further delineating insiders and outsiders in ways that challenge any easy characterization. While early wikis from Endy's lab at MIT had discussed visions of a "Synthetic Society" debated at "synthetic teas," an actual discussion of "risk assessment" at the 1.0 meeting was preceded by a request from Endy for all attending journalists to leave the room. A police officer was stationed in front of the auditorium doors. Claims of openness were similarly policed at the 3.0 meeting in 2007, where registrations and nametags were checked on entry to the plenary session, and a burly Swiss security guard was stationed at the front of the room during the first several talks, scowling intensely at the audience. It was clearly established that the conference would not suffer interruptions lightly.

This move was likely due in some measure to concerns about an activist response to the new field. During the third day of the 2.0 meeting at UC Berkeley in 2006, which was devoted to the broader issues and implications of synthetic biology, Endy and other conference organizers had faced a vigorous response from a coalition of more than two dozen public interest groups, believed to be led by the ETC Group based in Ottawa, Canada. An open letter from these organizations was presented to conference participants expressing concern not only that proper safeguards were not in place for such novel synthetic forms of research, but that synthetic biologists were conducting "closed-door" sessions and deciding future policy without broader participation by all stakeholders and a larger public, an echo, the activists said, of the famous 1975 biosafety meeting at Asilomar where the potential risks of novel efforts at genetic engineering were discussed. Such efforts, then and now, were characterized as inappropriate attempts at self-governance. Discussions centering around questions of security, risk, ethics, and community at the 2.0 meeting were clearly affected by the politics of the day. After an hour of discussion on the final afternoon, Tom Knight reflected both the general mood of the room, and the unsettled state of the newly emergent field: "We don't even know who is a synthetic biologist and who is not. It's probably a little premature to decide who's in and who isn't." Real-time discussions in a politically tense situation meant that any established categories of insider and outsider were rapidly being challenged.

An earlier plan to issue a public conference statement was shelved, with post-conference debates and suggestions to be hashed out on an open wiki instead, open to general amendment. When this was announced, a member of the audience immediately cried out in concern: "But that's open to the world!" Endy responded reassuringly but cryptically: "There will be a strong filter in place and we'll see what makes it in." This remarkable statement suggested that Endy was clearly no longer simply an "outsider." Rather, he was now helping to negotiate the place of the many newly arrived institutional "in-laws" who had to be accommodated and respectfully engaged with, even as they questioned Endy's vision of a happy marriage between engineering and biology. Seeking to raise social awareness, ethical considerations, and perhaps even legal injunction, these would-be in-laws would not forever keep their peace.

Indeed, the positioning of community activists with respect to emerging synthetic biology "insiders" continued to be negotiated with each passing flagship conference and interaction. At the 3.0 conference in Zürich, the ETC Group once again made its presence and views known, with a delightful illustrated poster drawing inspiration from the film *The Little Shop of Horrors*.[26] Following the experience at the 2.0 meeting, conference organizers at both the 3.0 meeting in Zurich and the 4.0 meeting, held in Hong Kong in 2008, offered sessions and panels that actively included community activists from the ETC Group and other civil society associations. The exclusion of several such groups almost entirely from attendance and participation at the 5.0 meeting in 2011 at Stanford—inciting the drafting of another open letter of concern—continued to further complicate questions of just who was inside and outside in synthetic biology.

THE OUTSIDER WELCOMED IN

Synthetic biology's potential drew national and international attention in rapid order, from industry and finance as well as from activists and protestors. With funding streams, corporate and political attention, and growing media coverage, synthetic biology had moved from a couple of outsiders' utopian visions to a full-scale intermixing of outsiders, outlaws, and in-laws all linked by circumstance. This complex set of strange bedfellows, involving scores of practicing scientists and engineers, as well as social scientists, policymakers, ethicists, civil society groups, lawyers, and DIY enthusiasts, further challenges any easy demarcation of "insider" and "outsider," or revolutionary outlaw and concerned in-law. Just who counted as a synthetic biologist or member of the synthetic biology community remained contested.[27]

Roger Brent would later identify Endy's approach as falling "outside the canon of . . . successful American biomedical research. [Synthetic biology] did not arise from the gifted experimental biologist well-funded by the NIH or the Howard Hughes Medical Institute, rising to the top of a hill and seeing beyond it to the next hill and then going to that next hill. These are clear offshoots of what would have been the predictable growth of the field."[28] As the field continued to mature, however, Endy was becoming part of a transdisciplinary inner circle of high-powered "thought leaders" who would sometimes jet to exotic locations and issue pronouncements on the future of synthetic biology, such as occurred in the case of the Ilulissat Statement in Greenland in 2007.[29]

But Endy's growing insider status was arguably nowhere more evident than in his increasingly frequent trips to Washington, DC, where he began to offer expertise and testimony, becoming if not quite a Beltway insider then at least a respected voice in relevant hearings. Endy became an ad hoc member of the Recombinant DNA Advisory Committee, a member of the Committee on Science, Technology, and Law at the U.S. National Academies, and was nominated to serve on the U.S. National Science Advisory Board for Biosecurity (where he had the final comment at a 2005 meeting, despite not being on the board). He also testified to the Congressional Committee on Energy and Commerce on the "Effects of Developments in Synthetic Genomics," on May 27, 2010—just hours before the representatives of British Petroleum were called in for questioning about the 2010 Gulf of Mexico oil spill. At nearly the same time, he was also one of a handful asked to testify before the Presidential Commission for the Study of Bioethical Issues in July 2010 regarding Craig Venter's work creating a minimal organism. Endy was slowly but surely transitioning from the outsider hacker and biological naïf with cool hobbies and a putatively revolutionary outlook, through the self-described "cat herder," central organizer, and key-player phase, to finally becoming a voice of transposed authority for the entire field, as the holder of the synthetic biology umbrella for ever-enlarging realms of nonscientific outsiders in need of expertise.

Hand in hand with this transition was Endy's switch from a more revolutionary rhetoric to a strategic downplaying of the novelty and risk of the new field when faced with powerbrokers in higher circles for whom such qualities might have raised concern.[30] Years of fraught encounters with larger publics and in-laws had evidently taught him to tone down the rhetoric. As conversations about the status and future of synthetic biology moved from conference sessions to higher circles of administrative power potentially interested in the field's economic potential, Endy apparently found it wiser to claim that the field was continuous with earlier (already regulated) efforts, and that any potential problems raised were common to all of biotechnology, not just synthetic biology.[31] As a diplomatic technique, this mellowing appeared to work, causing political decision makers to frequently echo his pronouncements. Suggesting that proper frameworks were already in place and that there was essentially "nothing new to see here" was a powerful rhetorical tool at a time when "outsiderness" could be and was sometimes construed as a potential national security threat. As the FBI became increasingly interested in the emergence of a hacker-driven "garage biology" driven by independent DIY enthusiasts, for example, it was increasingly strategic for Endy to play the insider rather than the outsider—to be the respectable "in-law" of a different sort, rather than the revolutionary outlaw. And yet the long-term value of his specific technical contributions seemed modest, even within the field of synthetic biology. As his tenure clock at MIT ticked loudly, and as affairs of the heart happily drew him westward and toward in-laws of the most traditional sort, Endy engineered a move to Stanford University in 2008, a relocation that would have clear professional consequences. No longer a mere visionary of an open source biology, Endy could now jumpstart his new vision of an bioindustrial fabrication laboratory, named BIOFAB, that would remake biology, and attempt to be both outsider and insider at once.

ENDY, FÊTED

Synthetic biology today remains an uneven collection of assorted approaches and schools, from standardized parts and metabolic engineering, to minimal genomes and cells, to origins of life research, with no clear center and no one dominant approach. There is no one scientific or engineering core to it, and it is as much a social phenomenon as an intellectual one. Seen in this light, Endy's compelling vision and his remarkable social accomplishments in developing a community in this new field cannot easily be overestimated. The attention given Endy is surely the result, at least in part, of his great charisma, the wealth of source material he has made available through public statements, interviews, and essays, and commonplace cultural narratives of disciplinary founding and formation (the quest for a "founding father"). And while it is clear that Endy's presence in the field has been instrumental to its development in a variety of ways, it is by no means clear that his vision of "synthetic biology" is even the dominant one in practice. Rather than seek to ensconce Endy as a "founding father" figure, therefore, it has seemed more instructive here to analyze Endy's path to synthetic biology and his rise to power within the newly emerging field as riven by ongoing negotiated shifts between "inside" and "outside." Nowhere was this transformation from outsider to insider better illustrated than at the conclusion of the 5.0 meeting at Stanford in June 2011. Amidst this gathering of putative "outsiders"—undergraduates, graduates, engineers, computer scientists, and more, all now "synthetic biologists"—Endy was given an impromptu and standing ovation during the awards banquet on the final evening of the conference for his many contributions to the field. Having successfully established a remarkable international movement for a new field traveling under the name of "synthetic biology," Endy had both entered into the halls of power and been duly received by the field he helped to synthesize from scratch. The outsider, now fêted, was in. He had arrived.

Campos, Luis. "That Was the Synthetic Biology That Was." In *Synthetic Biology: The Technoscience and Its Societal Consequences*. Edited by Markus Schmidt, Alexander Kelle, Agomoni Ganguli-Mitra, and Huib Vriend, 5–21. London: Springer Academic Publishing, 2009.

Campos, Luis. "The BioBrick™ Road." *Biosocieties* 7 (2012): 115–39.

Carlson, Robert H. *Biology Is Technology: The Promise, Peril, and New Business of Engineering Life*. Harvard University Press, 2010.

Endy, Drew. "Foundations for Engineering Biology." *Nature* 438 (November 24, 2005): 449–53. (Includes comic book *Adventures in Synthetic Biology*.)

The ETC Group. *Extreme Genetic Engineering: An Introduction to Synthetic Biology*. 2007. http://www.etcgroup.org/upload/publication/602/01/synbioreportweb.pdf

Keller, Evelyn Fox. "What Does Synthetic Biology Have to Do with Biology?" *BioSocieties* 4 (2009): 291–302.

NOTES

1. Endy, speaking at the BioBricks Foundation, "Technical & Legal Standards Workshop 2," UCSF, March 1, 2008.

2. All quotes from "Engineering Biology: A Talk with Drew Endy." http://www.edge.org/3rd_culture/endy08/endy08_index.html/

3. Kara Platoni, "Assembly Required," *Stanford Magazine* (July/August 2009). http://alumni.stanford.edu/get/page/magazine/article/?article_id=29598/

4. Ibid.

5. All quotes from "Engineering Biology."

6. Oliver Morton, "Life, Reinvented," *Wired* 13.01 (January 2005). http://www.wired.com/wired/archive/13.01/mit.html/

7. "Engineering Biology."

8. For more on the intellectual property dimensions of synthetic biology, see Luis Campos, "The BioBrick™ Road," *Biosocieties* 7 (2012): 115–39.

9. For more on the history of the naming of "synthetic biology," see Luis Campos, "That Was the Synthetic Biology That Was," in *Synthetic Biology: The Technoscience and Its Societal Implications*, eds. Markus Schmidt et al. (London: Springer, 2009), 5–21, 17–18.

10. Drew Endy and Adam Arkin, "A Standard Parts List for Biological Circuitry," DARPA white paper (October 1999), doi:1721.1/29794. http://openwetware.org/wiki/Endy:Reprints/

11. Alla Katsnelson, "Brick by Brick," *The Scientist* (February 1, 2009). http://www.the-scientist.com/?articles.view/articleNo/27099/title/Brick-by-Brick/

12. Morton, "Life Reinvented."

13. "Engineering Biology."

14. T. F. Knight, "Idempotent Vector Design for Standard Assembly of Biobricks." MIT Synthetic Biology Working Group Technical Reports (2003). http://hdl.handle.net/1721.1/21168/

15. Adam Bly, *Science Is Culture: Conversations at the New Intersection of Science + Society* (New York: Harper Perennial, 2010), 71–72.

16. Lee Silver, "Life, 2.0," *Newsweek* (June 3, 2007). http://www.thedailybeast.com/newsweek/2007/06/03/life-2-0.html

17. Peter Aldhous, "Redesigning Life: Meet the Bio-Hackers," *New Scientist* 2552 (May 20, 2006): 43–47, 46.

18. Chris Jones, "How to Make Life," *Esquire* (November 20, 2007). http://www.esquire.com/features/best-brightest-2007/synthbio1207/

19. "Designer Genes," *Good* (March 20, 2008). http://www.good.is/post/designer_genes/

20. Drew Endy, "Foundations for Engineering Biology," *Nature* 438 (November 24, 2005): 449–53.

21. J. B. Tucker and R. A. Zilinskas, "The Promise and Perils of Synthetic Biology," *New Atlantis* 12 (2006): 25–45.

22. For further details, see Campos, "The BioBrick™ Road."

23. Endy, "Foundations for Engineering Biology." Also available at http://www.nature.com/nature/comics/syntheticbiologycomic/

24. Platoni, "Assembly Required."

25. After Endy moved to Stanford, a new project at his recently inaugurated BIOFAB—funded to the tune of $10 million—had a rap-star-like "C-dog" component (referring to the "Central Dogma").

26. Ironically, the ETC Group's report, entitled "Extreme Genetic Engineering: An Introduction to Synthetic Biology," is now often referred to by practicing synthetic biologists as a clear exposition and sort of a basic primer to the field. The report is available at http://www.etcgroup.org/node/602/

27. Before being allowed to attend the 1.0 conference, I was interviewed by the conference organizers, who—among other things—wanted to make sure I would not be writing an exposé. At the time of the 2.0 meeting, I was at first told that the conference was oversubscribed and that there was no room for me to attend. I appealed the decision and attended. In 2007, at the 3.0 conference, I was awarded the conference prize for "best poster in social aspects."

28. "Roger Brent and the Alpha Project," *Ubiquity* (March 2004), doi:10.1145/985614.985615.

29. "Synthesizing the Future: A Vision for the Convergence of Synthetic Biology and Nanotechnology." http://www.research.cornell.edu/KIC/images/pdfs/ilulissat_statement.pdf/

30. Such a move had been encouraged in several previous conference discussions, where some participants had also suggested the difficulties inherent in the label "synthetic" (suggesting instead, among other labels, "happy shiny biology") and the search for the "killer app" (suggesting instead the "cure my grandmother app," to avoid semantic confusion).

31. In their 2007 report on "Extreme Genetic Engineering," the ETC Group had predicted that this sort of downplaying of the claims of synthetic biology would

become common: "Not 'business as usual': ETC Group notes that some synthetic biologists are beginning to shun the spotlight and may seek to avoid public scrutiny by asserting that it is impossible to clearly distinguish their work from earlier advances in recombinant DNA technology (genetic engineering). Because synbio is all part of the same toolbox, they argue, it simply isn't possible to compartmentalise their research for purposes of regulatory oversight. This refrain ('synthetic biology is really nothing new') is likely to be heard often in the coming months and years."

RICHARD C. LEWONTIN

EPILOGUE
THE PROBLEM WITH BOXES

The reader of *Outsider Scientists* might begin with the impression that the stories it relates are somehow examples of intellectual adventurism arising from an unrelated collection of idiosyncratic motivations on the part of unusually ambitious intellects. After all, no matter how intelligent one is, the road to fame in higher mathematics is rocky and poorly paved as compared with the superhighway to entrepreneurial success in genomics. While this may be true in some cases, to make such a generalization about mathematics and biology would be a mistake. There have also been, in contrast, extremely successful practitioners of other sciences who, heady with success, felt themselves equal to a quite new challenge and so entered biology. In a number of cases their optimism was not misplaced. A great deal of the early advance in molecular biology came from people like Max Delbrück who, after a distinguished career in physics, founded a major school of molecular genetics.

The entry into biology of a diverse set of investigators from various intellectual backgrounds is, in part, what we would expect from the growing phenomenological materialism that began in the mid-nineteenth century and that traces back to the *bête machine* of Descartes and *l'homme machine* of La Mettrie. If ecological communities are an expression of the activities of interacting species populations at multiple trophic levels, and species are collections of interbreeding and closely interacting organisms, and organisms are structured "machines" made up of communicating organs, and organs are made up of interacting cells, and so on down to the three-dimensional structures of macromolecules that determine metabolic activities but are subject to molecular noise and quantum uncertainty, then there are no rigidly defined ontological "boxes" into which biological phenomena fall, in principle. But then we have no clear-cut criteria for erecting the boundaries of epistemological boxes, either. Was the introduction of DNA sequencing into population genetics as a method for evaluating genetic variation in natural populations more influential as an epistemic or an ontological advance? It told us both how to find something out, as well as important facts about natural genetic variation.

A great deal can be learned about the epistemic history of "biology" from the academic structure of teaching and research at various times. While the study of the large molecules that are characteristic of organisms might appear to the uninitiated as an obvious example of "organic" chemistry, courses in Organic Chemistry when I was an undergraduate at Harvard in 1946–1951 were usually taught in the Chemistry Department, but Biochemistry belonged to the Biology Department. That is still the case. Although the active molecular cores of hemoglobin, myoglobin, cytochromes, and chlorophyll are all minor variants of porphyrin, not much was said about this extraordinarily important chemical structure in organic chemistry textbooks. As late as 1969, fifteen years after the appearance of Watson and Crick's paper on DNA, a leading textbook of organic chemistry devoted a mere page and a half of its 1,130 pages to DNA, while biology was consumed by an interest in that molecule.

But times have changed. A major epistemological consequence of the reduction of biological phenomena to macromolecular explanation has been the wholesale entry into biological science of academics identified institutionally as physical scientists. At Harvard there is now a "PhD Track in Engineering and Physical Biology" whose activities have included the organization of two symposia at which five out of nine presentations were by people identified as professors of physics, chemistry or engineering. Among the subjects of their presentations were, for example, "Photosynthesis: The details and (very) big picture," "Transition to collective behavior in developing eukaryotic cells," and "Optical control of protein expression and activity at the single cell level: Applications to morphogenesis in zebrafish."

An important consequence of what is now a common fluidity in epistemic level is that some phenomenon is attacked by a team of investigators, who come to understand it by progressively marching down the levels of analysis to an ultimate explanation. A striking example is the analysis of the appearance of resistance to phosphate insecticide in a sheep blowfly. The title of the article, "A single amino acid substitution converts a carboxylesterase to an organophosphate hydrolase and confers insecticide resistance on a blowfly,"[1] tells only the beginning of the story. Four of the authors are identified as coming from botany and zoology, and two from chemistry. They first found that a particular carboxylesterase in the resistant strain no longer had any esterase activity. They then isolated the enzyme and found that one amino acid within the active site at position 137 had mutated from glycine to aspartic acid. They then showed by in vitro tests that in two such natural and two induced mutations, the change in specificity from carboxylesterase to organophosphorus hydrolase occurred. Three-dimensional modeling of the enzyme protein shows that the amino acid substitution brings two other amino acid residues at the active site closer together so that they bind a water molecule, which then attacks the phosphorus atom. We pass step by step, in detail, from a biological effect to a single-atom effect. The analysis has moved from biology to chemistry to physics. In what epistemic box are we? How does the metaphor of the box help us to understand this way of doing science?

One answer is that we are still in the biological box, because the problem that gave rise to the sequence of investigations was a biological one, and the investigation was pursued further and further downward into chemical and then molecular structural levels only as far as was necessary to enable us to say, "Ah, that explains it," leaving nothing hanging. After all, it might have turned out that a gene conferring resistance to phosphate insecticides had entered the blowfly genome from a virus and by recombination been incorporated into the genome. Then a completely satisfactory explanation would have been framed in purely genetic terms, given our knowledge that such a molecular genetic mechanism is known to exist, and there would have been no motivation for the progressive exploration of deeper and deeper levels of causation. That is, the biological box is not defined solely by a collection of phenomena to be explained, but by the chains of sufficient explanation of those phenomena.

For a very long time, the study of living organisms was dominated by the consciousness of their immense diversity. The contrast between the bewildering variety of organisms that is the subject of biological description and the neat regular taxonomy of chemical elements in the periodic table seemed to demand a totally separate intellectual apparatus for the two realms of phenomena, even though they were admitted to be aspects of a single material universe. In the late nineteenth century, with the explanation of the diversity of seemingly continuous matter as the product of a small number of invisible particles, and the explanation of the diversity of living organisms as a manifestation of material "factors" passed from parent to offspring in the process of reproduction with mathematical regularity, the barriers between the phenomena of the living world and the contents of laboratory flasks were broken down. No deep epistemological break was now entailed in passing between explanations of living and dead. It should not surprise us that Pasteur's work on racemic forms of molecules and the importance to their properties of small differences in molecular structure should have been followed by his work on induced immunity to infection, a phenomenon that is based on small changes in the three-dimensional folding of molecules. Nor is it unexpected that Mendel would develop a simple factorial theory for heredity. Gregor Mendel was not primarily a monk who wandered into playing with the breeding of peas in the time between prayers. He was a student of physics at the Technical Hochschule in Ölmutz when he was recruited by C. F. Knapp, the scientifically active abbot of the Königenkloster in Brno. Knapp took him on precisely to carry out investigations on the basis of inherited variations in plants, conscious of the nature of Mendel's training. Mendel's eventual formulation of a factorial theory of heredity and the inferences he made based on quantitative observations were just what we might have expected from a student of physics at the time. Nor is it unexpected that his work on heredity should have remained unknown in the general scientific community for thirty-five years after its original publication. Mendel in his lifetime became an important person in Moravian science. He was a founder of the Brünn Society of Natural Science, of the Apiculture Society, and of the Austrian Meteorological Society. But both to his advantage and disadvantage, Moravia was peripheral to the influential centers of scientific activity in England, France, and Germany. Because he held a high position in a small and relatively isolated intellectual community, he had the freedom to pursue his own scientific course, but at the price of a long-delayed influence on world science.

We must be very careful of our understanding of the metaphor of "boxes" and of the intellectual histories of the scientists about which this book has been written. Robert MacArthur (chapter 11) was not originally a mathematician who then moved into the mathematical description and analysis of avian populations, of island biogeography, and of population ecology in general, but his brother was a physicist and his father a geneticist. MacArthur was interested in avian populations from his youth. He told the story of repeatedly going, as a boy, into the Vermont woods with his father and instructing his parent in the names of the birds as their calls were heard. Only when he reached adulthood did he learn that his father already knew all that and was humoring young Robert so as not to be seen as putting him down. Species assemblages and their biogeography were at the center of MacArthur's development as a scientist, and he was a student at Yale of G. Evelyn Hutchinson, the preeminent biogeographer and community ecologist of his generation, although not a maker of mathematical models. Yet as Hutchinson's student, MacArthur became a central figure in the group of community ecologists for whom mathematical formulations were a basic methodology in the search for generalizations about species assemblages.

When I arrived at Columbia in the Department of Zoology in the summer of 1951 to begin my graduate work in the laboratory of the preeminent geneticist of natural populations, Theodosius Dobzhansky, I got the idea of doing a radioactive labeling experiment on plant chromosomes that would determine when in the chromosomal division cycle the actual reproduction of the genetic material occurred. So I went upstairs and described my experiment to the chairman of the Botany Department, suggesting that I switch my affiliation. He assured me that it was a very interesting idea, but he said that Dobzhansky would never speak to him again if I switched over. He told me to go back downstairs. About a year afterward a post-doctoral fellow in the Botany Department, J. Herbert Taylor, announced the result of just such an experiment and launched a successful career on it. This was an experiment with a biological motivation but using methods from physics and chemistry. So how should we apply the metaphor of the box? DNA is absolutely part of the present biological box, but these experiments were really molecular chemistry. In this case, it is not that "outsiders coming into biology have expanded the scope or subject matter of biology" or that people "were trained with tools not considered to be part of the toolbox of a biologist," but that people who were in biology to start with, like Herbert Taylor and myself, were trained to consider physics and chemistry integral to our study of biology.

My first job, in 1954–1958, was at North Carolina State College of Agriculture and Engineering, where I was hired because the very active and powerful team of quantitative geneticists there (Ralph Comstock, Clark Cockerham, and others) felt the need for someone trained in classical population genetics. They understood that the formulations of quantitative genetics of selection were in many ways a differently parameterized parallel structure to the structure of Wrightian population genetics. At North Carolina State I acquired an outstanding graduate student, Ken-ichi Kojima, who was interested in putting linkage into models of artificial selection. In parallel, working together, we developed a simple set of equations for a two-locus, two-allele model of selection that was in the language of population genetics and contained the fitnesses of the nine genotypes and the recombination rate. (At about the same time, Motoo Kimura was working on and then published [1956] a paper on linkage and selection along similar lines, but with a different end in view and more in the Fisherian tradition.) Kojima and I then wrote a paper that finally appeared in 1960 and that was the explicit two-locus formulation of the genetical dynamical system for two loci each with two alleles and a parameter for the recombination rate, including free recombination $(r = .5)$, that would then apply to genes on different chromosomes.

Sometime in the late '50s, before I actually went to Australia on a year's leave in 1960, I had talked with Michael White, on one of his visits to the United States, about the inversion systems in the grasshopper, *Moraba scurra*. I realized that this was an exemplification of the two-locus theory and provided a chance to apply it to some observations in the field. This then led to the 1960 paper of Lewontin and White.[2]

This story more or less exemplifies how I have always worked. While many people start with observations in nature and develop theoretical tools to deal with them, which then may become part of the general theoretical apparatus of the field, my way of working has always been the reverse. I start by thinking about some general phenomenon (frequency-dependent selection, multilocus selection, selection in age-distributed continuously breeding life histories, etc.) and explore the theoretical dynamics of such systems. Given my limited mathematical abilities, a lot of that exploration over the years has been done using high-speed computer calculation and simulation, starting already when I was still at NC State and continuing up to the present. Then I come across observations made by others to which these theoretical formulations apply, and I apply them.

This theory-first, observation-second way of working even applied, in a way, to what would seem to be a purely observational corpus. In particular, my work on enzyme polymorphism with Jack Hubby came about in this way. The central problem that I inherited as a student was to determine the extent and nature of genetic variation in natural populations. For years I kept mentally constructing complicated experiments that would make it possible to identify the allelic variation at various loci, but it would have taken a lifetime of effort to apply these to a single locus, so I despaired of the possibility. Then I met Jack Hubby, and a person with a problem met a person with a method.

I suppose I could have stayed within the box of my disciplinary training and contributed to the general theoretical apparatus of population genetics. However, as I have already mentioned, my training was not limited to population genetics. Even my training in biology integrated physics and chemistry. As a graduate student, I also earned a master's degree in statistics that allowed me to develop the mathematical skills to pursue theoretical population genetics. Biology, for me, was a somewhat porous disciplinary box whose boundaries changed over time and when viewed from different perspectives.

The relationship between formal theoretical work and biological investigation has long been an ambiguous one and filled with tension. Most biologists are mathematically uninformed. Sir Solly Zuckerman, at one time a powerful figure in the British biological establishment, when asked what he did when he came across a mathematical formula in a publication, is reported to have replied, "I hum it." It is certainly true that Sir Ronald Fisher became one of the founders of modern population genetics entirely through his mathematical work, but his American counterpart, Sewall Wright, the son of a mathematician, began his scientific career breeding guinea pigs in order to investigate the heredity of continuously varying traits, and only later abandoned experimental work for mathematical theory.

It is extremely important that we understand that formal theory in biology is almost always, for its inventors, an attempt to find a way to pack the contents of the box, rather than a way of deciding what its contents will be. Moreover, irrespective of the use to which mathematical structures have been put, the technical level of successful and fruitful mathematics, as viewed from the standpoint of a professional mathematician, has always been very low. R. A. Fisher's "Statistical Methods for Research Workers"—although very original and tremendously influential in science as a whole in its creation of the method of maximum likelihood and the concept of analysis of variance—is easily taught to biology undergraduates, and it takes only a course in probability and some hard thought to be able to understand Sewall Wright's and Motoo Kimura's works on the stochastic theory of gene frequencies.

The contrast in the form between Fisher's version of population genetics and that of Wright and Kimura is a reflection of their paths of entry into evolutionary theory. Fisher's fundamental theorem of natural selection, framed in terms of "genetic variance," is precisely that expected of someone whose early career was in an experimental agricultural institution where continuously varying characters in plant and animal breeding are at the forefront of interest. Wright, on the other hand, from his experience with laboratory guinea pigs and morphological mutations, created a theoretical population genetics in terms of gene frequencies.

Mathematics of a higher order usually passes beyond real biology into the realm of pure formalism. In this volume, "Nicolas Rashevsky's Pencil-and-Paper Biology" illustrates the point. It was the view of Rashevsky and the school of mathematical biology that the properties of living organisms, from the cell to the population, species, and community levels could be generated by a rigorous mathematical and logical consideration of their structures and processes at the lowest level of assembly. Logic, then, and mathematics in particular, was a tool for deducing the properties of organisms and their assemblages from first principles. The fatal flaw in this axiomatic approach was that no need was felt to generate suggestions for perturbation experiments that could test predictions and their sensitivity to actual deviations from the rigorous a priori descriptions. In fact, among the members of Rashevsky's Committee on Mathematical Biology at the University of Chicago, there was a real hostility to a demand for empirical work. But until we try in practice, we do not know all the effective connections between things. We do not know what is in the box.

A problem that is not dealt with in this volume is the long-standing attempt to enlarge the biological box to include a variety of human historical, cultural, and individual behavioral phenomena. At a minimum this amounts to a description of what constitutes a human "nature" that is claimed to be characteristic of all normal members of the species as a consequence of the structure of the human central nervous system. More recently, this description is rationalized as a result of the operation of natural selection during the evolution of the species from its anthropoid ancestors. The most famous example of this enlargement of the box is the invention by a prominent "insider-outsider" expert on ant behavior, E. O. Wilson, of *sociobiology* as a general field of biological investigation of species, including a description of what is said to be a biologically determined human nature as a product of evolution by natural selection. Among the characteristics of human nature enumerated by sociobiology is a desire to be on the inside.

The invention of sociobiology illustrates the powerful effect of the entrance of Charles Darwin from outside the box of professional academic biology. The formulation of the general theory of evolution of organisms, with natural selection as its major motive force, has created a possibility for the explanation of any property of any species. One need only invent an "adaptive story" of how the possession of a heritable trait by individuals or groups might lead to their greater survival or reproduction and the consequent spread of the trait. Unfortunately, there is seldom the possibility of measuring the differential reproductive rates of real organisms in nature nor the heritability of their traits, so the claim of their establishment by natural selection is untestable. Ironically, the one species in which we have extensive records of age-specific mortality and fertility, causes of death, and detailed life histories and environments is *Homo sapiens,* yet we continue to make up stories out of our heads for the one occupant of the biological box about which we know the most.

NOTES

1. R. D. Newcomb et al., "A single amino acid substitution converts a carboxylesterase to an organophosphate hydrolase and confers insecticide resistance on a blowfly." *Proc. Nat. Acad. Sci. USA*, 94 (1997): 7464–7468.

2. The logic of the story is laid out in chapter 6 of R. C. Lewontin, *The Genetic Basis of Evolutionary Change* (New York: Columbia University Press, 1974), 273–281.

CONTRIBUTORS

LUIS CAMPOS
Department of History
University of New Mexico
Albuquerque, NM 87131
USA

MICHAEL R. DIETRICH
Department of Biological Sciences
Dartmouth College
Hanover, NH 03755
USA

W. TECUMSEH FITCH
Department of Cognitive Biology
University of Vienna
Vienna, Austria

SANDER GLIBOFF
Department of History and
Philosophy of Science
Indiana University
Bloomington, IN 47405
USA

OREN HARMAN
Program in Science, Technology,
and Society
Bar-Ilan University
Ramat Gan 52900
Israel

T. J. HORDER
Oxford University
Oxford OX1 3DW
England

EHUD LAMM
The Cohn Institute for the History and
Philosophy of Science and Ideas
Humanities Faculty
Tel Aviv University
Ramat Aviv, Tel Aviv 69978
Israel

RICHARD C. LEWONTIN
Museum of Comparative Zoology
Harvard University
Cambridge, MA 02138
USA

ERIKA LORRAINE MILAM
History Department
Princeton University
Princeton, NJ 08544
USA

MICHEL MORANGE
Institut d'Histoire et de Philosophie
des Science et des Techniques
75006 Paris
France

GREGORY J. MORGAN
College of Arts and Letters
Stevens Institute of Technology
Castle Point on Hudson
Hoboken, NJ 07030
USA

JAY ODENBAUGH
Department of Philosophy
Lewis and Clarke College
Portland, OR 97219
USA

MICHAEL RUSE
Department of Philosophy
Florida State University
Tallahassee, FL 32306
USA

SAHOTRA SARKAR
Department of Philosophy
University of Texas
Austin, TX 78712
USA

MAYA M. SHMAILOV
Program in Science, Technology,
and Society
Bar-Ilan University
Ramat Gan 52900
Israel

JONATHAN SIMON
Sciences et Société; Historicité,
Éducation et Pratiques
Université de Lyon
Université Lyon 1
69622 Villeurbanne Cedex
France

ROBERT A. SKIPPER, JR.
Department of Philosophy
University of Cincinnati
Cincinnati, OH 45221

HALLAM STEVENS
School of Humanities and
Social Sciences
Nanyang Technological University
Singapore 637332

WILLIAM C. SUMMERS
History of Medicine
Yale University
New Haven, CT 06520
USA

ALFRED I. TAUBER
Department of Philosophy
Boston University
Boston, MA 02215
USA

Index

Page references followed by *f* denote figures and photographs. References to endnotes give the endnote page number followed by n., the note number, and the text page of endnote citation in parentheses.

d'Herelle, Félix (cont.)
of patronage, 12; personal traits of, 10, 63, 66, 72–73; philosophy of, 72; photograph of, 61*f*; protobiology, 62, 71
Diamond, Jared (*Ecology and Evolution of Communities*), 181
DIAS (Dublin Institute for Advanced Study), 93, 95–96
Dickinson, Roscoe, 111
diphtheria, 261
Dirac, Paul A. M., 93
disciplines: biology as, 6–7; as category, 3, 6; movement between, 3–4, 7; subdisciplines, 4
DNA structure, Pauling and, 118
Dobzhansky, Theodosius, 17, 195n. 10 (183), 306, 352–353
dominance, 37
Doppler, Christian, 34
Dorset, Marion, 69
Dounce, A. L., 102
Dublin Institute for Advanced Study (DIAS), 93, 95–96
DuBridge, Lee, 119, 120
duck-billed platypus, 1, 22
Duclaux, Émile, 51, 53, 55
Duke University, 130
Dumas, Jean-Baptiste, 47, 52
Dunn, L. C., 38–39
dynamic sufficiency, 325

EC (embryonal carcinoma) cell lines, 280, 281, 282
Eccles, John, 17, 307, 316
École normale, Pasteur and, 46–47
ecology: community, 184, 189–190, 192–193; MacArthur and, 181–194; patterns in, 183, 186, 190–194; population, 183, 188, 193; systems, 187
Ecology and Evolution of Communities (Cody and Diamond), 181
Edelman, Gerald, 6
Ehrlich, Paul, 261, 272n. 12 (261)
Einstein, Albert, 95
Eisley, Loren, 27, 29
Elements of Physical Biology (Lotka), 164
Ellis, Derek, 232

embryological development, Jacob's studies on, 279–284
embryology, Metchnikoff and, 260, 263–265
embryonal carcinoma (EC) cell lines, 280, 281, 282
embryonic layers, 264–265
embryonic stem (ES) cells, 282–283
emergent properties, 184
Eminent Victorians (Strachey), 86
Endy, Drew, 9, 21, 331–348; *Adventures in Synthetic Biology*, 339, 340; analogy use by, 18; collaborators, 14; detractors of, 17; early life and career of, 333–335; ethos of cool and, 339; European backlash toward, 340–341; funding difficulties, 12; insider status of, 343–345; International Genetically Engineered Machines (iGEM) competition, 338–339; open source biology and, 333, 334, 336; personal characteristics of, 10; photograph of, 332*f*; simplification by, 17; standardized biological parts and, 334–339; Synthetic Biology 1.0 conference, 331–333, 336–338, 342; training of, 11; translation of concepts by, 16
enlightened absolutism, 30–31
entropy, negative, 98, 100, 101
enzyme polymorphism, Lewontin and, 354
Ephrussi, Boris, 280
Erewhon (Butler), 75, 76, 84
errors, theory of, 151
ES (embryonic stem) cells, 282–283
Esquire (magazine), 339
ESS (evolutionarily stable strategy), 313, 323–325
essentialism, 246
ETC Group, 342–343, 347–348n. 31 (344)
eugenics, R. A. Fisher and, 147, 149–150, 152, 157
Eugenics Education Society, 152
Eugenics Review, 152
evolution: Butler and, 75–87; cell lines and, 265–266; conservation of developmental genes, 282; correlation in, 153; cultural, 79, 85; Elaine

Morgan and, 223–233; eugenics and, 149; exaptation, 214, 215; of family and fatherhood, 317, 321; Fisher and, 153–157; game theory and, 322–325; history of evolutionary thinking, 81–82; Lamarck and, 81, 86; of language, 202, 210, 213–216, 217; machine, 77–78, 79; "Mendelism," 27; Metchnikoff and, 263–269; molecular, 119–120; molecular clock and, 120; multilevel selection, 319–320, 325; natural selection and, 78–79, 81, 86–87, 120, 149, 154–157, 215, 356; neutral theory of, 120; organismal identity and, 269–270; popular writing on, 15; Price equation, 318–320, 325; sequence comparison and, 136; from a statistical genetics point of view, 153–157; teleology in, 305

Evolution, Old and New (Butler), 75, 80

Evolution: The Modern Synthesis (Huxley), 87

evolutionarily stable strategy (ESS), 313, 323–325

evolutionary developmental biology ("evo-devo"), 202, 215, 216

evolutionary selection, Hull and, 241, 244, 247, 250

evolutionary synthesis, 7

Evolution in Changing Environments (Levins), 188

Ewald, Paul, 96

Ewens, Warren, 155

exaptation, 214, 215

experimental agriculture, Fisher and, 153

externalism, David Hull and, 247

F9 antigen, 282

Fathers and Sons (Gosse), 86

feedback. *See* negative feedback

feminist anthropology. *See* Morgan, Elaine

fermentation: d'Herelle and, 63–64, 66–67; Pasteur and, 49–51

Fisher, R. A. (Ronald Aylmer), 8, 147–160; analysis of variance and, 153;

on antagonism toward outside views, 10; eugenics and, 147, 149–150, 152, 157; evolution and, 153–157; *The Genetical Theory of Natural Selection*, 12, 154–155, 157; George Price compared to, 326; institutional support for, 13; maximum likelihood method of, 151, 152; as outsider, 147; patrons of, 11–12, 152, 157–158; personal characteristics of, 10; photograph of, 148*f*; population genetics and, 354–355; probability and, 150–151; Rothamsted Experimental Station, 13; *Statistical Methods for Research Workers*, 153, 157, 355; training of, 11; variance and, 153–154, 155, 355

Fitch, Tecumseh, 216

fitness, 155

fluorescence imaging, George Price and, 315

Ford Foundation, 118

formal language theory, 205, 207–208

Foucault, Michel, 284

Franklin, Rosalind, 118

free will, Schrödinger and, 99

French Republican educational system, 46, 58n. 2 (46)

Frimmel, Franz, 95

fundamental niche, 186

funding bodies, 12

Gaertner, Carl Friedrich, 36

Galison, Peter, 3

Galtier, Pierre Victor, 55

Galton, Francis, 149

game theory: evolutionary stable strategy, 313, 323–325; George Price and, 322–325

Gamow, George, 102

gases, theory of, 155–156

gastrula, 264

Geison, Geralde, 48

GenBank, 13, 18, 128, 137–139

gene frequencies, 355

General Chemistry (Pauling), 118

generative grammar/syntax, 204, 207–208

171; as outsider, 161, 167–169, 175–176; photograph of, 162*f*; relational biology and, 173–174; rhetorical style of, 168; at University of Chicago, 161, 169–173; Warren Weaver support of, 12; Westinghouse Research Laboratories and, 161, 164–165, 168–169

realized niche, 186

recessiveness, 37

recombination, by outsiders, 286–287

Registry of Standard Biological Parts, 338–339

Reiner, John, 170

relational biology, 173–174

replicon model, 280

Robinson, Art, 121, 122

Robot, 31, 42n. 8 (31)

Rockefeller Foundation: Pauling and, 112–113, 115; Rashevsky and, 169; Warren Weaver and, 12, 112–113, 169–172

Romanes, George, 15, 82

Rosen, Robert, 161–162

Rosenblueth, Arturo, 13, 14, 293, 295–298, 305–306

Ross, Ronald, 261

Rothamsted Experimental Station, 152–153

Roux, Emile, 55, 68, 271n. 6 (260)

Ruse, Michael, 238, 247, 249

Russell, Bertrand, 206, 207

Salmon, Daniel, 66

Salmonella cholerasuis, 69

Samuelson, Paul, 316

Sartor Resartus (Carlyle), 84

Saussure, Ferdinand de, 206

The Scars of Evolution (Morgan), 232

Schatz, Carol (wife of Noam Chomsky), 204, 212

Schmidt, Markus, 340

Schmitt, Stéphane, 284

Schrödinger, Erwin, 8, 93–109; collaborators, 14; color vision and, 95; free will and, 99; genetic code and, 98, 102; genetics and, 95–104; inventiveness of, 19; legacy of, 100–103; obstacles faced by, 20; as outsider, 103–105; personal

characteristics of, 9–10, 106n. 10 (96); photograph of, 94*f*; terminology transport, 5; thermodynamics of living systems, 97–98, 100; training of, 11; Vedanta metaphysics and, 95, 99, 103; *What is Life?*, 95–105, 134

Schrödinger equation, 93, 112

Schwann, Theodor, 59n. 7 (49)

Schwartz, Samuel, 315

science: philosophy and, 238–250; as the study of patterns, 190

Science as a Process (Hull), 240–242

scientific change/progress, David Hull and, 242, 244–246

scientific method, 238, 255n. 17 (240)

scientific orthodoxy, structural elements enforcing, 174–175

scientific theory, abduction and, 211–212

Scott, Donald, 167

Scriven, Michael, 238

Seed (magazine), 339

segregation, rule of, 38

selection: kin, 321; linkage and artificial selection, 353; multilevel, 319–320, 325; natural, 78–79, 81, 86, 87, 120, 149, 154–157, 215, 356; Price equation and, 318–320, 325; sexual, 322

"self," 263, 265, 268–269, 273n. 23 (265)

The Selfish Gene (Dawkins), 325

self-reproducing automata, 291–305, 301*f*, 303*f*, 304*f*, 306–307

self-reproduction, 295, 298

sequence analysis, 136–138

serum therapy, 261

Seven Years' War (1756–1763), 30

sexual selection, 322

Seyfarth, Robert, 232

Shannon, Claude E., 15, 17, 306, 314, 316, 318–320

Shaw, George Bernard, 86

sheep breeding, 33

Sherrington, C. S., 103, 104

Shockley, William Bradford, Jr., 17, 315

sickle cell anemia, 116, 118

sickle cell hemoglobin, 116

Silesia, 30

silkworms, disease in, 52